William F. Maag Library
Youngstown State University

MODERN ASPECTS OF ELECTROCHEMISTRY

No. 37

Modern Aspects of Electrochemistry

Modern Aspects of Electrochemistry, No. 36:

- The study of electrochemical nuclear magnetic resonance (EC-NMR); *in situ* both sides of the electrochemical interface via the simultaneous use of ^{13}C and ^{193}Pt NMR.
- Recent impressive advances in the use of rigorous ab initio quantum chemical calculations in electrochemistry.
- Fundamentals of ab initio calculations, including density functional theory (DFT) methods, help to understand several key aspects of fuel cell electrocatalysis at the molecular level.
- The development of the most important macroscopic and statistical thermodynamic models that describe adsorption phenomena on electrodes.
- Electrochemical promotion and recent advances of novel monolithic designs for practical utilization.
- New methods for CT analysis and an explanation for the existing discrepancy in Li diffusivity values obtained by the diffusion control CT analysis and other methods.

Modern Aspects of Electrochemistry, No. 35:

- Impedance spectroscopy with specific applications to electrode processes involving hydrogen.
- Fundamentals and contemporary applications of electroless metal deposition.
- The development of computational electrochemistry and its application to electrochemical kinetics.
- Transition of properties of molten salts to those of aqueous solutions.
- Limitations of the Born Theory in applications to solvent polarization by ions and its extensions to treatment of kinetics of ionic reactions.

LIST OF CONTRIBUTORS

M. ABRAHAM
Chemistry Department
University of Montreal
Montreal, Quebec, Canada

M.-C. ABRAHAM
Chemistry Department
University of Montreal
Montreal, Quebec, Canada

FAISAL M. AL-FAQEER
Saudi Aramco E-7600
Dhahran 31311, Saudi Arabia

C. L. BRIANT
Division of Engineering
Brown University
182 Hope Street, Box D
Providence, RI 02912, USA

CHRISTOS COMNINELLIS
Unit of Electrochemical Engineering
Institute of Chemical and Biological Process
 Science
School of Basic Sciences, Swiss Federal
 Institute of Technology
CH-1015 Lausanne, Switzerland

GYÖRGY FÓTI
Unit of Electrochemical
 Engineering,
Institute of Chemical and
 Biological Process Science
School of Basic Sciences, Swiss
 Federal Institute of
 Technology
CH-1015 Lausanne, Switzerland

K. HEMMES
Fac Technology Policy and
 Management of TUDelft
Jaffalaan 5
2628 BX DELFT
The Netherlands

HOWARD W. PICKERING
Department of Materials Science
 and Engineering
Pennsylvania State University
University Park, PA 16802

ZOLTÁN NAGY
Argonne National Laboratory
Materials Science Division
Argonne, Illinois, 60439, USA

A Continuation Order Plan is available for this series. A continuation order will bring delivery of each new volume immediately upon publication. Volumes are billed only upon actual shipment. For further information please contact the publisher.

MODERN ASPECTS OF ELECTROCHEMISTRY

No. 37

Edited by

RALPH E. WHITE
University of South Carolina
Columbia, South Carolina

B. E. CONWAY
University of Ottawa
Ottawa, Ontario, Canada

C. G. VAYENAS
University of Patras
Patras, Greece

and

MARIA E. GAMBOA-ADELCO
Managing Editor
Superior, Colorado

Kluwer Academic / Plenum Publishers
New York, Boston, Dordrecht, London, Moscow

The Library of Congress cataloged the first volume of this title as follows:

Modern aspects of electrochemistry, no. [1]
    Washington Butterworths, 1954–
    v. illus., 23 cm.
    No. 1–2 issued as Modern aspects series of chemistry.
    Editors: No 1– J.O'M. Bockris (with B. E. Conway, No. 3–)
    Imprint varies: no. 1, New York, Academic Press.—No. 2, London, Butterworths.
    1. Electrochemisty—Collected works. I. Bockris, John O'M. ed. II. Conway, B. E. ed.
(Series: Modern aspects series of chemistry)

QD552.M6   54-12732 re

ISBN: 0-306-48226-6

© 2004 Kluwer Academic / Plenum Publishers, New York
233 Spring Street, New York, New York 10013

http://www.wkap.nl/

10   9   8   7   6   5   4   3   2   1

A C.I.P. record for this book is available from the Library of Congress

All rights reserved

No part of this work may be reproduced, stored in a retrieval system, or transmitted in any form or by any means, electronic, mechanical, photocopying, microfilming, recording, or otherwise, without written permission from the Publisher, with the exception of any material supplied specifically for the purpose of being entered and executed on a computer system, for exclusive use by the purchaser of the work

Permissions for books published in Europe: *permissions@wkap.nl*
Permissions for books published in the United States of America:
permissions@wkap.com

Printed in the United States of America

Preface

This volume of Modern Aspects contains seven chapters. The major topics covered in the first six chapters of this volume include fundamentals of solid state electrochemistry; kinetics of electrochemical hydrogen entry into metals and alloys; oxidation of organics; fuel cells; electrode kinetics of trace-anion catalysis; nano structural analysis. The last chapter is a corrected version of chapter four from Volume 35.

Faisal M. Al-faqeer and Howard W. Pickering begin the first chapter by going back to 1864 and Cailletet who found that some hydrogen evolved and was absorbed by iron when it was immersed in dilute sulfuric acid. The absorption of hydrogen into metals and alloys can lead to catastrophic failures of structures. They discuss the kinetics of electrochemical hydrogen entry into metals and alloys.

In chapter three, Clyde L. Briant reviews the electrochemistry, corrosion and hydrogen embrittlement of unalloyed titanium. He begins by reviewing the basic electrochemistry and general corrosion of titanium. He also discusses pitting and galvanostatic corrosion followed by a review of hydrogen embrittlement emphasizing the formation of hydrides and the effect of these on titanium's mechanical properties.

Christos Comninellis and György Fóti discuss the oxidative electrochemical processes of organics in chapter three. They begin by defining direct and indirect electrochemical oxidation of organics. They introduce a model that allows them to distinguish between active (strong) and non-active (weak) anodes. Different classes of organic compounds are used for kinetic models of organic oxidation at active and non-active type anodes.

In chapter five, Kas Hemmes presents a comprehensive discussion of fuel cells and Carnot engines; Nernst law; analytical fuel cell modeling; reversible losses and Nernst loss; and irreversible losses, multistage oxidation and equipartition of driving forces. He compares two analytical models with experimental results on a 110 cm^2 Molten Carbonate Fuel Cell. He also discusses fuel cell configurations and geometries, and numerical fuel cell modeling of various cells such as the Alkaline Fuel Cell, Polymer Electrolyte Membrane Fuel Cell, Direct Methanol Fuel Cell, Phosphoric Acid Fuel Cell, Molten Carbonate Fuel Cell, and Solid Oxide Fuel cell. He ends with a discussion of new developments and applications of fuel cells: fuel cells in trigeneration systems; coal/biomass fuel cell systems; indirect carbon fuel cells; and direct carbon fuel cells.

Zoltán Nagy discusses in chapter five the catalytic effect of trace anions in outer-sphere heterogeneous charge-transfer reactions. He begins with the observations of H. Gerischer and continues with the observations of D. C. Johnson and E. W. Rosnick. The measurements of J. Weber, Z. Samec, and V. Marecek are explored as well. A discussion of the observations of F. R. Smith and J. T. Wadden and the work of Z. Nagy and L. A. Curtis, and H. C. Hung and Z. Nagy are included. He states that it is reasonable to conclude that trace-anion catalysis of outer-sphere charge-transfer reactions may be a very general phenomenon.

Chapter six by M. Abraham and M. C. Abraham discusses the results of experimental and theoretical investigations on bridging electrolyte-water systems as to thermodynamic and transport properties of aqueous and organic systems. This chapter is a corrected version of chapter four in Volume 35. In the course of editing Volume 35, the authors provided a final set of edited and corrected diskettes. However, in conversion to CD format for the final pagination and indexing stage, using supposedly the same software, many errors and transpositions of diagrams occurred in the final production stage of the chapter. The Editors and Publisher find it necessary and appropriate to reprint this corrected chapter in its entirety in this volume. We apologize to the authors and our readers for any inconvenience this may have caused.

R. E. White

University of South Carolina
Columbia, South Carolina

B. E. Conway

University of Ottawa
Ottawa, Ontario, Canada

C. G. Vayenas

University of Patras
Patras, Greece

Contents

Chapter 1

KINETICS OF ELECTROCHEMICAL HYDROGEN ENTRY INTO METALS AND ALLOYS

Faisal M. Al-Faqeer* and Howard W. Pickering

I. Introduction	1
II. Hydrogen Evolution Reaction Mechanisms	3
III. Hydrogen Degradation in Metals and Alloys	5
IV. Devanathan-Strachurski Electrochemical Permeation Cell	9
1. Electrochemical Hydrogen Permeation Cell	10
2. Metal Membrane Preparation	12
3. Solution Preparation	12
4. Experimental Procedure	12
5. Directly Obtained Quantities	13
6. Hydrogen Diffusivity	14
7. Hydrogen Concentration	17
8. Trapping Effect	18
9. Difficulties	20
10. Advantages of the Technique	22
V. IPZ Analyses	22
1. The Original IPZ Analysis	22
2. Membrane Thickness Effect	29
3. Calculation of θ_H and Rate Constants of the HER from Polarization Data	30
VI. The Generalized IPZ Analysis	32
VII. The New IPZA Analysis	40
VIII. Analysis of Permeation Data under Corrosion Conditions	46
IX. Conclusions	48

List of Symbols 49
References 50

Chapter 2

THE ELECTROCHEMISTRY, CORROSION AND HYDROGEN EMBRITTLEMENT OF TITANIUM

C.L. Briant

I. Introduction 55
II. Electrochemistry and General Corrosion 58
III. Galvanic Corrosion 69
IV. Pitting Corrosion 72
V. Hydrogen Embrittlement of Titanium 75
VI. Summary 84
References 84

Chapter 3

ELECTROCHEMICAL OXIDATION OF ORGANICS ON IRIDIUM OXIDE AND SYNTHETIC DIAMOND BASED ELECTRODES

György Fóti and Christos Comninellis

I. Introduction 87
II. Electrochemical Oxidation of Organics in Aqueous Media Using Active Anodes 91
   1. Oxidation of Organics on Ti/IrO$_2$ Anodes in the Potential Region before O$_2$ Evolution 93
   2. Oxidation of Organics on Ti/IrO$_2$ Anodes in the Potential Region of O$_2$ Evolution 96
   3. Kinetic Model of Organic Oxidation on Ti/IrO$_2$ Anodes   103
III. Electrochemical Oxidation of Organics in Aqueous Media Using Non-Active Anodes 108
   1. Oxidation of Organics on p-Si/BDD Anode in the Potential Region before O$_2$ Evolution 110

2. Oxidation of Organics on p-Si/BDD Anode in the
     Potential Region of O_2 Evolution 113
  3. Modeling of Organics Incineration on p-Si/BDD Anode 116
     (*i*) Electrolysis Under Current Control ($j < j_{\text{lim}}$) 119
     (*ii*) Electrolysis Under Mass Transport ($j > j_{\text{lim}}$) 122
     (*iii*) Experimental Verification 123
IV. Conclusion 126
    List of Symbols 128
    References 129

Chapter 4

FUEL CELLS

K. Hemmes

I. Introduction 131
II. Fuel Cells and Carnot Engines 132
   1. Fuel Cells versus Carnot Engines 132
   2. Fuel Cells Combined with Carnot Engines 136
      (*i*) FC-CE Systems Having an Overall FC Reaction
            with $\Delta S < 0$ (Reversible) 136
      (*ii*) FC-CE Systems Having an Overall FC Reaction
            with $\Delta S < 0$ (Irreversible) 138
      (*iii*) FC-CE Systems Having an Overall FC Reaction
            with $\Delta S = 0$ (Reversible) 141
      (*iv*) FC-CE Systems Having an Overall FC Reaction
            with $\Delta S = 0$ (Irreversible) 143
      (*v*) FC-CE Systems Having an Overall FC Reaction
            with $\Delta S > 0$ (Reversible) 144
      (*vi*) FC-CE Systems Having an Overall FC Reaction
            with $\Delta S > 0$ (Irreversible) 145
III. Nernst Law 146
   1. Derivation of Nernst Law 146
   2. MCFC with Diluted Fuel Gas 150
   3. Pure Hydrogen as a Fuel 151
IV. Analytical Fuel Cell Modeling 156
   1. Analytical versus Numerical Modeling 156

2. Analytical FC Model by Division in Hypothetical Sub-Cells	160
3. Analytical FC Model Using Differential Equations	166
V. Reversible Losses and Nernst Loss	170
1. Nernst Loss	170
2. Minimizing Nernst Loss by Lowering the Operating Temperature	174
3. Minimizing Nernst Loss by Decreasing the Utilization	174
4. Minimizing Nernst Loss by Decreasing the OCV by Gas Recycling	175
5. Minimizing Nernst Loss by Changing the Relation $V_{eq}(u)$	175
6. Can We Minimize the Nernst Loss by Changing the Relation $V_{eq}(x)$?	177
7. FC-CE Systems Including Nernst Loss	179
8. Modeling the Initial Dip in the Nernst Potential	181
VI. Irreversible Losses, Multistage Oxidation and Equipartition of Driving Forces	183
1. Irreversible Losses	183
2. Multistage Oxidation	185
3. General Formulation of the Equipartition Principle	187
4. Equipartition of Entropy Production in a Non-isothermal Fuel Cell	189
VII. Further Analysis	190
1. Maximum Power versus Maximum Efficiency	190
2. Fuel Cell under Potentiostatic Control	191
3. Fuel Cell Operated at Constant Load Resistance	192
4. Fuel Cell under Galvanostatic Control	193
5. Heat Effects in Fuel Cells	194
(i) Heat Dissipation Due to Irreversible Losses	194
(ii) Heat Dissipation Due to Reversible Processes	195
VIII. Comparison of Two Analytical Models with Experimental Results on a 110 cm^2 MCFC	197
1. Experimental Results	197
2. The Simple and Extended Analytical Fuel Cell Model	198
3. The Simple Analytical Fuel Cell Model Compared with Experimental Results	202
4. The Extended Analytical Fuel Cell Model Compared with Experimental Results	204

5. Conclusions and Applications of the Analytical Models to Other Fuel Cell Types	208
IX. Fuel Cell Configurations and Goemetries	209
1. Disk Shaped Fuel Cell Geometry	210
2. Tubular Fuel Cell Geometry	212
3. 1 in/1 Out or Single-Chamber Fuel Cell Concept	216
4. 2 in/1 Out or Dead-End Mode Concept	217
5. 3 in/3 Out Concept; MCFC with a Separate CO_2 Supply	218
X. Numerical Fuel Cell Modeling	223
1. Introduction	223
2. AFC: Alkaline Fuel Cell	223
3. PEMFC: Polymer Electrolyte Membrane Fuel Cell	225
4. DMFC: Direct Methanol Fuel Cell	228
5. PAFC: Phosphoric Acid Fuel Cell	229
6. MCFC: Molten Carbonate Fuel Cell	230
7. SOFC: Solid Oxide Fuel Cell	231
XI. New Developments and Applications	232
1. Fuel Cells in Trigeneration Systems	232
2. Coal/Biomass Fuel Cell Systems	233
3. Indirect Carbon Fuel Cell (IDCFC)	234
4. Direct Carbon Fuel Cell (DCFC)	239
List of Symbols	244
References	247

Chapter 5

TRACE-ANION CATALYSIS OF OUTER-SPHERE HETEROGENEOUS CHARGE-TRANSFER REACTIONS

Zoltán Nagy

I. Introduction	253
II. Summary of Observations	254
III. The Ferrous/Ferric Reaction	256
1. Gerischer	257
2. Johnson and Resnick	257
3. Weber et al.	258

4. Smith and Wadden		259
5. Nagy et al.		260
6. Hung and Nagy		260
7. Discussion		264
IV. Concluding Remarks		267
References		268

Chapter 6

THERMODYNAMIC AND TRANSPORT PROPERTIES OF BRIDGING ELECTROLYTE-WATER SYSTEMS

Maurice Abraham and Marie-Christine Abraham

I. Introduction	271
II. Thermodynamic Properties	273
1. Adsorption Theory of Electrolytes	276
(*i*) Component Activities	276
(*ii*) Characteristic Features of Water Activity Coefficient Curves	287
(*iii*) Excess Properties	292
(*iv*) Temperature Dependence of the Adsorption Parameters	299
2. Approaches Related to Regular Solution Theories	300
(*i*) Approach with Mole Fraction on an Ionized Basis	300
(*ii*) Approach with Mole Fraction on an Un-Ionized Basis	304
3. Surface Properties	305
4. Other Approaches and Observations	309
(*i*) Water Vapor Pressure	309
(*ii*) Molar Volume	310
(*iii*) Substitution of an Organic Substance to Water	311
III. Transport Properties	312
1. Activation Energy for Viscous Flow	313
2. Transition State Theory of Viscosity	315
3. Free Volume for Viscous Flow	325
4. Equation for Fluidity with Apparent Parameter	329
5. Activation Energy for Electrical Conductance	334

6. Transition State Thoery of Electrical Conductance .. 337
7. Free Volume for Electrical Conductance 341
8. Equation for Equivalent Electrical Conductance with Apparent Parameter 343
9. Relation Between Viscosity and Electrical Conductance 346
10. Other Approaches and Observations 349
    (*i*) Logarithmic Equation for Equivalent Electrical Conductance 349
    (*ii*) Power-Law Equations for Viscosity and Electrical Conductivity 349
    (*iii*) Substitution of an Organic Substance to Water 350
References 353

INDEX 359

Kinetics of Electrochemical Hydrogen Entry into Metals and Alloys

Faisal M. Al-Faqeer* and Howard W. Pickering

Department of Materials Science and Engineering, The Pennsylvania State University, University Park, PA 16802, U.S.A

I. INTRODUCTION

In 1864, Cailletet found that when an iron sample was immersed in dilute sulfuric acid, some hydrogen evolved and some was adsorbed by the iron.[1-3] In 1922, Bodenstein found that application of a cathodic current would result in an increase in the amount of hydrogen entering iron.[4] His results showed that the amount of hydrogen permeating an iron membrane was directly proportional to the square root of the applied current density.[2-4] These two observations demonstrated that some of the hydrogen atoms produced electrochemically during corrosion of iron may enter the metal lattice and permeate through the metal and that the quantity of hydrogen entering the metal can be increased by applying a cathodic current. The entry of the electrolytic hydrogen into metal is related to the hydrogen evolution reaction.

---

*Present Address: Saudi Aramco E-7600, Dhahran 31311, Saudi Arabia

Modern Aspects of Electrochemistry, Number 37, Edited by Ralph E. White *et al.* Kluwer Academic / Plenum Publishers, New York, 2004.

Atomic hydrogen has a uniquely high diffusivity being able to rapidly diffuse through some metals at room temperature. The diffusivity of hydrogen within iron, with its rather loosely packed bcc crystal lattice, is extremely high. The diffusion coefficient of hydrogen in fully annealed iron is of the order of 10^{-5} cm^2 s^{-1} at room temperature.[5,6] This is about 12 orders of magnitude greater than that of other interstitials such as carbon and nitrogen. The high mobility of hydrogen atoms into metals is commonly attributed to the small size of its atom.[7]

The entry of these hydrogen atoms into the metal lattice is an undesirable reaction because of the damage that the absorbed hydrogen causes to the physical and mechanical properties of the metal.[3,7-10] Internal bursts and blisters may be realized during corrosion or cathodic polarization if large amounts of hydrogen collect in a localized area.[3,8,9] The absorbed hydrogen causes hydrogen embrittlement and, as a result, possible catastrophic failure of structures in service.[3,8,11]

There are many sources of absorbed hydrogen in metals. Hydrogen can enter metals and alloys during cleaning, pickling, welding or electroplating processes, to name a few. Or, it may be picked up from the service environment as a result of cathodic protection or corrosion.[3,5] In all of these processes, the hydrogen evolution reaction (HER) is a possible cathodic reaction with atomic hydrogen as a byproduct.[3,6,12-18] It is usually an unavoidable electrode reaction proceeding simultaneously with these processes.

Mechanistic analyses have been developed which evaluate the kinetics of hydrogen entry into metals via the hydrogen evolution (HER) and absorption (HAR) reactions. These analyses can determine the individual rate constants of the hydrogen evolution reaction, k_1 and k_2, (from the steady state permeation data or the polarization curve) and the kinetic-diffusion rate constant, k, the hydrogen surface coverage, θ_H, and the hydrogen concentration in the charging side of the membrane, C^o, (from permeation data). The surface coverage of a second adsorbate, θ_A, can also be evaluated. Example applications of these analyses are given for characterizing the effects of hydrogen sulfide and hexamethylenetetramine (HMTA) on the HER and HAR. They are known as The IPZ and IPZA (Iyer, Pickering, Zamanzadeh, Al-Faqeer) analyses. The surface coverage's of HMTA were determined using the newer IPZA analysis, which takes into account competitive adsorption under Langmuir isotherm. Good agreement was obtained with its coverage's determined by two other independent methods, Electrochemical Quartz Crystal Microbalance (EQCM) and the estimated corrosion current density (i_{corr}), which is a powerful documentation of the overall IPZ framework.

II. HYDROGEN EVOLUTION REACTION MECHANISMS

The hydrogen evolution reaction has long been known. The reaction looks rather simple at first glance; however, a thorough understanding of the mechanism was not achieved until after 1950 through the pioneering work by Tafel, Volmer, Horiuti, Frumkin, and others. It is one of the most frequently studied electrochemical reactions because it is one of the simplest and most fundamentally important reactions in electrochemistry. It has received great attention from corrosion engineers and scientists because of its crucial role in the corrosion of many metals in acid media.[10] The reaction is also of basic importance in the related problem of hydrogen embrittlement, since it controls the hydrogen entrance into metals from aqueous solution.[10] This reaction has been studied extensively on various metals over a period of many years. The HER and permeation mechanisms on pure Fe are fairly clear as a result of this work.

Two basic mechanisms have been accepted for the hydrogen evolution reaction. They are the discharge, chemical recombination (Volmer-Tafel) mechanism[3,5,6,12,14,19-31] and the discharge, electrochemical desorption (Volmer-Heyrovsky) mechanism.[5,6,10,25, 26, 31-34] Both mechanisms of hydrogen evolution involve an adsorbed hydrogen atom on the metal surface (M-H_{ads}). Hydrogen enters the metal through this species.[2,5,6]

The HER is not a single step reaction, but rather proceeds in two steps,[3,5,6,12-14,19-21,23-28] with only one intermediate reaction.[35] The first step is now fairly well understood to take place by a single electron transfer.[3,5,6,10,12-14,19-21,23-28,36] This step involves a proton discharge on the surface of metals to form an adsorbed hydrogen atom. Hydrogen absorption occurs via adsorbed hydrogen atoms which are produced during the proton discharge process.[9,37] The discharge step on iron is,

$$H^+ + e^- + Fe \xrightarrow{k_1} Fe\text{-}H_{ads} \qquad (1)$$

In a corrosive environment, the majority of the adsorbed atomic hydrogen on a metal surface (M-H_{ads}) will combine and form molecular hydrogen, which accumulates and bubbles off the surface. This combination of the adsorbed hydrogen atoms is the second step of the HER.[3,5,6,10,12-14,19-21,23-28,36] This step can take place through an atom-atom combination as suggested by the chemical recombination mechanism (Volmer-Tafel)[3,5,6,12,14,24-28] or through an ion-atom reaction as suggested by the electrochemical recombination mechanism (Volmer-Heyrovsky).[5,6,10,25,26,32] These two reactions on an iron surface are, respectively,

$$Fe-H_{ads} + Fe-H_{ads} \xrightarrow{k_2} 2Fe + H_2 \qquad (2)$$

$$Fe-H_{ads} + H^+ + e^- \xrightarrow{k_3} Fe + H_2 \qquad (3)$$

Some of the adsorbed hydrogen on the metal surface is absorbed inside the metal (H_{abs}) through the hydrogen absorption reaction (HAR). The HAR is a side reaction that occurs in parallel with reactions (2) and (3) of the HER. On iron, it is,[10,24,32-34,38]

$$Fe-H_{ads} \underset{k_{des}}{\overset{k_{abs}}{\rightleftarrows}} Fe + H_{abs} \qquad (4)$$

The electrolyte/metal interface has a pronounced effect on the rate at which the hydrogen evolution and absorption reactions take place. Species present in the environment have a profound effect on the quantity of hydrogen that can enter the metal. A few solution species of practical concern in different industries include: sulfide in crude oil and sour gas, residual traces of arsenic in acids, cyanide in plating solutions, and inhibitors added to pickling solutions. In all of these cases, the sulfide, cyanide, etc., cause hydrogen related problems that would not exist in their absence.[8,14,20,21,24,33,34,39-41]

III. HYDROGEN DEGRADATION IN METALS AND ALLOYS

Hydrogen embrittlement occurs as a result of several sequential events. Hydrogen adsorption on the metal surface from the aqueous phase occurs through the first step of the HER as shown in reaction (1). This is followed by its absorption into the metal lattice through the HAR as shown in reaction (4),[42] by diffusion to specific sites and additional events that lead to hydrogen embrittlement, the latter events being beyond the scope of this paper. Hydrogen accumulation in steel and other alloys increases the probability of cracking. There are different terminologies used to describe the failure of vessels in service. Examples include sulfide hydrogen embrittlement, stress corrosion cracking (SCC), hydrogen blistering, hydrogen induced cracking (HIC), step wise cracking (SWC), stress oriented hydrogen induced cracking (SOHIC) and delayed fracture.

Different industries suffer from hydrogen related failures. For example in the oil and gas industries, Exxon reported that 172 of the 505 vessels inspected in their 19 refineries contained cracks, and 40% of them required replacement.[43] Amoco Canada found that 48% of their inspected vessels had cracks, with 3% of these vessels requiring replacement and most of them requiring repairs.[43] Syncrude Canada reported that 50% to 60% of the inspected vessels contained cracks and 5 of them were replaced, while many required repairs.[43] An industry survey done by American Petroleum Institute (API) in 1998 showed that 25% of all inspected vessels contained cracks and 33% of them required replacement.[43] Other industries such as paper and pulp are seriously affected as well.

In these examples of hydrogen degradation of metals and alloys, cracking is preceded by hydrogen atoms entering the metal and occupying interstitial sites. However, steel containing interstitial hydrogen is not always damaged. It almost always loses ductility (hydrogen embrittlement), but cracking usually takes place only under conditions of sufficiently high applied or residual tensile stress and can occur during hydrogen absorption or subsequently. Failures of this kind are called hydrogen stress cracking or hydrogen cracking. The cracks tend to be mostly transgranular.[44]

Hydrogen embrittlement is a severe environmental type of failure. When hydrogen is present, materials fail at load levels that are very low compared to those that a hydrogen-free material can withstand. The result is usually catastrophic fracture which occurs unexpectedly, sometimes after many years of service.[3,8,11] The incident of the catastrophic failure of a monoethanoamine (MEA) tower in one of the oil companies is an example. This incident caused 17 deaths and cost the company $100 million.[43]

Hydrogen embrittlement is a primary cause of premature failure of metal components. The effect of hydrogen on the mechanical behavior of metals has been investigated and mitigated by different approaches that include both thermo-mechanical treatments and materials selection. Hydrogen embrittlement is thought to be affected by a material's strength, microstructure, composition, nature and density of trapping sites, content of hydrogen, and temperature.[3,8,11]

Hydrogen induced cracking or HIC is not strongly influenced by stress. HIC initiates at elongated nonmetallic inclusions (often MnS) in the base metal. Cracks linking the blisters formed at inclusions may follow a stair-step pattern through the wall. The term step-wise cracking (SWC) describes this feature compared to the crack path observed for sulfide stress cracking (SSC). For soft steel in which stresses are applied, the fracture morphology appears to be between HIC and SSC and is sometimes referred to as stress-oriented hydrogen induced cracking (SOHIC). Most but not all cracking that occurs in the petroleum industry is HIC, but this term is more widely used in the petroleum industry to describe all of the cracking of steel that occurs in the presence of hydrogen.[40]

In HIC the major cathodic reaction in the corrosion of steel is the HER. As for the other types of hydrogen degradation, the adsorption and absorption of hydrogen into the steel occurs as described above. However, some of the absorbed hydrogen according to reaction (4), diffuse and accumulate within atomic-scale void spaces, e.g., at non-metallic inclusions.[12,40] Then, the hydrogen atoms combine and form molecular hydrogen in this void space according to reaction (2).[40] Molecular hydrogen is too large to diffuse through the metal, so it is confined to this space (trapped) and forms a small volume of H_2 gas within this space in the steel. The H_2 gas continues to form as long as the HER is occurring at the steel surface. As a result, the pressure of the H_2 gas in the void increases.[12,40] It could increase and approach the value of the hydrogen fugacity existing at the HER charging sites on the outer surface, but before this happens the steel's yield stress is often exceeded. This causes deformation of the steel in the vicinity of the void and an upheaval or blister to form on the metal surface. Strictly speaking, a blister refers to the protrusion seen at the surface, which only forms during the pressure buildup but remains thereafter as a visible protrusion on the surface of the steel. It is important to note that although blister formation can only occur during the HER, crack propagation due to hydrogen can occur long after the hydrogen is introduced into the steel by the HER. Solution pH, the amount of hydrogen entering the steel, temperature, steel composition, processing and microstructure of the steel are the major factors that affect HIC.[40]

The steel's composition has a profound effect on HIC resistance. A reduction in sulfur content increases resistance by reducing the number of nonmetallic inclusions that can collect hydrogen from which HIC could initiate. Calcium treatment has proved beneficial in reducing the extent of HIC because it controls the sulfide-inclusion shape. Calcium-treated steels generally have rounded sulfide inclusions rather than deleterious elongated inclusions (stringers).[40] The addition of copper decreases both the corrosion rate of steel in H_2S and the tendency for HIC, provided the solution pH is greater than 5. At pH < 5, the protective iron-sulfide film no longer is present and copper is not beneficial.[40,44] A molybdenum coating on a nickel substrate showed an inhibiting effect on hydrogen absorption.[45]

HIC is also dependant on steel manufacturing, heat treatment, and strength. These factors have a strong relationship with the microstructure. If HIC is anticipated, then steel that is resistant to this form of hydrogen attack should be used. Generally, seamless pipe is resistant to HIC because the manufacturing process limits the development of elongated sulfide inclusions. On the other hand, both electric resistance and submerged arc welded pipes have exhibited HIC. These pipes are manufactured from strip and plate that are flat rolled, which increases the tendency to form elongated inclusions. However, careful control of the steel composition, processing and manufacturing processes can increase the resistance of these types of pipes to the point that they can be used satisfactorily in sour service.[44]

Any of the forms of hydrogen damage that do not require the simultaneous occurrence of the HER can produce what is referred to as delayed fracture. One characteristic of this type of damage is the delay time for the appearance of cracks after stress is applied. The delay time is only slightly dependent on stress level, but it decreases with increase in hardness or tensile strength. A critical minimum stress exists, below which delayed cracking will not take place at any time. The critical stress decreases with an increase in hydrogen concentration. The delay in fracture apparently results because of the time required for hydrogen to diffuse to specific areas near a crack nucleus until the concentration reaches a damaging level. A sharp notch in a steel surface favors plastic deformation at its base, and hence, lowers the critical minimum stress and shortens the delay time.[46]

Hydrogen can exist in metals in two forms: namely, trapped hydrogen and mobile or diffusible hydrogen. Pressouyre[47] found that hydrogen diffusing in a lattice interacts with lattice imperfections. Many of these defects interact with hydrogen more strongly than the lattice does, which results in attraction and concentration of the hydrogen at these sites. Since hydrogen atoms become bound to these defect sites, they are called hydrogen traps.[47] Traps include dislocations, grain boundaries, voids and particle matrix interfaces.[47] Some of these traps act as sinks because they prevent hydrogen from reaching a potential flaw and growing to a critical concentration. Traps that are sinks could be beneficial by reducing the susceptibility to hydrogen embrittlement and subsequent cracking or premature failure. Sinks are classified as irreversible traps.[47,48]

Some traps can be crack initiating sites depending on the trap capacity and whether the critical combination of stress and concentration of hydrogen is achieved. Reversible traps may also function as sources, initially storing hydrogen and then providing hydrogen to a crack site, thus accelerating the hydrogen degradation process. A reversible trap can be defined as one at which hydrogen has a short residence time at the temperature of interest with a relatively low interaction energy.[47,48]

The two important effects of trapping are the apparent increase in hydrogen solubility and the apparent decrease in the diffusivity of hydrogen. An iron base alloy absorbs hydrogen to the solubility limit of the host lattice, in addition to the hydrogen needed to occupy the traps. Thus, the apparent solubility or the total hydrogen concentration may be significantly greater than the lattice solubility. Trapping increases the apparent solubility. A trapped atom must acquire energy substantially greater than the lattice migration energy to escape the trap and contribute to measured transport parameters. Consequently, in the presence of trapping, the apparent diffusivity will be less than the lattice diffusivity.[49]

The mechanisms by which alloying elements and/or additives in the electrolyte inhibit the hydrogen absorption step and subsequent damage to equipment are not well understood. This has been in part because of the difficulty in measuring the kinetic (reaction rate constants) and thermodynamic (coverage) parameters of both, the HER and HAR, and in part due to the difficulty in analyzing the interaction of absorbed hydrogen with the defect structure of the metal. The development of the Iyer-Pickering-Zamenzadeh (IPZ) analysis[50] in 1989 was a major impetus in successfully determining these parameters of the HER and HAR in many metal/electrolyte systems. The analysis assumes the discharge chemical (Volmer-Tafel) recombination mechanism for the HER and develops some relations which can be used to determine the discharge reaction rate constant (k_1), recombination reaction rate constant (k_2), hydrogen surface coverage (θ_H) and the kinetic diffusion constant (k).[10,19-21,23,24,30,36,50-52] The latter includes the absorption and desorption reaction rate constants (k_{abs} and k_{des}, respectively), hydrogen diffusivity (D), and the membrane thickness (L).

IV. DEVANATHAN - STRACHURSKI ELECTROCHEMICAL PERMEATION CELL

The electrochemical hydrogen permeation cell and technique were developed by Devanathan and Stachurski in 1962 to study the diffusion of hydrogen in metals.[5,25,26] In this technique, the specimen is placed between two glass cells, one used for charging the specimen with hydrogen and the other for measuring the hydrogen flux permeating through the membrane.

It was initially used to study hydrogen diffusion through palladium.[25] The method was then used by many researchers to investigate the diffusion of hydrogen through other metals.[6,9,10,19,20,22-24,29,32-34,36-38,42,50,53-75] This technique is considered as the most sensitive and accurate for measuring the hydrogen flux through metals with time because small currents that result from oxidation of hydrogen at the exit surface of the membrane can be very accurately measured.[70] The technique is simple and very useful for the study of hydrogen absorption into different metals, alloys and coatings in the presence or absence of additives to the aqueous solution. It has also been used to study the trapping effect and hydrogen defect interactions,[48,60,76] and the kinetics of carbide precipitation during tempering of low carbon martensite at 453K.[60] Thus, the electrochemical method of hydrogen permeation through metal membranes is a very important technique for the study of the diffusion and absorption of hydrogen in metals.

1. Electrochemical Hydrogen Permeation Cell

The typical electrochemical hydrogen permeation cell is composed of two identical compartments, each being a standard 3-electrode cell separated by the metal sheet.[5,25,26] Each surface of the metal functions as the working electrode in the respective cell. The metal sheet separates the electrolyte solution in each compartment from each other. In the charging compartment, hydrogen ions are discharged on the entry surface (charging side) of the sheet electrode and form adsorbed hydrogen atoms on the surface according to reaction (1). The majority of the adsorbed hydrogen atoms will combine to form the H_2 molecule according to either reaction (2) or reaction (3), and eventually enter the solution and evolve as gas bubbles A usually small fraction of the hydrogen atoms that are adsorbed on the entry surface become absorbed inside the electrode according to reaction (4) and diffuse through the metal thickness to the other (exit) surface of the metal electrode.[2,21,54,70,76]

The exit surface of the metal sheet is anodically polarized with the electrical circuitry of the exit compartment, which is completely separated from the electrical circuitry of the entry (charging) compartment. With the exit surface of the sheet under anodic polarization, arriving hydrogen atoms, which are diffusing through the metal from the entry surface under their concentration gradient, are immediately oxidized to hydrogen ions and enter the electrolyte in the exit compartment. The anodic current density in the exit compartment at steady state is a measure of the steady state permeation of hydrogen through the metal.[2,54,70,76] These opposite surfaces of the metal sheet, which are the working electrodes of the two compartments, have a specific exposed area to the electrolyte, 0.8 cm^2 in the experiments reported below, and the sheet is on the order of $\leq$ 1mm thick depending on the metal and conditions. Each compartment also has its own reference electrode and counter electrode. A schematic of the permeation cell is shown in Fig. 1.

In the charging (left) side of the cell in Fig. 1, a cathodic current is applied to the metal membrane using a potentiostat which reduces the protons to atomic hydrogen that is adsorbed on the membrane's input

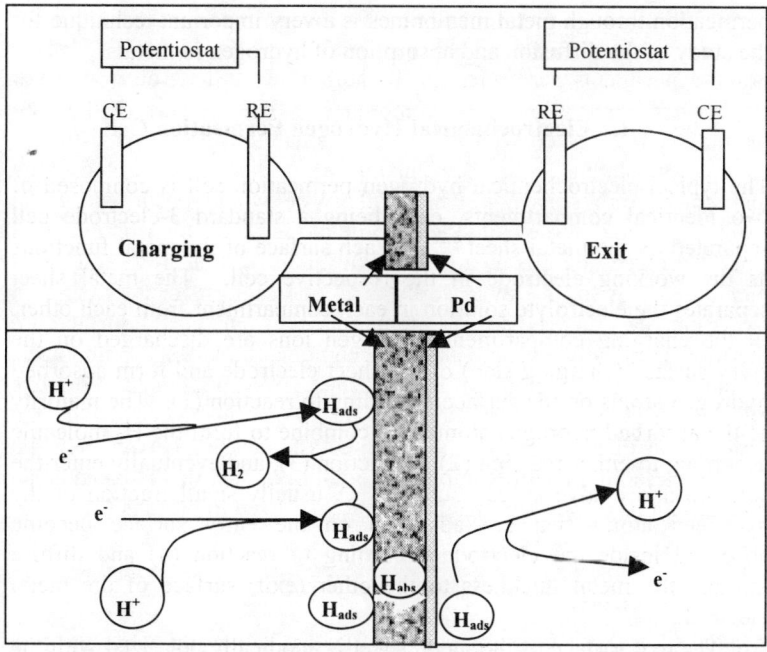

Figure 1. Electrochemical Hydrogen Permeation Cell. (Reprinted from: F. M. Al-Faqeer, PhD Thesis, *New Advancements in the Analysis Procedures of the Electrochemical Hydrogen Permeation Experimental Data*, The Pennsylvania State University, University Park, 2002, p. 75.[77])

(charging) surface. In the exit (right) compartment of the cell, a computerized potentiostat with CMS100 Software is used to apply a potential, typically 0.15 V (Hg/HgO), to the exit surface of the membrane, which is in contact with an alkaline solution (0.1N NaOH), in order to oxidize the hydrogen arriving at this exit surface to the proton.

The hydrogen permeation current transient is measured until a steady state is achieved. Values of the steady state hydrogen permeation current are obtained for different applied charging currents. The electrode potential is measured against a reference electrode, e.g., a saturated calomel electrode (SCE), at the charging side of the cell at the different cathodic charging currents.

2. Metal Membrane Preparation

Some experiments performed by the authors were done on iron, 1020 steel and Cr low alloy steel membranes. The exit side of the iron membrane was coated with a very thin palladium (Pd) layer using an electroless plating solution (Palamerse). This layer protects the membrane from corrosion which otherwise would occur, since the applied potential at the exit side is noble of the reversible iron potential and, therefore, oxidizing to the exit side of the iron and steel membranes. Also, the layer works as a good catalyst for the hydrogen oxidation reaction. Hydrogen atoms adsorbed on the Pd surface are thermodynamically unstable and their combination is fast. Without the Pd surface, even a high overpotential is not always sufficient to oxidize all the arriving hydrogen atoms to protons. This can lead to the observation of hydrogen bubbles at the exit surface. However, when using a palladium coating, bubbles are not seen, and this is taken as an indication that all the hydrogen atoms are oxidized to protons.[61]

3. Solution Preparation

Analytical grade chemicals and double distilled water were used to prepare the electrolytic solutions in this research. The solutions were pre-electrolyzed for 24 hours at 2 mA. The purpose of the pre-electrolysis is to remove the impurities, which can affect the quality of the permeation data collected. Then, the solutions were de-aerated with hydrogen for 2 hours before the experiments. The pH of the solution was measured before and after each experiment and no significant change was found.

4. Experimental Procedure

The iron or steel sheet was fitted between the two compartments in the hydrogen electrochemical permeation cell after it had been polished on both surfaces and coated with palladium on one surface. The sheet was installed so that the palladium-coated side was the exit side of the membrane. An alkaline solution (0.1N NaOH) was introduced in the exit compartment of the cell. An oxidizing potential of 0.15V vs. Hg/HgO was applied in the exit (right) compartment of Fig. 1 in order to reduce the background current to below 1 μA before starting the permeation experiment.[10,66]

Then, with the potentiostat turned on and set at a specific cathodic current value so that the charging (left) side of the iron or steel sample would be cathodically protected, the charging solution was added to the charging (left) compartment. Cathodic protection of the iron or steel surface in the left compartment prevented its corrosion when it was exposed to the charging solution. Experiments were started at the highest charging current to be used in the series of galvanic charging experiments. This procedure maximized and stabilized the film reduction for subsequent charging at lower currents. This procedure prevented a further film reduction process at the lower cathodic currents. Also, hydrogen traps will become saturated and stabilized at the highest current. Similarly, when hydrogen charging was done potentiostatically, permeation measurements were carried out by starting at the highest hydrogen over voltage and proceeding to lower over voltages. It should be noted that the highest cathodic current or over voltage was below the critical over voltage at which the irreversible damage such as cracks occur in the material.[10]

At each applied cathodic current (or over voltage), steady state hydrogen over voltage (or cathodic current) and both transient and steady state permeation fluxes was monitored. When the permeation current no longer changed with time and the steady state was reached for a specific charging current, the cathodic charging current was decreased to a new value and the permeation current was measured with time until it reached a new steady state. This stepwise procedure was repeated at many decreasing current densities. This was done in order to collect enough data to be analyzed using available analysis techniques such as the IPZ analysis. Fig. 2 shows a typical output from the permeation experiment of different charging current densities of 1.25 to 0.125 mA cm^{-2}.

5. Directly Obtained Quantities

Two quantities can be calculated from the permeation experiment with no need of modeling or any theoretical analysis. These two quantities are the hydrogen diffusivity in the metal membrane and the hydrogen concentration in the membrane at the charging side.[10,21,25,62,76]

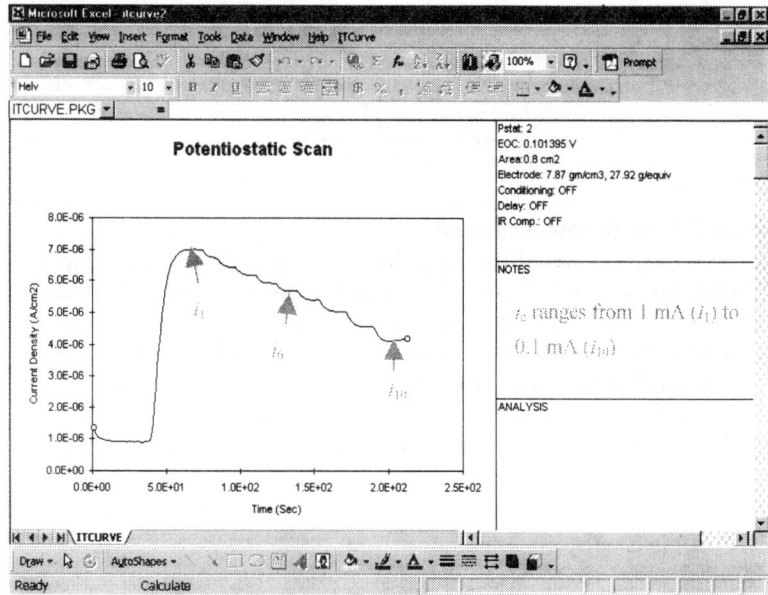

Figure 2. Typical permeation curve with different steps of the charging current. (Reprinted from: F. M. Al-Faqeer, PhD Thesis, *New Advancements in the Analysis Procedures of the Electrochemical Hydrogen Permeation Experimental Data*, The Pennsylvania State University, University Park, 2002, p. 75.[77])

6. Hydrogen Diffusivity

By monitoring the hydrogen permeation flux (in terms of the oxidation current in the anodic cell) as a function of time, extensive knowledge of the transport properties can be obtained. An important measurable parameter of the transient permeation curve is the half rise time and break through time. The lattice diffusivity (in the absence of trapping) or the effective diffusivity (in the presence of trapping) can be obtained from either of these times using different equations.[10,14,21,25,76]

The permeation experiment is the most commonly used method for measuring hydrogen diffusivity in iron and some other metals. The half rise time, $t_{1/2}$, which is the time required for the permeation flux to attain half of the steady state level is obtained directly from the measured permeation transient. Then, the hydrogen diffusivity can be obtained using,[25,62]

$$D = \frac{L^2}{0.142 \ t_{1/2}} \quad (5)$$

where L is the membrane thickness.

Using Eq. (5) the diffusivity of hydrogen on fully annealed iron was evaluated to be 4 x 10^{-5} cm^2 s^{-1}.[68] Figure 3 shows the half rise time for the permeation transient in this research for fully annealed, commercially pure iron used in this research. It gives a value for the hydrogen diffusivity of 2 x 10^{-5} cm^2 s^{-1}.

An alternative method is to measure the breakthrough time, t_b. The breakthrough time is the time it takes the hydrogen atom to diffuse

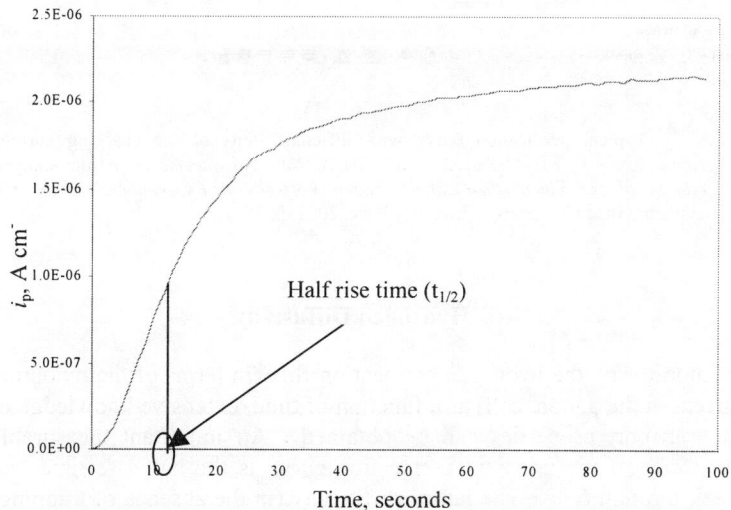

Figure 3. Half rise time from the permeation transient for the commercially pure iron used in this research at a charging current of 1.25 mA cm^{-2}. (Reprinted from F. M. Al-Faqeer, PhD Thesis, *New Advancements in the Analysis Procedures of the Electrochemical Hydrogen Permeation Experimental Data*, The Pennsylvania State University, University Park, 2002, p 75.[77])

through the membrane from the charging to the exit side. It can be calculated using Eq. (6):[25,76]

$$t_b = \frac{L^2}{15.3 \, D} \tag{6}$$

Figure 4 shows the calculation of the breakthrough time from the permeation transient. Using Eq. (6) the diffusion coefficient can then be estimated. There is no essential difference between the two methods.

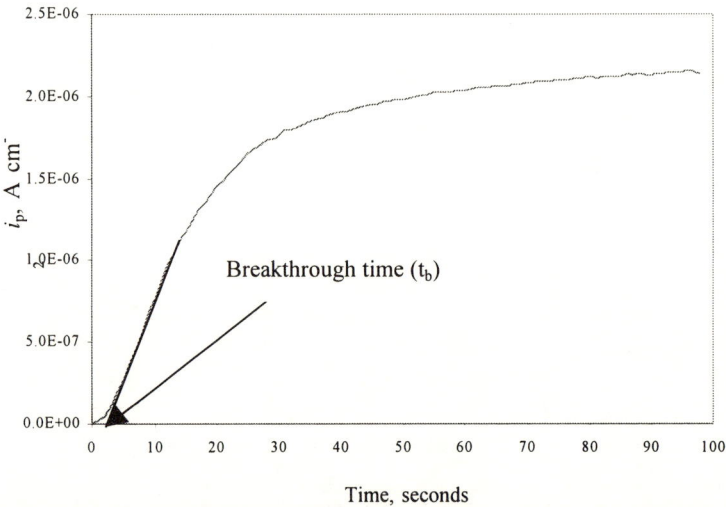

Figure 4. Breakthrough time from the permeation transient for the commercially pure iron used in this research at a charging current of 1.25 mA cm^{-2}. (Reprinted from: F. M. Al-Faqeer, PhD Thesis, *New Advancements in the Analysis Procedures of the Electrochemical Hydrogen Permeation Experimental Data*, The Pennsylvania State University, University Park, 2002, p. 75.[77])

7. Hydrogen Concentration

The hydrogen concentration inside the membrane at the charging (left) side in Fig. 1 can be calculated from the steady state permeation rate.[5,25,26,76] The concentration profile of the atomic hydrogen in the metal membrane between the charging and exit sides at steady state is shown in Fig. 5 for a concentration independent diffusivity, D. The flux of the hydrogen atoms passing through the metal membrane using Fick's first law is,

$$J = -D\frac{dC}{dx} \qquad (7)$$

where J is the steady state flux, D is the concentration independent hydrogen diffusivity, C is the concentration of hydrogen and x is the distance with $x = 0$ at the entry (charging) surface and $x = L$ at the exit surface.

Using Eqs. (8) and (9) as the boundary conditions,

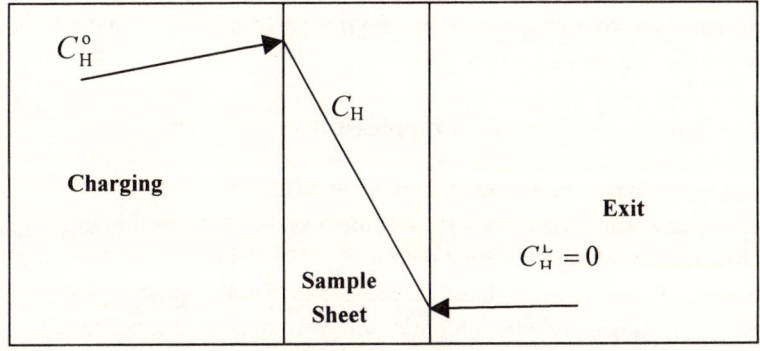

Figure 5. Concentration profile of hydrogen at steady state permeation of hydrogen within the membrane and a hydrogen diffusivity, which is independent of the hydrogen concentration.

$$C(x=0) = C_H^o \tag{8}$$

$$C(x=L) = C_H^L = 0 \tag{9}$$

The flux is obtained as,

$$J = D\frac{C_H^o}{L} \tag{10}$$

Substituting the current for flux, J (= i/nF), the steady state hydrogen permeation current density, i_∞, is,[31]

$$i_\infty = FD\frac{C_H^o}{L} \tag{11}$$

where F is the Faraday constant and i_∞ is the steady state hydrogen permeation current density.

Equation (11) can be used to estimate the absorbed hydrogen concentration at the charging side of the membrane, C_H^o, using the measured steady state permeation rate if the hydrogen diffusivity in the membrane is known.

8. Trapping Effect

The diffusivity measured at room temperature, though typically showing a lot of scatter, is still below the estimated values by extrapolation of high temperature data.[66,78] The smaller diffusivity was related in part to trapping.[78] Darken and Smith[79] investigated the hydrogen transport through cold worked steel and attributed the delayed hydrogen transport in the permeation transient to trapping. Permeability values are not influenced by trapping since the permeability is a steady state characterization of the lattice, i.e., Equation (11) holds in the presence or absence of traps. The definition of the steady state requires that the trapped and lattice hydrogen populations be in local equilibrium. Transport is then controlled exclusively by lattice parameters for annealed and deformed iron and any differences in the lattice diffusivity values for the annealed and cold worker steels. In spite of the longer time lag in the permeation transient measurements for cold worked steel, the steady state fluxes are different only to the extent the lattice diffusivities are different.[80]

There are two types of traps, non-saturable (irreversible) traps and saturable traps (reversible). Saturable traps are dislocations, impurity atoms, and internal interfaces where logic suggests that their capacity for hydrogen is finite. Stevens, et al.,[81] studied the trapping effect by determining the permeation rise time for both first and second polarizations. The second polarization in each of their experiments was identical to the first and was applied following an outgassing of the sample. They suggested that the type of traps present in each structure can be deduced from comparison of successive polarizations. The first polarization transient gives information on the concurrent behavior of irreversible and reversible traps; subsequent polarizations predominately involve the re-saturation of reversible traps. A substantial increase in the second apparent diffusivity relative to the first one indicates the presence of a significant irreversible trap density. The decrease in this relative difference at higher tempering temperature is consistent with the decrease in available carbide/matrix surface area that accompanies the coarsening process. These results are consistent with a strong influence of irreversible trapping in reducing the mobility and the bulk transport of hydrogen in these alloys.[81]

Zakroczymski[82] used the comparison of multiple charging/discharging runs to evaluate the amount of trapped hydrogen in irreversible traps. He assumed that during the first charging, both reversible and irreversible traps are active, while during a successive transient, only reversible traps are effective. On the other hand, reversible trapping is revealed clearly in the permeation decay transient after removing the hydrogen source at the entry side of the membrane.

If a steady state permeation current density is obtained, either there are no traps or all traps are filled (saturated), i.e., i_p continues to increase with time while traps are filling, but ideally i_∞ (traps) = i_∞ (without traps) if the lattice diffusivity is the same in both situations.

However, the lattice diffusivity is obtained from the breakthrough time only in the case of a trap free material, where it gives a (lower) "effective" diffusivity if traps are present. Similarly, if the half rise time or half decay time (after i_∞ is reached) is used to obtain D, the lattice and effective D are obtained in the absence and presence of traps, respectively. Hence, the value of C° is, in principle, obtained from Eq. (11) with or without traps if the lattice diffusivity (which would have to be independently known or measured for the trap-free materials) is used. In practice, this analysis may not hold up, e.g., i_∞ (traps) ≠ i_∞ (without traps).

9. Difficulties

The electrochemical hydrogen permeation technique is a very sensitive technique and getting very good reproducible results requires overcoming many difficulties. The main difficulties of using this technique and possible ways to overcome them are discussed here.

It is very important to choose an appropriate membrane thickness (L) for this type of experiment.[10] The upper limit for L is controlled by the time to attain a steady state condition. If the metal sample is too thick, the time to perform an experiment is impractically long. It can take a very long time to reach steady state (e.g., 2-3 days).[10,76] Then, other problems may be encountered. These include adsorption of impurities at the entry surface (cathode), change in concentration (especially pH) of the catholyte and background corrosion in the anodic cell. Therefore, it is always advisable to keep the time to attain steady state well within 24 hours. The thickness can be estimated if the diffusivity is known, $L_e = 2\sqrt{D_L^e t}$, where D_L^e is the estimated diffusivity of hydrogen and t is the time.[10]

If the membrane is too thin, the rate determining step of the permeation may be a surface reaction instead of diffusion of hydrogen through the membrane.[3,10,76] The applicability of the rate determining diffusion can be tested by carrying out a number of measurements of permeation at the same overpotential and different thicknesses.[76] Extremely thin membranes will be impractical to prepare or use. Also, grain boundary effects become dominant in these membranes. Grain boundaries may short circuit hydrogen diffusion and also localize the hydrogen adsorption and absorption process. This means that the hydrogen diffusivity value evaluated from the transient permeation measurement and the adsorption and absorption characteristic parameters evaluated from the steady state permeation will not be representative of the bulk material. Therefore, the lower limit of L should be at least five to ten times the average grain diameter.[10]

The metallurgical condition of the membrane is also an important consideration. Initial experiments are better carried out on well-annealed, high purity specimens in which traps for hydrogen are less of a problem so that the hydrogen diffusivity obtained from Eq. (5) or Eq. (6) will be closer to the true lattice hydrogen diffusivity. Subsequent experiments can be carried out on specimens that are quenched or tempered or having controlled additions of impurities or alloying elements and/or cold work so these effects can be carefully studied.[10]

The surface of the membrane should be flat, well polished and free of films. Oxide or other corrosion product films are easily formed on many metallic surfaces, and may affect the kinetics of the hydrogen entry and transport in metals and alloys.[10] If there is film on the evolution (entry) side, some of the charging current will be spent reducing it. If significant surface film reduction has occurred, the breakthrough time will be longer than normal since time will occur before the hydrogen diffusion commences. Also, the calculated diffusion coefficient will be lower than normal.[10,76]

Anodic dissolution of the metal on the exit side may take place. Usually the potential applied at the exit side is oxidizing to the membrane surface. One solution is to plate out a thin layer of palladium onto the anodic side. This layer protects the iron from dissolution and offers negligible resistance against the throughput of hydrogen if sufficiently thin.[10,76]

The membrane should not have internal damage before or during the experiments. Internal damage in the membrane can occur during the permeation experiment if the charging current or overpotential is very high. In this case, hydrogen charging will be high and concentration of the absorbed hydrogen will exceed the critical one at which damage will occur. This greatly affects the reproducibility of results.[10]

Impurities in the solution have a big effect on the results and their reproducibility. It is important to keep all impurities in the solution to the lowest possible level because many impurities can adsorb on the surface and produce erroneous results. Therefore, pre-electrolysis is required to clean the catholyte of all impurities in order to overcome this problem.[10,76,83-89]

10. Advantages of the Technique

The electrochemical hydrogen permeation method has many advantages. The first advantage of this system is that it is very simple and reasonably inexpensive to set up. The system requires no calibration before or after the experiments. The technique is very sensitive and therefore very low hydrogen permeation fluxes or currents can be measured.[76]

V. IPZ ANALYSES

Hydrogen absorption and evolution reactions on metals have been studied extensively in the literature after the development of the electrochemical hydrogen permeation cell.[2,3,5-12,14,19,20,22-26,28-34,36-39,42,50-52,54-71,76,82,90-103] Correlation of the hydrogen absorption reaction with the hydrogen evolution reaction was carried out in different studies.[3,6,10,19,20,22-24,26,29,30,32,36,38,50-52,66,76,90] Iyer, et, al., developed the IPZ analysis in 1989 to evaluate the steady state hydrogen permeation data and describe the kinetics of the hydrogen evolution and absorption reactions.[50] The analysis was also able to evaluate the hydrogen surface coverage, which is an important parameter that determines the amount of hydrogen that enters into metals.

1. The Original IPZ Analysis

The original IPZ analysis developed a set of equations, which relate the charging, i_c, recombination, i_r, and steady state hydrogen permeation, i_∞, current densities and the charging overpotential, η.

The original IPZ analysis assumes that the hydrogen evolution reaction (HER) proceeds according to a coupled discharge-recombination (Volmer-Tafel) mechanism (Eqs. 1 and 2). Langmuir conditions were assumed for the hydrogen surface coverage. The analysis also assumes that the charging current density is the sum of the recombination and steady state permeation current densities:[10,36,50]

$$i_c = i_r + i_\infty \tag{12}$$

The original analysis does not consider the adsorption of other species on the surface.

The hydrogen absorption reaction will take place as a side reaction of the HER. Some of the adsorbed hydrogen atoms will absorb (H_{abs}) into the metal lattice and quickly establish the back equilibria (Eq. 4).

The original IPZ analysis expresses the current density of the first step of the HER (reaction 1) using the Langmuir adsorption isotherm as follows[10,19,20,22-24,29,30,36,50]

$$i_c = Fk_1C_{H^+}(1-\theta_H)\exp(-a\alpha\eta) \tag{1}$$

where C_{H^+} is the hydrogen ion concentration, a = F/RT where F, R and T have their usual meanings, α is the transfer coefficient and η is the hydrogen overpotential ($\eta = E - E_{rev}$, where E_{rev} is the equilibrium potential of the HER). The physical parameters, k_1, C_{H^+} and α in Eq. (13) are presumed to be constants for a given system under charge transfer control of the HER, i.e., only i_c and θ_H increase when η is increased.

Equation (13) was rearranged in the analysis to,

$$i_c\exp(a\alpha\eta) = i_o'(1-\theta_H) \tag{14}$$

where $i_c\exp(a\alpha\eta)$ is referred to as the charging function and is solely determined by conditions at the charging side of the membrane, and i_o' is given by,

$$i_o' = Fk_1C_{H^+} = i_o/(1-\Delta_H^e) = i_o \text{ for } \Delta_H^e \ll 1 \tag{15}$$

where i_o is the exchange current density of the hydrogen evolution reaction and Δ_H^e is the equilibrium hydrogen surface coverage ($\Delta_H^e \approx 0$).

The rate of the recombination step of the hydrogen evolution reaction (reaction 2) was given by the analysis as[10,19,20,22-24,29,30,36,50]

$$i_r = Fk_2\theta_H^2 \tag{16}$$

Equation (11) gives the steady state hydrogen permeation current density.

Kim and Wilde,[38] showed that the steady state hydrogen permeation current, i_∞, is also given by,

$$i_\infty = F k_{abs} \Delta_H - F k_{des} C_H^o \tag{17}$$

The IPZ analysis uses Eq. (17) in the development of its main relations.[10,19,20,22-24,29,30,36,50] Substituting C_H^o from Eq. (11) into Eq. (17) gives

$$i_\infty = F k_{abs} \Delta_H - F k_{des} \frac{L i_\infty}{FD} \tag{18}$$

Rearranging Eq. (18) one obtains,[10,24,50]

$$i_\infty = F \frac{k_{abs}}{1 + \dfrac{L k_{des}}{D}} \Delta_H \tag{19}$$

The kinetic-diffusion constant, k, is expressed in terms of the adsorption and absorption rate constants as,[10,19,20,24,50,63]

$$k = \frac{k_{abs}}{1 + k_{des} \dfrac{L}{D}} \tag{20}$$

The dimensionless group ($k_{des} L / D$) can be interpreted as the relative rate of the reverse absorption and the diffusion process. Equation (19) can be simplified using the kinetic-diffusion constant, k, to,

$$i_\infty = F\, k\, \Delta_H \tag{21}$$

Equation (21) relates the steady state hydrogen permeation current density to the hydrogen surface coverage, θ_H. Equations (20) and (21) are two major relations in the IPZ analysis.[10,19,20,24,50]

Substituting the hydrogen surface coverage from Eq. (16) into Eq. (21) and rearranging gives,[10,19,20,24,50]

$$i_\infty = k \sqrt{\frac{F}{k_2}} \sqrt{i_r} \qquad (22)$$

Equation (22) shows the relation between the steady state hydrogen permeation current density, i_∞, and the square root of the hydrogen recombination current density, $\sqrt{i_r}$. This relation was previously observed experimentally by Dafft, et al., in 1979.[31] The relation should be a straight line passing through the origin with a slope of $k\sqrt{F/k_2}$ for the application of the IPZ analysis. This relation is schematically illustrated in Fig. 6.

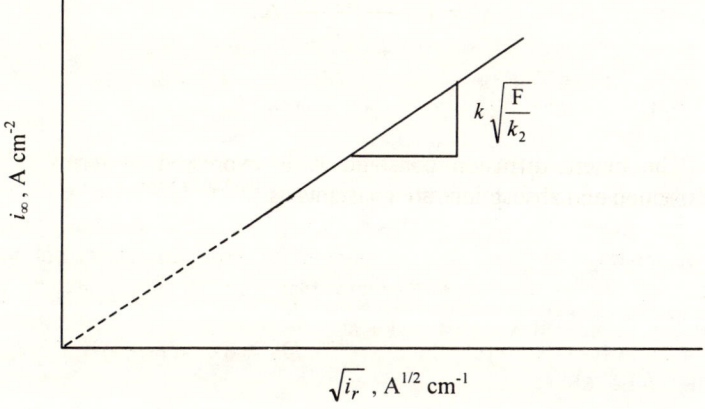

Figure 6. A Schematic Illustration of Eq. (22). (Reprinted from: M. H. Abd Elhamid, B. G. Ateya and H. W. Pickering, *J. Electrochem. Soc.*, **144** (199)L58.[22])

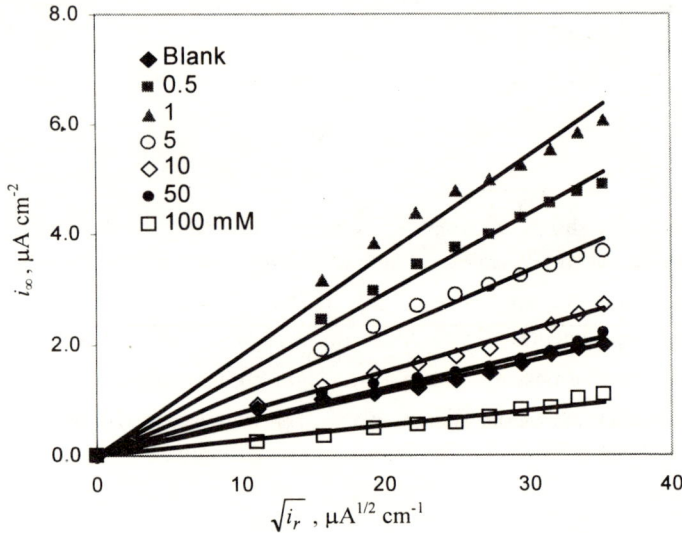

Figure 7. The relation between i_∞ and $\sqrt{i_r}$ for iron in 1M Na_2SO_4 (pH 2) with different HMTA concentrations. (Repinted from: F. M. Al-Faqeer, K. G. Weil and H. W. Pickering, *J. Electrochem. Soc.*, in press. [104])

If the experimental data do not show a straight-line relation passing by the origin, the IPZ analysis cannot be used to analyze the experimental data since one or more of its requirements is not met. Figure 7 shows experimental data that approximately satisfy the requirements of Eq. (22).[104]

Similarly, substituting the hydrogen surface coverage, θ_H, from Eq. (21) into Eq. (14) gives

$$i_c \exp(a\alpha\eta) = i_o(1 - \frac{1}{Fk} i_\infty) \qquad (23)$$

Equation (23) shows the relation between the charging function, $i_c\exp(a\alpha\eta)$ and the steady state hydrogen permeation current density,

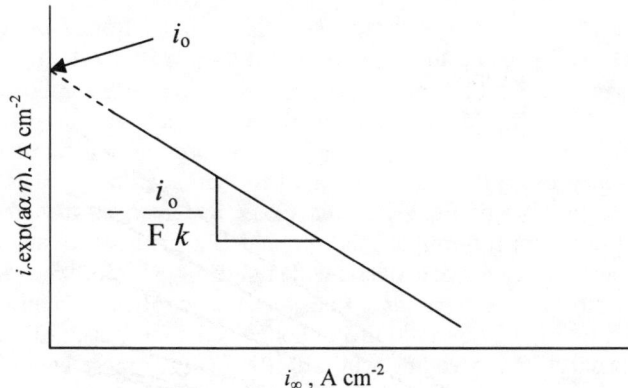

Figure 8. A schematic illustration of Eq. (23). (Reprinted from: F. M. Al-Faqeer and H. W. Pickering, *J. Electrochem. Soc.*, **148** (2001) E248.[23])

i_∞. This relation should be a straight line with a slope of $-i_o'/Fk$ and an intercept of i_o. This is another requirement of the original IPZ analysis. A schematic representation of the relation is shown in Fig. 8. Figure 9 and Table 1 show experimental data for different HMTA concentrations that approximately satisfy the requirements of Eq. (22).[19,104]

The original IPZ analysis revealed the major relations, (20), (21), (22) and (23), between the various variables in the hydrogen permeation system. The kinetic-diffusion constant, k, and i_o can be found from the slope and intercept of Eq. (23). The recombination reaction rate constant, k_2, can be evaluated using the estimated kinetic-diffusion constant, k, and the slope of Eq. (22).

Table 1
Values of i_o, $-i_o/Fk$, k (Eq. 23) and the % Surface Control for Iron Membranes of Different Thickness in 0.1N H2SO4 + 0.9N Na2SO4 at Room Temperature (pH 1.8)[19]

L, mm	i_o, A cm^{-2}	$-i_o/Fk$	k, mol cm^{-2} s^{-1}	Surface control, %
0.10	2.0x10^{-6}	0.32	6.6x10^{-11}	38
0.25	2.0x10^{-6}	0.40	5.4x10^{-11}	20
0.50	2.0x10^{-6}	1.20	2.1x10^{-11}	10
0.85	2.0x10^{-6}	1.45	1.4x10^{-11}	7

The discharge rate constant, k_1, can be found from the exchange current density using the formula, $i_o = Fk_1C_H$. The hydrogen surface coverage, θ_H, can be evaluated using the kinetic-diffusion constant, k, and Eq. (21). Therefore, these kinetic and thermodynamic parameters of both the hydrogen evolution and absorption reactions can be determined using the IPZ analysis when the experimental data satisfy the requirements of Eqs. (22) and (23). See Iyer, et. al.[50] for some values obtained by applying the original IPZ analysis to steady state hydrogen permeation data over a wide range of pH for three different iron/electrolyte systems and one nickel/electrolyte system. It is important to note, however, that although satisfying Eqs. (22) and (23) is necessary for proceeding with the IPZ analysis, these conditions do not say anything about whether or not the parametric values obtained from the analysis will be the true values. Thus, comparisons

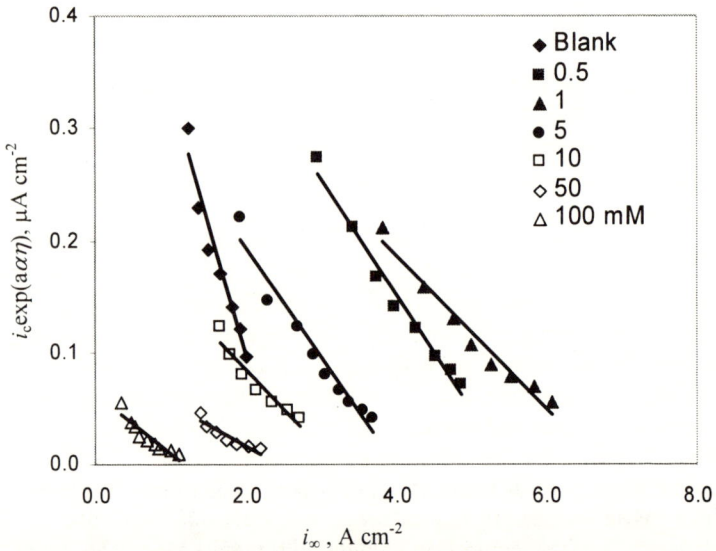

Figure 9. The relation between the charging function, $i_c\exp(a\alpha\eta)$, and i_∞ for iron in 1M Na_2SO_4 (pH 2) with different HMTA concentrations. (Repinted from: F. M. Al-Faqeer, K. G. Weil and H. W. Pickering, *Journal of Electrochemical Society*, in press. [104])

with values obtained by other independent methods are needed but are generally unavailable. One exception is for values of the coverage of the surface by a second adsorbate, θ_A, in addition to θ_H for hydrogen. The comparisons for θ_A using three different independent techniques are favorable and are discussed below in connection with the new IPZA analysis.

2. Membrane Thickness Effect

The IPZ analysis can be used to evaluate the absorption and desorption rate constants, k_{abs} and k_{des}. Rearranging Eq. (20) can give a relation between the kinetic-diffusion constant, k, and the membrane thickness, L.[10,19,50,63]

$$\frac{1}{k} = \frac{1}{k_{abs}} + \frac{L}{D}\frac{k_{des}}{k_{abs}} \qquad (24)$$

The contribution of the first term to the sum of the two terms on the right side of Eq. (24) is a measure of the amount of surface control during the absorption-permeation process.[19]

Permeation experiments need to be done at different metal membrane thicknesses, L, in order to use Eq. (24). Once the kinetic-diffusion constants, k, are obtained for different membrane thicknesses, L, Eq. (24) can be used to obtain k_{abs} and k_{des}. Equation (24) shows a straight line between the reciprocal of the kinetic-diffusion constant, $1/k$, and the membrane thickness, L. The slope of this relation will give the value of the desorption constant, k_{des}. The intercept will give the absorption rate constant, k_{abs}. This relation is illustrated in Fig. 10.

From the slope and intercept, $k_{abs} = 2 \times 10^{-10}$ mol cm^{-2} s^{-1} and $k_{des} = 1.9 \times 10^{-3}$ cm s^{-1} for the values of k obtained from the IPZ analysis for iron membranes of different thickness given in Table 1. The extent to which these values of k_{abs} and k_{des} may vary with L due to the significant surface control in the thinner membranes is a point of further discussion. These values are compared to literature values in Abd Elhamid, et. al.,[19] e.g., the literature gives $k_{des} = 10^{-3}$ to 10^{-2} cm s^{-1} over a wide range of pH.[26,105,106] For k_{abs}, only estimates could be obtained from related quantities which yielded values of $k_{abs} = 10^{-8}$ to 10^{-9} mol cm^{-2} s^{-1}.[38,105,107]

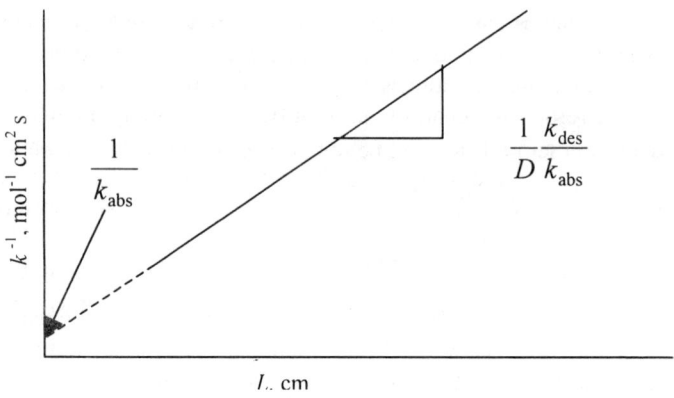

Figure 10. A schematic illustration of Eq. (24). (Reprinted from F. M. Al-Faqeer and H. W. Pickering, in: *International Conference on Hydrogen Effects on Materials Behavior and Corrosion Deformation Interactions*, Ed. by N. R. Moody, TMS, Warrensville, PA.)[24]

Abd Elhamid, et al.,[19] also shows in Fig. 5 of their paper that θ_H does not depend within the experimental error on the thickness of the membrane, L. Thus, θ_H may strongly depend only on the overpotential, as is expected for certain mechanisms of hydrogen evolution and Langmuir adsorption conditions, e.g., at $\theta_H \ll 1$.[2,38,108] Since i_∞ increases as L decreases,[19] i_r decreases at constant i_c in accord with Eq. (12) ($i_c = i_r + i_\infty$). It follows from Eq. (16) ($i_r = Fk_2\theta_H^2$) that k_2 also decreases as L decreases, although the qualitative understanding of such an interaction needs further consideration. Unfortunately, no results were given in Abd Elhamid, et. al. paper[19] regarding a k_2 vs. L relationship.

3. Calculation of θ_H and Rate Constants of the HER from Polarization Data

Abd Elhamid, et al.[30] showed that the hydrogen surface coverage, θ_H, and the rate constants of the hydrogen discharge and recombination reactions, k_1 and k_2, respectively, can be obtained by modifying the original IPZ analysis and applying the spin-off analysis procedure to the polarization curve of the HER. They assumed in their study that the permeation current density, i_∞, is very low in the substrate material, e.g., as in the case of some metals like copper, and iron in solutions free of species that promote the absorption of hydrogen such as H_2S or As_2O_3. Therefore, the charging current density, i_c, will be equal to the recombination current density, i_r, so that $i_r = i_c = Fk_1\theta_H^2$. This analysis has the beauty of requiring only the most basic of electrochemical equipment found in most corrosion and electrochemical laboratories, namely the potentiostat.

Abd Elhamid, et al.[30] developed Eq. (25) that shows a straight line relation between $i_c\exp(a\alpha\eta)$ and $\sqrt{i_c}$. A schematic illustration is shown in Fig. 11. They used this relation to calculate the exchange current density, i_o, and k_2. Then, k_1 was calculated from the exchange current density using $i_o = Fk_1C_{H^+}$. The hydrogen surface coverage, θ_H, was calculated from $\theta_H = \sqrt{i_r}/\sqrt{Fk_2}$:

$$i_c\exp(a\alpha\eta) = i_o(1 - \frac{\sqrt{i_c}}{\sqrt{Fk_2}}) \tag{25}$$

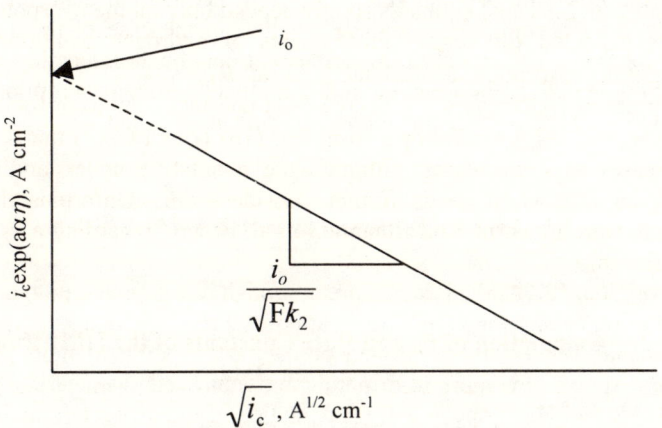

Figure 11. A schematic illustration of Eq. (25). (Reprinted with the permission of M. A. V. Devanathan and Z. Stachurski, *Proceedings of the Royal Society*, **A270** (1962) 90.[25])

Table 2
Rate Constants and Transfer Coefficient for HER on Iron and Copper in 0.1N H_2SO_4 + 0.9N Na_2SO_4 at Room Temperature Obtained by the Polarization Curve Method. Comparison with Values Obtained from Permeation Data Using the IPZ Analysis are Also Shown for Iron in the Same Solution[30]

Analysis	i_o, A cm^{-2}		k_2, mol cm^{-2} s^{-1}		K_1, cm s^{-1}		α	
	Fe	Cu	Fe	Cu	Fe	Cu	Fe	Cu
IPZ	2.0×10^{-6}	-	2.0×10^{-8}	-	1.3×10^{-6}	-	0.5	-
Polarization Data	2.0×10^{-6}	8.0×10^{-8}	3.0×10^{-8}	1.8×10^{-7}	1.3×10^{-6}	5.0×10^{-8}	0.5	0.5

The values of k_1 and k_2 obtained by applying this simplified analysis method to the measured polarization curves for copper and iron in 0.1 N H_2SO_4 - 0.9 N Na_2SO_4 are given in Table 2.[30] The hydrogen coverage, θ_H, was also obtained as a function of overpotential for these systems by this simplified method. A good agreement was found for these same variables when compared with values obtained by application of the original IPZ analysis to permeation data, e.g., see Table 2 for a comparison with IPZ values in the case of permeation data for iron. The extent of the agreement may be an indication of the extent to which the important assumption that hydrogen absorption was negligible in the polarization curve experiments was met.

VI. THE GENERALIZED IPZ ANALYSIS

The original IPZ analysis was a breakthrough development in hydrogen adsorption, absorption and permeation studies. The analysis, where applicable, renders the hydrogen permeation experiment more useful since it shows how more information and improved characterization of the HAR and HER, specifically, calculation of the different kinetic and thermodynamic parameters of these reactions, can be obtained from the permeation measurements. However, the original IPZ analysis has some limitations, which include the assumption of the Langmuir adsorption isotherm for the adsorbed hydrogen atoms. This isotherm is usually applicable at low values of coverage ($\theta < 0.1$) and at coverages close to unity ($\theta > 0.9$).[23,24,27,102,109-111] This assumption was actually found to be not so limiting since in some instances the original IPZ analysis was found to be applicable also for intermediate values of hydrogen surface coverage of hydrogen on iron surfaces.[19,20,30,50] According to this isotherm, the surface is uniform and all sites are equivalent. This isotherm also assumes no interaction between the adsorbed hydrogen atoms on the surface.

There were some metal/electrolyte systems where the IPZ requirements were not met and hence the original analysis could not be used to evaluate the different parameters of the hydrogen evolution and adsorption reactions. Figure 12 shows that the zero-intercept requirement of Eq. (22) was not met in the presence of hydrogen sulfide in acidic media.[24]

This illustrated the need to modify the analysis to be more general and applicable. One idea was to apply conditions for the Frumkin adsorption isotherm instead of those of the Langmuir isotherm, in spite of the fact that there is no independent evidence that the Frumkin approach is a better physical description than the Langmuir approach for hydrogen adsorption under these conditions. The Frumkin isotherm assumes the surface to be heterogeneous and that the adsorbed hydrogen atoms will interact. It adds an interaction parameter to the kinetic rate expressions, and thus another parameter to be evaluated by the IPZ analysis.

It has been shown by Frumkin for hydrogen layers adsorbed on a metallic electrode that in the case of medium coverages of the surface, the best approximation is to assume the activity of the adsorbed hydrogen to be directly proportional in a wide interval to the quantity $e^{f\theta}$, where f is a constant.[27]

The Frumkin adsorption isotherm assumes that the free energy of adsorption of a species decreases with coverage according to the following equation,[23,24,27,109-111]

$$\Delta G_\theta^0 = \Delta G_0^0 + f\,R\,T\theta \tag{26}$$

where f is a dimensionless factor that describes the deviation from the ideal Langmuir behavior. The fRT quantity is sometimes called the rate of change of the standard free energy of adsorption. ΔG_θ^0 and ΔG_0^0 are the standard free energies of adsorption at a certain coverage, θ, and

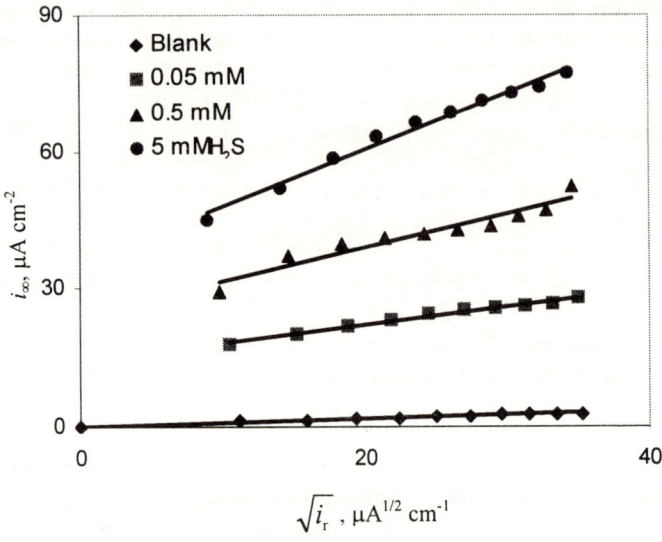

Figure 12. The relation between i_∞ and $\sqrt{i_r}$ obtained at different hydrogen sulfide concentrations at pH 2 illustrating a strongly non-zero intercept when the solution contains sulphide ion. (Reprinted from F. M. Al-Faqeer and H. W. Pickering, in: *International Conference on Hydrogen Effects on Materials Behavior and Corrosion Deformation Interactions*, Ed. by N. R. Moody, TMS, Warrensville, PA, to be published.[24])

at zero coverage, $\theta = 0$, respectively, R is the gas constant and T is the absolute temperature. The Langmuir adsorption isotherm assumes that the standard free energy of adsorption is independent of coverage (i.e., $f = 0$).

Iyer, et al.,[36] had some success in applying the Frumkin adsorption isotherm to the IPZ analysis, but they had to assume a value for f since the developed relations were not able to calculate this parameter. Recently, Al-Faqeer, et al.[23] were able to modify the original IPZ analysis using the Frumkin adsorption isotherm that evaluates all parameters including f. In other words, using the modified relations it is not necessary to assume a value for f. This modification is discussed in detail below.

Kinetics of Electrochemical Hydrogen Entry into Metals and Alloys

Writing the rate equations in view of the Frumkin adsorption conditions, one has for the rate of proton discharge and hydrogen evolution,[10,23,24,27,36]

$$i_c = i_o(1 - \theta_H)\exp(-a\alpha\eta)\exp(-\alpha f \theta_H) \tag{27}$$

$$i_r = Fk_2\theta_H^2 \exp(2\alpha f \theta_H) \tag{28}$$

Inserting θ_H from Eq. (21) into Eq. (28),[23,24]

$$i_r = Fk_2\left(\frac{i_\infty}{Fk}\right)^2 \exp\left(2\alpha f \frac{i_\infty}{Fk}\right) \tag{29}$$

Taking the square root of Eq. (29) and rearranging,[23,24]

$$\sqrt{i_r} = \sqrt{\frac{k_2}{F}} * \frac{i_\infty}{k} * \exp\left(\alpha f \frac{i_\infty}{Fk}\right) \tag{30}$$

Dividing both sides of Eq. (30) by i_∞ and taking the natural logarithm (ln),[23,24]

$$\ln\left(\frac{\sqrt{i_r}}{i_\infty}\right) = \ln\left(\sqrt{\frac{k_2}{F}} * \frac{1}{k}\right) + \frac{\alpha f}{Fk}i_\infty \tag{31}$$

Equation 31, which was also derived by Iyer, et al.,[10,36] in a similar form, but with different slope and intercept, can be used for determining the rate constants when the value of f is known or can otherwise be estimated. According to this relation, a plot of $\ln\left(\sqrt{i_r}/i_\infty\right)$ vs. i_∞ should show a straight line with a slope of $\alpha f/Fk$ and an intercept of $\ln\left(\sqrt{k_2/F}/k\right)$, see Fig. 13.

Recently, Al-Faqeer, et. al.,[23] arrived at another relationship which, when applied to the steady state permeation data, can yield the rate constants without a prior knowledge of the value of f. In fact, the

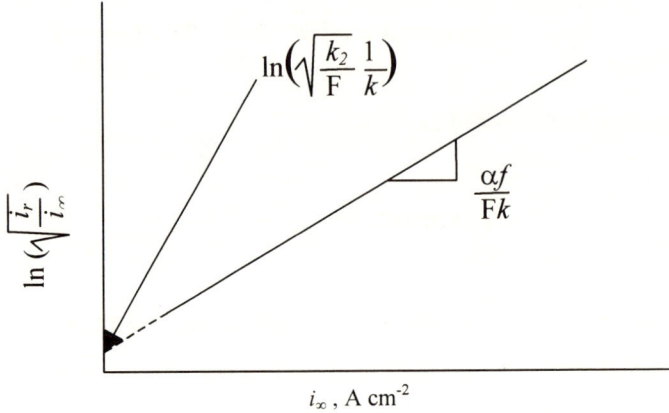

Figure 13. A schematic Illustration of Eq. (31).

value of f is a result of the analysis.[23,24] This derivation is discussed in detail below.

Rearranging Eq. (27) and solving for $\exp(\alpha f \theta_H)$,

$$\exp(\alpha f \theta_H) = \frac{i_o'(1-\theta_H)}{i_c \exp(a\,\alpha\eta)} \quad (32)$$

Taking the square root of Eq. (28),

$$\sqrt{i_r} = \sqrt{F\,k_2}\,\theta_H \exp(\alpha f \theta_H) \quad (33)$$

Substituting $\exp(\alpha f \theta_H)$ from Eq. (32) into Eq. (33),

$$\sqrt{i_r} = \sqrt{F\,k_2} * \theta_H * \frac{i_o(1-\theta_H)}{i_c \exp(a\,\alpha\eta)} \quad (34)$$

Rearranging and inserting for θ_H from Eq. (21) into Eq. (34),

Kinetics of Electrochemical Hydrogen Entry into Metals and Alloys

$$\sqrt{i_r}\; i_c \exp(a\alpha\eta) = \sqrt{Fk_2} * \frac{i_\infty}{Fk} * i_o (1 - \frac{i_\infty}{Fk}) \qquad (35)$$

Dividing both sides of Eq. (35) by i_∞,

$$\frac{\sqrt{i_r}}{i_\infty}\; i_c \exp(a\alpha\eta) = \sqrt{\frac{k_2}{F}} * \frac{i_o'}{k}(1 - \frac{1}{Fk} i_\infty) \qquad (36)$$

Plots of $(\sqrt{i_r}/i_\infty) i_c \exp(a\alpha\eta)$ vs. i_∞ should give a straight line with a slope of $-\sqrt{k_2/F} * i_o/Fk^2$, and an intercept of $\sqrt{k_2/F} * i_o/k$. This relation is schematically shown in Fig. 14.

Equations (31) and (36) can be simplified to Eqs. (22) and (23) if the Langmuir adsorption conditions are assumed instead of the Frumkin adsorption conditions. In other words, this generalized IPZ analysis reduces to the original Langmuir IPZ analysis if the standard free energy of adsorption is assumed to be independent of the hydrogen

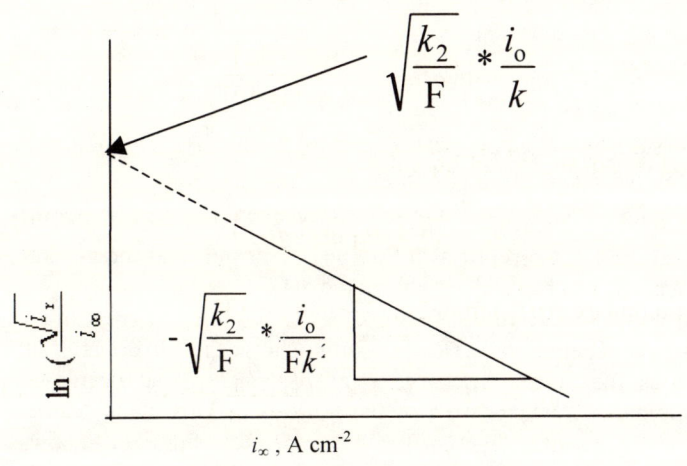

Figure 14. A schematic illustration of Eq. (36).

coverage, i.e., $f = 0$. Thus, the original IPZ analysis is a special case of the generalized IPZ analysis, in the same way the Langmuir isotherm is a special case of the Frumkin isotherm.

The procedure for calculating the unknowns using the generalized IPZ analysis is:

1. From the slope and intercept of Eq. (36), k can be calculated.
2. From the slope of Eq. (31) and knowing k from step 1, f can be calculated, after using the charging (Tafel) data to obtain the value of α.
3. From the intercept of Eq. (31) and knowing k from step 1, k_2 can be calculated.
4. From the intercept of Eq. (36) and knowing k from step 1 and k_2 from step 3, i_o can be calculated.
5. From Eq. (21) and knowing k from step 1, the hydrogen surface coverage (θ_H) can be calculated.

This procedure was used successfully to evaluate the different rate constants, k_1 and k_2, the kinetic-diffusion constant, k, the hydrogen surface coverage, θ_H, exchange current density, i_o, and f for metal/electrolyte systems that did not satisfy the original IPZ requirements. The parameter f was found to be constant for each of these systems.[21,23,24] Experimental data in the presence of hydrogen sulfide will be used as an example of these metal/electrolyte systems as indicated by Fig. 12.

Figure 15 shows the relation between $\ln(\sqrt{i_r}/i_\infty)$ and the measured steady state hydrogen permeation current density, i_∞. The figure shows straight lines in accord with Eq. (31).

Figure 16 shows linear relations between $(\sqrt{i_r}/i_\infty)i_c \exp(a\alpha\eta)$ and the measured i_∞ in accord with Eq. (36). This indicates the generalized IPZ analysis can be used to analyze the data.[24]

Table 3 shows values for i_o, k_1, k_2, k and f at different hydrogen sulfide concentrations. These results show that hydrogen sulfide increases the proton discharge rate constant, k_1, and decreases the recombination rate constant of the HER, k_2. This could lead to an increase in the hydrogen surface coverage, θ_H. Figure 17 shows the relations between θ_H, and cathodic potential, E. It reveals that the θ_H increases with the increase in both hydrogen sulfide concentration and

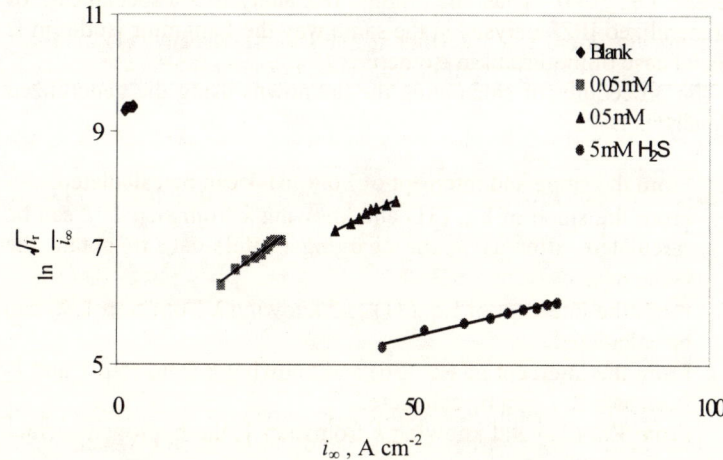

Figure 15. The relation between ln $(\sqrt{i_r}/i_\infty)$ and i_∞ obtained at different H_2S concentrations at pH 2. (Reprinted from F. M. Al-Faqeer and H. W. Pickering, in: *International Conference on Hydrogen Effects on Materials Behavior and Corrosion Deformation Interactions*, Ed. N. R. Moody, TMS, Warrensville, PA, to be published.)

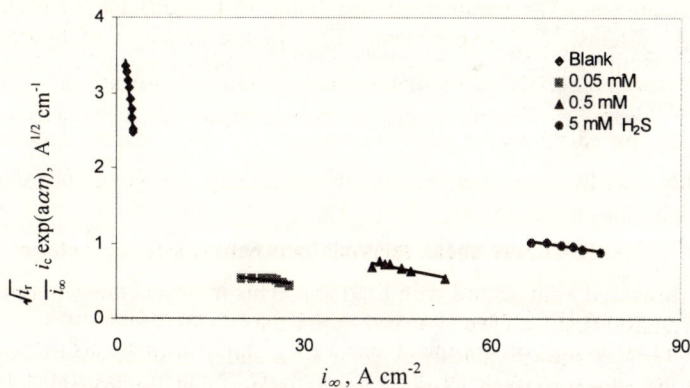

Figure 16. Data The relation between $(\sqrt{i_r}/i_\infty)i_c \exp(a\alpha\eta)$ and i_∞ obtained at different hydrogen sulfide concentrations at pH 2.(Reprinted from F. M. Al-Faqeer and H. W. Pickering, in: *International Conference on Hydrogen Effects on Materials Behavior and Corrosion Deformation Interactions*, Ed. N. R. Moody, TMS, Warrensville, PA, to be published.[24])

Table 3
Values of k, k_2, i_o, k_1 and f Obtained on Iron Membrane at Different Hydrogen Sulfide Concentrations at pH 2[24]

H$_2$S Content	k mol cm^{-2} s^{-1}	k_2 mol cm^{-2} s^{-1}	i_o µA cm^{-2}	k_1 cm s^{-1}	f
Blank	7.02 x 10^{-11}	4.96 x 10^{-8}	0.421	4.36 x 10^{-7}	0.49
50 µM	6.61 x 10^{-10}	2.82 x 10^{-9}	2.78	2.89 x 10^{-6}	8.94
500 µM	9.62 x 10^{-10}	2.66 x $^{-9}$	9.22	9.56 x 10^{-6}	8.87
5 mM	1.43 x 10-9	9.24 x 10^{-10}	26.1	2.70 x 10-5	8.26

the potential in the cathodic direction, in agreement with earlier work, e.g., Iyer, et al.,[36] for hydrogen sulfide and McBreen and Genshaw[2] and Kim and Wilde[38] for overpotential.

Table 3 also shows that hydrogen sulfide affects f. The value of f increased from almost zero (0.49) in the absence of hydrogen sulfide to nearly 8.7 in the presence of hydrogen sulfide. This indicates that the standard free energy of adsorption is coverage dependent, and that the Frumkin adsorption isotherm was correctly applied.

The increase in θ_H and f, dominate over the decrease in the hydrogen recombination rate constant of the HER, k_2, which leads to an increase in the hydrogen evolution reaction rate. However, the increase in the hydrogen absorption reaction rate is attributed to increases in θ_H and k.

VII. THE NEW IPZA ANALYSIS

The original IPZ analysis assumes that hydrogen atoms are the only adsorbed species on the metal surface.[10,50] The IPZ analysis, however, has been used to evaluate the hydrogen evolution reaction (HER) and the hydrogen absorption reaction (HAR) in the presence of organic and inorganic inhibitors such as benzotriazole (BTA)[112] and iodide ions[20] under conditions where these inhibiting species might compete with hydrogen for the available adsorption sites on the metal surface.

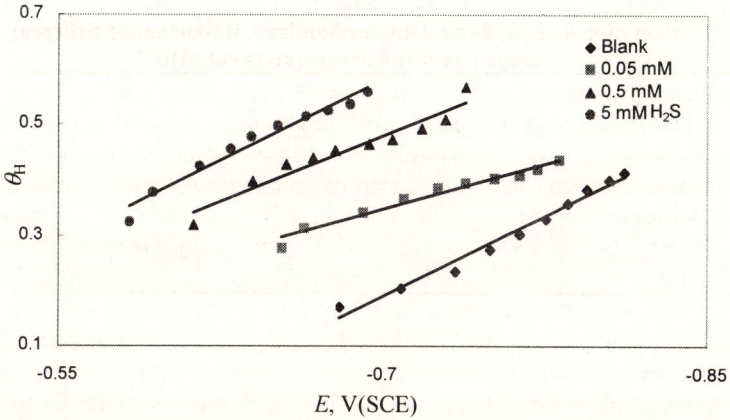

Figure 17. The relations between θ_H and E obtained at different hydrogen sulfide concentrations at pH 2. (Reprinted from: F. M. Al-Faqeer and H. W. Pickering, in: *International Conference on Hydrogen Effects on Materials Behavior and Corrosion Deformation Interactions*, Ed. by: N. R. Moody, TMS, Warrensville, PA, to be published.[24])

Most organic inhibitors of corrosion under open circuit conditions function by adsorbing on the surface. The adsorption of inhibitors is influenced by the nature and the surface charge of the metal, by the type of the aggressive electrolyte and by the chemical structure of the inhibitors.[113] In the case of hydrogen permeation studies where the surface is under cathodic polarization, if these inhibitors are present in the electrolyte, they could adsorb on the membrane's surface and in turn decrease the free sites available for hydrogen adsorption. Therefore, the calculated hydrogen surface coverage using the original IPZ analysis, which does not take into account the presence of a second adsorbing species (besides hydrogen), will be larger than the actual hydrogen surface coverage.

Thus, the need to include the second adsorbate coverage in the original IPZ analysis of the hydrogen permeation data is clear. Through a careful study of the original IPZ analysis, Al-Faqeer, et al.,[77,104,114] were able to modify the analysis to include the coverage of the second adsorbate (competitive adsorption). This analysis procedure was named IPZA (Iyer/Pickering/Zamanzadeh/Al-Faqeer).[104,114] The new IPZA analysis was used to evaluate all the relevant parameters including the coverage of the second adsorbate, which were hexamethylenetetramine and iodide ions in these experiments.

The IPZA analysis, like the IPZ analyses, assumes that the HER proceeds according to a coupled discharge-recombination (Volmer-Tafel) mechanism. Langmuir adsorption conditions are assumed for the hydrogen surface coverage in the new IPZA analysis[77,104,114] while the IPZ analyses are available for both Langmuir and Frumkin-Temkin adsorption conditions.[21,23,24] The IPZA analysis similarly assumes that the charging current density is the sum of the recombination and steady state permeation current densities ($i_c = i_r + i_\infty$). The IPZA analysis, unlike the IPZ analyses, recognizes that a portion of the surface may be covered by solution species other than hydrogen as a result of a competitive adsorption.

Writing the rate equations in view of the Langmuir adsorption isotherm as was assumed by the original IPZ analysis[10,50] and considering the surface coverage of the second adsorbate on the membrane surface, θ_A, one has for the discharge step of the HER,

$$i_c = Fk_1 C_{H+} (1-\theta_A - \theta_H) \exp(-a\alpha\eta) \tag{37}$$

Rearranging, Eq. (37) one has,

$$i_c \exp(a\alpha\eta) = i_o' (1-\theta_A-\theta_H) \tag{38}$$

For the recombination step, Eq. (16) of the original IPZ analysis is taken to be still valid, i.e., it is assumed not to be affected by the surface coverage of the second adsorbate, since it is the recombination of two hydrogen atoms, which results in this current. Also Eq. (22) of the original IPZ analysis is assumed not to be affected by the surface coverage of the second adsorbate.

Substituting the hydrogen coverage, θ_H, from Eq. (22) into Eq. (38) and taking $\theta_H^c \approx 0$ gives,

$$i_c \exp(a\alpha\eta) = i_o (1-\theta_A - \frac{1}{Fk} i_\infty) \tag{39}$$

Rearranging Eq. (39),

$$i_c \exp(a\alpha\eta) = i_o(1-\theta_A) - \frac{i_o}{Fk} i_\infty \qquad (40)$$

Equation (40) shows the relation between the charging function, $i_c\exp(a\alpha\eta)$ and the steady state hydrogen permeation current density, i_∞. This relation should be a straight line with a slope of $-i_o/Fk$ and an intercept of $i_o(1-\theta_A)$. This is similar to the relation of the original IPZ (Eq. 23) but different since it includes the surface coverage of the second adsorbate.

This added another parameter to the usual ones evaluated by the original IPZ analysis. There are two parameters in the intercept of Eq. (40): the exchange current density, i_o, and the surface coverage of the second adsorbate, θ_A. De-convolution of these two parameters is a complication. However, if one of these parameters can be estimated or determined using another relation or another technique, all other kinetic and thermodynamic parameters of the HER and HAR can be evaluated using Eq. (22) and the modified relation, Eq. (40).

One of these parameters is the exchange current density of the HER, i_o, which is usually evaluated using the Tafel part of the polarization curve. This information is part of the electrochemical hydrogen permeation technique. The cathodic charging current and the overpotential are measured at the charging side of the iron sample. Therefore, the construction of the Tafel plot to evaluate the exchange current density can be done with data collected from the hydrogen permeation experiment with no additional work. The cathodic Tafel relation is,[111,115]

$$\eta = \frac{2.303}{a\alpha} \log i_o - \frac{2.303}{a\alpha} \log i_c \qquad (41)$$

Plotting η vs. $\log(i_c)$ will give a straight line with a slope of $-2.303/aa$, and an intercept of $(2.303/aa)\log i_o$. From the slope, one can determine the transfer coefficient, α. In fact, the original IPZ analysis uses the Tafel plots to determine the value of α. Then, using the determined value of α and the intercept of Eq. (41), the exchange current density can be determined.

The IPZA procedure to evaluate the different parameters is summarized as:

1. From the slope of the Eq. (41), α can be determined.
2. From the intercept of Eq. (41) and knowing α from step 1 above, i_o can be determined.
3. Knowing i_o and the solution pH, k_1 can be determined using Eq. (15).
4. From the slope of Eq. (40) and knowing i_o, the kinetic-diffusion constant, k, can be determined.
5. From the intercept of Eq. (40) and knowing i_o, the surface coverage of the second adsorbate, θ_A, can be determined.
6. From the slope of Eq. (22) and knowing k, the recombination reaction rate constant, k_2, can be determined.
7. From Eq. (21) and knowing k, the hydrogen surface coverage, θ_H, at different steady state hydrogen permeation current densities, i_∞, can be determined.

This shows that all kinetic and thermodynamic parameters of the hydrogen evolution and absorption reactions can be determined using the IPZA analysis, including the surface coverage of a second adsorbent, θ_A.

Experimental data in the presence of HMTA will be used here to demonstrate the use of the IPZA to calculate the different parameters including the surface coverage of HMTA, θ_{HMTA}.[104] Figure 7 and 9 indicate that the data satisfy the requirement of Eqs. (22) and (40), respectively. Figure 18 shows the relation between the applied hydrogen charging current density, i_c, and the electrode potential, E, at the charging surface of the iron membrane. This figure is needed in order to determine the transfer coefficient, α, and the exchange current density, i_o. The above procedure was applied to Figs. 7, 9 and 18 to determine all parameters.

Table 4 shows values for i_o, k_1, k_2, and k determined by the IPZA analysis at different HMTA concentrations. Figure 19 shows the relations between θ_H, and overpotential, η. θ_{HMTA} was determined using two other techniques, the electrochemical quartz crystal microbalance, EQCM, and the estimated corrosion current densities (potentiodynamic technique).[104] These results are presented in Fig. 20.

Figure 20 shows a good agreement of the determined coverage's using the different techniques. A similar agreement was found for

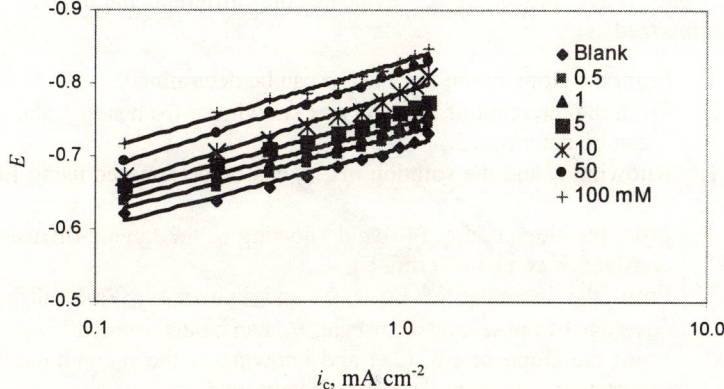

Figure 18. The relation between the log of the charging current density (i_c) and the electrode potential (E) on the charging surface of the iron membrane for different HMTA concentrations on iron in 1M Na_2SO_4 at pH 2. (Repinted from F. M. Al-Faqeer, K. G. Weil and H. W. Pickering, *J. Electrochem. Soc.*, in press. [104])

iodide ions.[114] The HMTA surface coverage's determined by the EQCM are larger than those calculated by the IPZA analysis or from the corrosion current densities. This could be due to the surface roughness of the electroplated iron.[104]

Table 4
Values of k, k_2, i_o and k_1 Obtained on Iron Membrane at Different Hexamethylenetetramine (HMTA) Concentrations at pH 2[104]

HMTA Content	$10^{11} k$ mol cm^{-2} s^{-1}	$10^7 k_2$ mol cm^{-2} s^{-1}	i_o μA cm^{-2}	$10^7 k_1$ cm s^{-1}
Blank	2.37	0.17	0.67	6.9
0.5 mM	5.95	0.16	0.59	6.1
1 mM	7.32	0.16	0.48	5.0
5 mM	5.45	0.24	0.43	4.4
10 mM	5.41	0.49	0.35	3.6
50 mM	5.19	0.77	0.21	2.2
100 mM	4.27	2.24	0.17	1.8

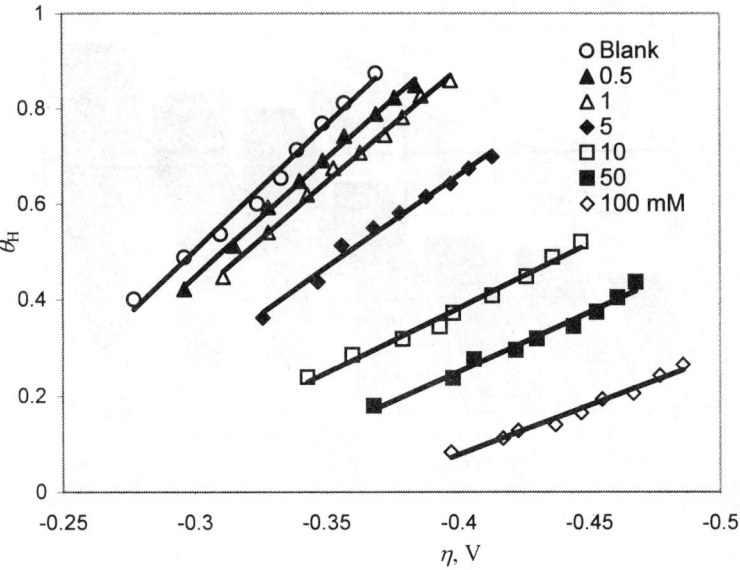

Figure 19. Hydrogen surface coverage, θ_H, at different HMTA concentrations and electrode overpotential. (Repinted from F. M. Al-Faqeer, K. G. Weil and H. W. Pickering, *Journal of Electrochemical Society*, in press. [104])

VIII. ANALYSIS OF PERMEATION DATA UNDER CORROSION CONDITIONS

All IPZ analyses including the new IPZA analysis treat hydrogen permeation data under cathodic charging conditions. Ramasubramanian, et al.,[63] developed an IPZ analysis to characterize the hydrogen permeation data under corroding conditions. The analysis requires the knowledge of the corrosion rate of the specific metal or alloy in solutions of different acidity. The analysis is capable of determining the rate constants of the hydrogen discharge, recombination and adsorption reactions.

Kinetics of Electrochemical Hydrogen Entry into Metals and Alloys

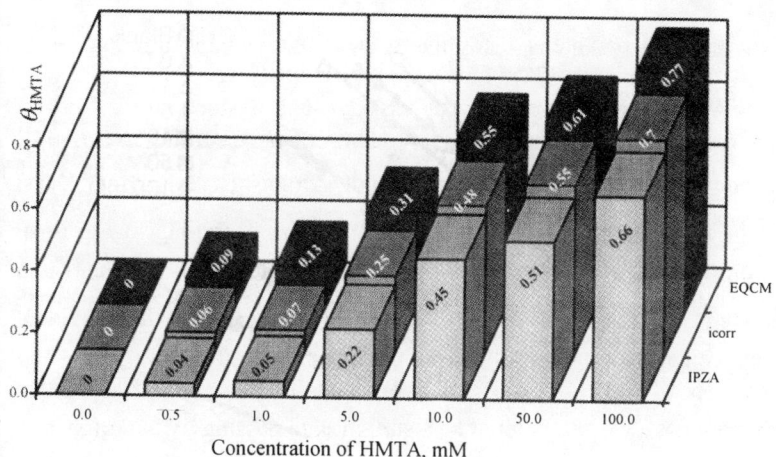

Figure 20. HMTA surface coverage on iron as a function of its concentration in 1M Na_2SO_4 at pH 2 by three independent measuring techniques. (Repinted from F. M. Al-Faqeer, K. G. Weil and H. W. Pickering, Journal of Electrochemical Society, in press.[104])

They developed relations for the coupled discharge-electrochemical recombination (Eqs. 1 and 3) and the coupled discharge-chemical recombination mechanisms (Eqs. 1 and 2). For the discharge-chemical recombination mechanism, the following relations were developed,

$$i_\infty = k \sqrt{\frac{F}{k_2}} \sqrt{i_r} \qquad (42)$$

$$i_c \exp(a\alpha\eta) = Fk_1(1-\theta_H)C_{H^+}^p \qquad (43)$$

$$i_c \exp(a\alpha\eta)C_{H^+}^{p-1} = Fk_1 - \frac{k_1}{k} i_\infty \qquad (44)$$

where p is a parameter determining the effective concentration for i_c. Equation (42) is Eq. (22) of the original IPZ analyses.

A plot of log $[i_c\exp(a\alpha\eta)]$ vs. C_{H^+} has a slope of p and an intercept of log (Fk_1) assuming low values of θ_H according to Eq. (43). From the intercept, k_1 can be calculated. A plot of $i_c \exp(a\alpha\eta)C_{H^+}^{p-1}$ vs. i_∞ has a slope of Fk_1 and an intercept of k_1/k according to Eq. (44). From this step, k can be calculated. Now, a plot of i_∞ vs. $\sqrt{i_r}$ has a slope of $k\sqrt{F/k_2}$ according to Eq. (42). This shows that the individual rate constants for the HER can be determined from the slope and intercept of Eqs. (42), (43) and (44). For this model to be applicable the plot of i_∞ vs. $\sqrt{i_r}$ should be a straight line passing by the origin and the value of k_1 calculated from the intercept of Eq. (43) and the slope of Eq. (44) should be similar.

Similar equations were developed for the coupled discharge-electrochemical recombination mechanism. More details can be found in Ramasubramanian, et al.[63]

IX. CONCLUSIONS

The kinetics of hydrogen entry into metals has been studied using different analyses. The original IPZ analysis has been applied to steady state permeation data to yield the individual rate constants of the hydrogen evolution reaction, k_1 and k_2, the kinetic-diffusion constant, k, the hydrogen surface coverage, θ_H, and the hydrogen concentration in the charging side, C^o. It was also shown how θ_H, k_1 and k_2 could be determined from the measured HER polarization curve for metal/electrolyte systems of low hydrogen absorption into the metal.

The IPZ analysis was modified assuming the validity of the Frumkin adsorption isotherm where a general analysis was developed. The experimental data in the presence of hydrogen sulfide fit the relations derived in the case of the generalized IPZ analysis. However, there is no independent evidence that the Frumkin approach is a better physical description than the Langmuir approach for hydrogen adsorption under these conditions. Furthermore, the adsorption of the sulfur species was ignored; therefore, any effect that they may have had on the available hydrogen adsorption sites on the surface was ignored too, for the sake of an approximation.

The new IPZA was developed to yield the surface coverage of a second adsorbate in addition to all other parameters. The IPZA analysis assumed the validity of the Langmuir adsorption isotherm. The analysis was used to analyze experimental data in the presence of HMTA. The surface coverage of HMTA was successfully determined with a good agreement using two different independent methods. These are the EQCM and i_{corr} (polarization technique). A similar agreement was found for iodide ions.[114]

ACKNOWLEDGEMENT

We thank K. G. Weil, B. G. Ateya and M. Abd Elhamid for helpful discussions. The authors gratefully acknowledge financial support by the U. S. Steel Corporation, Saudi Aramco and the National Science Foundation, DMR 96-12303 and INT 9724698.

LIST OF SYMBOLS

a	constant, F/RT, V^{-1}
C_H	hydrogen ion concentration in the electrolyte, mol L^{-1}
C_H	hydrogen concentration within the metal membrane, mol cm^{-3}
C_H^o	hydrogen concentration in the metal membrane at the charging surface, mol cm^{-3}
C_H^L	hydrogen concentration in the metal membrane at the exit surface, mol cm^{-3}
D	hydrogen diffusion coefficient, $cm^2\ s^{-1}$
F	Faraday constant, 96484.6 C mol^{-1}
i_c	charging current density, A cm^{-2}
i_p	transient hydrogen permeation current density, A cm^{-2}
i_∞	steady state hydrogen permeation current density, A cm^{-2}
i_o	exchange current density, A cm^{-2}
i_o'	$i_o/(1-\theta_H^c)$, A cm^{-2}
i_r	hydrogen recombination current density, A cm^{-2}
k_{abs}	absorption rate constant, mol $(cm^2\ s)^{-1}$
k_{des}	desorption rate constant, cm s^{-1}
k	kinetic-diffusion constant, mol $cm^{-2}\ s^{-1}$

k_1	discharge rate constant, cm s^{-1}
k_2	recombination rate constant, mol (cm^2 s)$^{-1}$
L	membrane thickness, cm
R	gas constant, 8.314 J (mol K)$^{-1}$
T	absolute temperature, K
$t_{1/2}$	time when i_p reaches one-half of i_∞ (half rise time), s
t_b	time for the initial hydrogen atoms to reach the exit surface (breakthrough time), s
x	distance perpendicular to the metal membrane surfaces which increase from zero at the charging surface to L at the exit surface, cm
α	transfer coefficient, dimensionless
η	overpotential of the HER, V
θ_H	hydrogen surface coverage, dimensionless
θ_H^e	hydrogen surface coverage at equilibrium, dimensionless
θ_A	coverage of the second adsorbate, dimensionless
θ_{HMTA}	HMTA surface coverage, dimensionless

REFERENCES

[1] L. Cailletet, *Compte Rendu*, **58**, (1864) 327.
[2] J. McBreen and M. A. Genshaw, in *Fundamental Aspects of Stress Corrosion Cracking*, Ed. by R. W. Steahle, A. J. Forty and D. van-Rooyen, NACE International, Houston, TX 1969, p. 51.
[3] J. J. DeLuccia, in *Hydrogen Embrittlement: Prevention and Control*, Ed. by L. Raymond, ASTM, Philadelphia, PA, 1988, p. 17.
[4] M. Bodenstein, Zeitschrift fur *Electrochemie*, **28**, (1922)517.
[5] M. A. V. Devanathan, Z. Stachurski and W. Beck, *Journal of the Electrochemical Society*, **110**, (1963) 886.
[6] J. O. M. Bockris, J. McBreen and L. Nanis, *Journal of the Electrochemical Society*, **112**, (1965) 1025.
[7] M. Smialowski, in *Stress Corrosion Cracking and Hydrogen Embrittlement of Iron Base Alloys*, Eds.R. W. Steahle, J. Hochmann, R. D. McCright and J. E. Slater, 405, NACE International, Houston, TX, 1977.
[8] R. D. McCright, in *Stress Corrosion Cracking and Hydrogen Embrittlement of Iron Base Alloys*, Ed. by R. W. Steahle, J. Hochmann, R. D. McCright and J. E. Slater, 306, NACE International, Houston, TX, 1977.
[9] N. Ohnaka and Y. Furutani, *Corrosion*, **46**, (1990) 129.
[10] R. N. Iyer and H. W. Pickering, Annual Review of Materials Science, **20**, 299 (1990).
[11] A. Taha and P. Sofronis, *Engineering Fracture Mechanics*, **68**, (2001) 803.
[12] M. A. Fullenwider, Hydrogen Entry and Action in Metals, Pergamon Press, New York 1983, p. 125.
[13] M. Smialowski, *Hydrogen in Steel : Effect of Hydrogen on Iron and Steel During Production, Fabrication, and Use*, Pergamon Press, Oxford , NY, 1962, p. 452.
[14] J. Flis, in *Corrosion of Metals and Hydrogen-Related Phenomena*, Elsevier Science Publication Co., Amsterdam, 1991, p. 241.

[15] G. B. Nielsen, J. E. T. Andersen and J. C. Reeve, in *Modern Aspects of Electrochemistry*, Vol. 31, Ed. by R. E. White, B. E. Conway and J. O. M. Bockris, Plenum Press, New York, 1997, p. 251.
[16] M. Enyo, in *Modern Aspects of Electrochemistry*, Vol. 11, Ed. by J. O. M. Bockris and B. E. Conway, Plenum Press, New York, 1975, p. 251.
[17] B. G. Pound, in *Modern Aspects of Electrochemistry*, Vol. 25, Ed. by J. O. M. Bockris, B. E. Conway and R. E. White, Plenum Press, New York, 1993, p. 63.
[18] T. Mizuno and M. Enyo, in *Modern Aspects of Electrochemistry*, Vol. 30, Ed. by R. E. White, B. E. Conway and J. O. M. Bockris, Plenum Press, New York, 1996, p. 451.
[19] M. H. Abd Elhamid, B. G. Ateya and H. W. Pickering, *Journal of the Electrochemical Society*, **147** (2000) 2959.
[20] M. H. Abd Elhamid, B. G. Ateya and H. W. Pickering, *Journal of the Electrochemical Society*, **147** (2000) 2258.
[21] F. M. Al-Faqeer, MS Thesis, *Effect of Sulfur Containing Compounds on Hydrogen Absorption by Iron*, The Pennsylvania State University, University Park, 1999.
[22] M. H. Abd Elhamid, B. G. Ateya and H. W. Pickering, *Journal of the Electrochemical Society*, **144** (1997) L58.
[23] F. M. Al-Faqeer and H. W. Pickering, *Journal of the Electrochemical Society*, **148** (2001) E248.
[24] F. M. Al-Faqeer and H. W. Pickering, in *International Conference on Hydrogen Effects on Materials Behavior and Corrosion Deformation Interactions*, Ed. by N. R. Moody, TMS, Warrensville, PA (to be published).
[25] M. A. V. Devanathan and Z. Stachurski, *Proceedings of the Royal Society*, **A270** (1962) 90.
[26] M. A. V. Devanathan and Z. Stachurski, Journal of the Electrochemical Society, **111** (1964) 619.
[27] A. Frumkin and N. Aladjalova, *Acta Physicochimica URSS*, **19** (1944) 1.
[28] M. R. Gennero de Chialvo and A. C. Chialvo, *Journal of Electroanalytical Chemistry*, **388** (1995) 215.
[29] M. H. Abd Elhamid, B. G. Ateya, K. G. Weil and H. W. Pickering, *Corrosion*, **57** (2001) 428.
[30] M. H. Abd Elhamid, B. G. Ateya, K. G. Weil and H. W. Pickering, *Journal of the Electrochemical Society*, **147** (2000) 2148.
[31] E. G. Dafft, K. Bohnenkamp and H. J. Engell, *Corrosion Science*, **19** (1979) 591.
[32] J. O. M. Bockris, J. L. Carbajal, B. R. Scharifker and Chandrasekaran, *Journal of the Electrochemical Society*, **134** (1987) 1957.
[33] G. Jerkiewicz, J. J. Borodzinski, W. Chrzanowski and B. E. Conway, *Journal of the Electrochemical Society (USA)*, **142** (1995) 3755.
[34] S. Y. Qian, B. E. Conway and G. Jerkiewicz, in *Hydrogen at Surface and Interface Proceedings of the International Symposium*, Vol. 2000-16, Ed. by G. Jerkiewicz, J. M. Feliu and B. N. Popov, , The Electrochemical Society, INC., Pennington, 2000, p. 98.
[35] N. Krstajic, M. Popovic, B. Grgur, M. Vojnovic and D. Sepa, *Journal of Electroanalytical Chemistry*, **512** (2001) 16 .
[36] R. N. Iyer, I. Takeuchi, M. Zamanzadeh and H. W. Pickering, *Corrosion*, **46** (1990) 460.
[37] C. Kato, H. J. Grabke, B. Egert and G. Panzner, *Corrosion Science*, **24** (1984) 591.
[38] C. D. Kim and B. E. Wilde, *Journal of the Electrochemical Society*, **118** (1971) 202.
[39] W. Beck, A. L. Glass and E. Taylor, *Journal of the Electrochemical Society*, **112**, (1965). 53

[40] B. D. Craig, *Sour-Gas Design Considerations*, Society of Petroleum Engineering, Richardson, TX, 1993, p. 25.
[41] D. A. Jones, *Principles and Prevention of Corrosion*, Macmillan Publishing Company, New York, 1992, p. 248.
[42] H. A. Duarte, D. M. See, B. N. Popov and R. E. White, *Corrosion*, **54** (1998)187.
[43] P. F. Timmins, *Solutions to Hydrogen Attack in Steel*, ASM International, Materials Park, OH ,1997, p. 3.
[44] B. D. Craig, *Corrosion*, **40** (1984) 471.
[45] A. Mezin, J. Lepage and P. B. Abel, *Thin Solid Films*, **272** (1996) 124.
[46] H. H. Uhlig and R. W. Revie, *Corrosion and Corrosion Control: An Introduction to Corrosion Science and Engineering*, John Wiley & Sons, New York, 1985, p. 142.
[47] G. M. Pressouyre, *Acta Met.*, **28** (1980) 895.
[48] G. M. Pressouyre and I. M. Bernstein, *Corrosion Science*, **18** (1978) 819.
[49] H. H. Johnson, *Met. Trans. A*, **19A** (1988) 2372.
[50] R. N. Iyer, H. W. Pickering and M. Zamanzadeh, *Journal of the Electrochemical Society*, **136** (1989) 2463.
[51] R. N. Iyer, H. W. Pickering and M. Zamanzadeh, *Journal of the Electrochemical Society*, **134** (1987) C424.
[52] R. N. Iyer, H. W. Pickering and M. Zamanzadeh, *Scripta Metallurgica*, **22** (1988). 911
[53] A. M. Allam, B. G. Ateya and H. W. Pickering, *Corrosion*, **53** (1997) 284.
[54] C. Azevedo, P. S. A. Bezerra, F. Esteves, C. J. B. M. Joia and O. R. Mattos, *Electrochimica Acta*, **44** (1999) 4431.
[55] K. Banerjee and U. K. Chatterjee, *Scripta Materialia*, **44** (2001) 213.
[56] D. H. Coleman, G. Zheng, B. N. Popov and R. E. White, *Journal of the Electrochemical Society*, **143** (1996) 1871.
[57] H. A. Duarte, D. M. See, B. N. Popov and R. E. White, *Journal of the Electrochemical Society*, **144** (1997) 2313.
[58] A. Durairajan, A. Krishniyer, B. S. Haran, R. E. White and B. N. Popov, *Corrosion*, **56** (2000) 283.
[59] A. Durairajan, R. E. White and B. N. Popov, in *Hydrogen at Surface and Interface Proceedings of the International Symposium*, Vol. 2000-16, G. Jerkiewicz, Ed. by J. M. Feliu and B. N. Popov, The Electrochemical Society, INC., Pennington, 2000, p. 72.
[60] M. I. Luppo and J. Ovejero-Garcia, *Materials Characterization*, **40** (1998) 189.
[61] P. Manolatos, M. Jerome and J. Galland, *Electrochimica Acta*, **40** (1995) 867.
[62] J. McBreen, L. Nanis and W. Beck, *Journal of the Electrochemical Society*, **113** (1966) 1218.
[63] M. Ramasubramanian, B. N. Popov and R. E. White, *Journal of the Electrochemical Society*, **145** (1998) 1907.
[64] A. Sehgal, B. G. Ateya and H. W. Pickering, *J. Electroch. Soc.*, **142** (1995) L198.
[65] T. L. Ward and T. Dao, *Journal of Membrane Science*, **153** (1999) 211.
[66] E. Wu, J. Electroch. Soc., **134** (1987) 2126.
[67] T.-H. Yang and S.-I. Pyun, *Journal of Electroanalytical Chemistry*, **414** (1996) 127.
[68] T.-Y. Zhang and Y.-P. Zheng, *Acta Materialia*, **46** (1998). 5023.
[69] Y.-P. Zheng and T.-Y. Zhang, *Acta Materialia*, **46** (1998). 5035.
[70] M. J. Danielson, *Corrosion Science*, **44** 829 (2002).
[71] E. Owczarek and T. Zakroczymski, *Acta Materialia*, **48** 3059 (2000).
[72] T. Zakroczymski, Z. Szklarska-Smialowska and M. Smialowski, *Werkstoffe und Korrosion*, **26** (1975) 617.
[73] S. S. Chatterjee, B. G. Ateya and H. W. Pickering, *Met. Trans. A*, **9A** (1978) 389.

[74]M. Zamanzadeh, A. M. Allam, C. Kato, B. G. Ateya and H. W. Pickering, *J. Electrochem. Soc.*, **129** (1982) 284.
[75]M. Zamanzadeh, A. M. Allam, H. W. Pickering and G. K. Hubler, *J. Electrochem. Soc.*, **127** (1980) 1688.
[76]J. O. M. Bockris, in *Stress Corrosion Cracking and Hydrogen Embrittlement of Iron Base Alloys*, Ed. by R. W. Steahle, J. Hochmann, R. D. McCright and J. E. Slater, 286, NACE, Houston, Texas, Unieux-Firminy, France, 1977.
[77]F. M. Al-Faqeer, PhD Thesis, *New Advancements in the Analysis Procedures of the Electrochemical Hydrogen Permeation Experimental Data*, The Pennsylvania State University, University Park, 2002, p. 75.
[78]H. H. Johnson and R. W. Lin, in *Hydrogen Effects in Metals*, Ed. by I. M. Bernstein and A. W. Thompson, The Metallurgical Society of AIME, Warrendale, PA, 1981, p. 3.
[79]L. S. Darken and R. P. Smith, Corrosion, **5** (1949) 1.
[80]A. J. Kumnick and H. H. Johnson, *Met. Trans. A*, **6A** (1975) 1087.
[81]M. F. Stevens, I. M. Bernstein and W. A. McInteer, in *Hydrogen Effects in Metals*, Ed. by I. M. Bernstein and A. W. Thompson, The Metallurigical Society of AIME, Warendale, PA, 1981, p. 341.
[82]T. Zakroczymski, *J. Electroanal. Chem.*, **475** (1999) 82.
[83]K. Arai, F. Kusu, K. Ohe and K. Takamura, *Electrochimica Acta*, **42** (1997) 2493.
[84]M. R. Deakin, T. T. Li and O. R. Melroy, *J. Electroanal. Chem.*, **243** (1988) 343.
[85]H. Ehahoun, C. Gabrielli, M. Keddam, H. Perrot, Y. Cètre and L. Diguetb, *J. Electrochem. Soc.*, **148** (2001) B333.
[86]L. Han-Wei, H. Uchida and M. Watanabe, *J. Electroanal. Chem.*, **413** (1996) 131.
[87]P. Kern and D. Landolt, *Journal of the Electrochemical Society*, **148** (2001) B228.
[88]P. Kern and D. Landolt, *Journal of Electroanalytical Chemistry*, **500** (2001) 170.
[89]P. Kern and D. Landolt, *Journal of the Electrochemical Society*, **147** (2000) 318.
[90]J. O. M. Bockris and P. K. Subramanyan, *Electrochimica Acta*, **16** (1971) 2169.
[91]J. L. Crolet and M. R. Bonis, in *Corrosion 2001*, , NACE, Houston, TX , 2001, Paper 067.
[92]J. W. Davenport, G. J. Dienes and R. A. Johnson, *Physical Review B*, **25** 2165 (1982).
[93]J. Flis, T. Zakroczymski, V. Kleshnya1, T. Kobiela and R. Dus, *Electrochimica Acta*, **44** (1999) 3989.
[94]M. Garet, A. M. Brass, C. Haut and F. Guttierez-Solana, *Corrosion Science*, **40** (1998) 1073.
[95]L. A. C. J. Garcia, C. J. B. M. Joia, E. M. Cardoso and O. R. Mattos, *Electrochimica Acta*, **46** (2001) 3879.
[96]R. P. Hu, P. Manolatos, M. Jerome, M. Meyer and J. Galland, *Corrosion Science*, **40** (1998) 619.
[97]A. Kawashima, K. Hashimoto and S. Shimodaira, *Corrosion*, **32** (1976) 321.
[98]R. S. Lillard, D. G. Enos and J. R. Scully, *Corrosion (USA)*, **56** (2000) 1119.
[99]L. Nanis and T. K. Namboodhiri, *J. Electrochem. Soc.*, **119** (1972) 691.
[100]R. W. Revie, in Modern Aspects of Electrochemistry, Vol. 26, Ed. by B. E. Conway, J. O. M. Bockris and R. E. White, , Plenum Press, New York, 1994, p. 234.
[101]M. R. Gennero de Chialvo and A. C. Chialvo, *J. Electroanal. Chem.*, **415** (1996) 97.
[102]M. R. Gennero de Chialvo and A. C. Chialvo, *Electrochimica Acta*, **44** (1998) 841.
[103]M. R. Gennero de Chialvo and A. C. Chialvo, J.Electroanal. Chem., **448** (1998) 87.
[104]F. M. Al-Faqeer, K. G. Weil and H. W. Pickering, Journal of Electrochemical Society, (In Press).
[105]Y. T. Zhang, Y. P. Zheng and Q. Y. Wu, *J. Electrochem. Soc*, **146** (1999) 1741.

[106] E. A. Maleeva, K. S. Pedan and V. N. Kudryavtsev, *Russ. J. Electrochem.*, **30** (1994) 1228.
[107] B. K. Subramanyan, in *Comprehensive Treatise of Electrochemistry*, Vol. 4, Ed. by J. O. M. Bockris, B. E. Conway and R. E. White, , Plenum Press, New York, 1981, p. 411.
[108] J. O. M. Bockris and R. K. Reddy, *Modern Electrochemistry*, Plenum Press, New York, NY, 1971.
[109] M. Temkin and V. Pyzhev, *Acta Physicochimica URSS*, **12** (1940) 327.
[110] E. Gileadi and B. E. Conway, in *Modern Aspects of Electrochemistry*, Vol. 3, Ed. by J. O. M. Bockris and B. E. Conway, Butterworths, Washington, 1964, p. 347.
[111] E. Gileadi, *Electrode Kinetics for Chemists, Chemical Engineers, and Materials Scientists*, Wiley-VCH, New York, 1993, p. 118.
[112] M. H. Abd Elhamid, PhD Thesis, *Mechanistic Analysis of the Hydrogen Evolution and Absorption Reactions on Iron*, , The Pennsylvania State University, University Park, 2000, p. 89.
[113] S. Kertit and B. Hammouti, *Applied Surface Science*, **93** (1996) 59.
[114] F. M. Al-Faqeer, K. G. Weil and H. W. Pickering, *Electrochemica Acta*, (In Press).
[115] R. H. Hausler, *Corrosion*, **33** (1977) 117.

2

The Electrochemistry, Corrosion and Hydrogen Embrittlement of Titanium

C.L. Briant

Division of Engineering, Brown University, Providence, RI 02912

I. INTRODUCTION

Titanium has the reputation of being one of the most corrosion resistant metals. It can be used in many acidic and basic environments and not suffer attack. Only exposure to hydrofluoric acid and concentrated boiling mineral acids cause significant corrosion. The reason for this general immunity is that titanium forms a tenacious oxide film on its surface. Figure 1 shows the Pourbaix diagram for titanium, and one notes the large range of pH and electrochemical potential over which an oxide is stable.[1] This film is usually ductile, but even if it is broken it can be easily repaired because most solutions contain sufficient oxygen to allow it to reform. As a result of this characteristic, the greatest concern about the use of titanium occurs in situations in which the titanium oxide films can be reduced or in which the degrading species can penetrate the oxide film. An example of the first situation would be crevice corrosion. An example of the second would be pitting. A third example, which could be in either category, is hydrogen embrittlement of titanium; in some cases the hydrogen reduces the film,

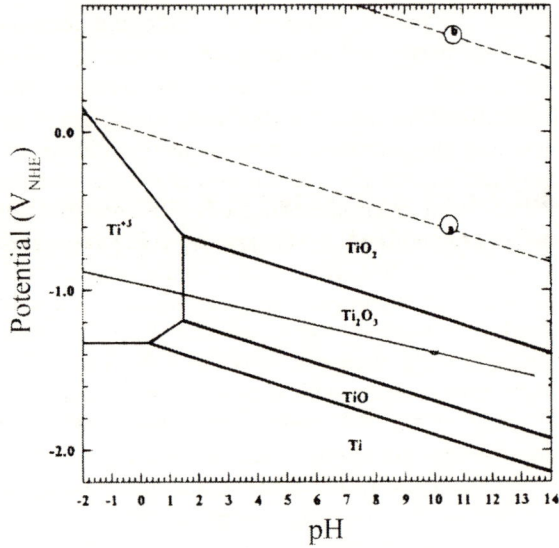

Figure 1. The Pourbaix diagram for titanium.(Reprinted from Ref. 1 with permission from The Electrochemical Society.)

but in others atomic hydrogen diffuses through the film and enters the metal.

Because of this general immunity of titanium to environmental attack and because this metal has low density and a relatively high tensile strength, it has been employed in a number of demanding engineering applications. For example, it has been used extensively for piping and heat exchangers in many desalination plants.[2-6] Another use is in the chemical processing industry where titanium vessels, heat exchangers, and piping systems are used to handle aggressive chemical compounds.[6-8] As a final example, titanium is often used in environments where bio-fouling might occur.[9]

At room temperature titanium exists as a hexagonal-close-packed metal. Its melting temperature is 1667 °C, but at 840 °C it undergoes an allotropic phase transformation from the hexagonal-close-packed α-phase that is stable at temperatures below this value to the body-centered-cubic β-phase, which is the stable phase between this temperature and the melting temperature. Alloying additions stabilize one of these two phases and are grouped accordingly. Elements that stabilize the α-phase include aluminum, tin, oxygen, nitrogen, and carbon; β-stabilizers include molybdenum, vanadium, niobium, and iron.[10] These alloying additions can significantly affect mechanical and corrosion properties, and they also allow engineers to tailor an alloy composition so that it meets specific needs. References 11 and 12 give background information on a number of frequently used titanium alloys. However, in this review, we shall deal exclusively with titanium to which no intentional alloying additions have been made. This material is classified by grade and Table 1 shows typical compositions and yield strengths. Note that the interstitial level and yield strength both increase with increasing grade number. Of these materials, grade 2 is the most commonly used. Since the electrochemistry does not strongly depend on the grade of titanium we will not be specific about which grade of titanium is being discussed in this paper.

In this review we describe the basic electrochemistry of titanium, the corrosion of titanium in specific environments, and hydrogen embrittlement of titanium. In particular, we shall pay special attention to the behavior of titanium in sea water (or NaCl solutions), since there is great interest in using titanium in applications where it is exposed to this environment.[2-5,13-18] The organization of the review will be as follows. First we shall consider the basic electrochemistry and general

Table 1
Typical Yield Strengths, Ultimate Tensile Strengths, and Maximum Impurity Contents for Grades 1-4 Titanium
(Composition Values in Weight Percent)

Grade	Yield Strength (MPa)	Ultimate Tensile Strength (MPa)	O	Fe	H	C	N
1	170	240	0.18	0.2	0.015	0.08	0.03
2	275	345	0.25	0.3	0.015	0.08	0.03
3	380	450	0.35	0.3	0.015	0.08	0.05
4	485	550	0.4	0.5	---	0.1	0.05

corrosion. Then we will discuss pitting and galvanic corrosion. Finally, we will review in some detail hydrogen embrittlement of this material, with emphasis on the formation of hydrides within the material and the effect of hydrogen on its mechanical properties.

II. ELECTROCHEMISTRY AND GENERAL CORROSION

Figure 2 shows a typical polarization curve for titanium in hydrogen saturated 20% H_2SO_4 at room temperature.[19] The reversible hydrogen potential is -0.21 V_{SCE}. A small active region is observed, but at a critical current density of 140 µA/cm^2 passivity begins. The surface maintains this passive state, which occurs through the formation of

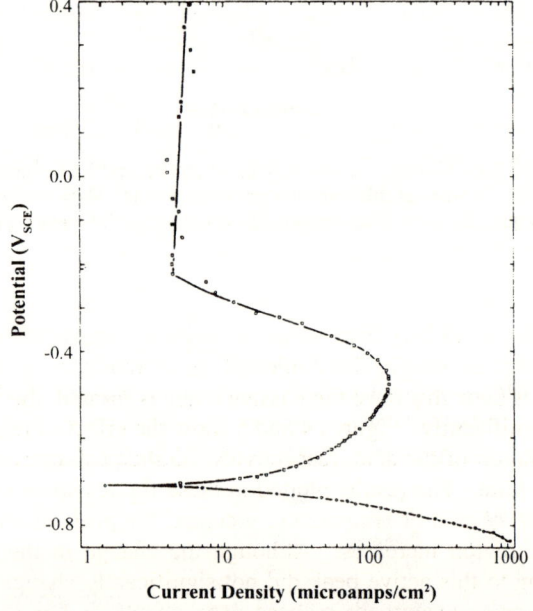

Figure 2. The polarization curve for titanium in 20% sulfuric acid at room temperature. Reprinted from Reference 19 with permission from The Electrochemical Society.

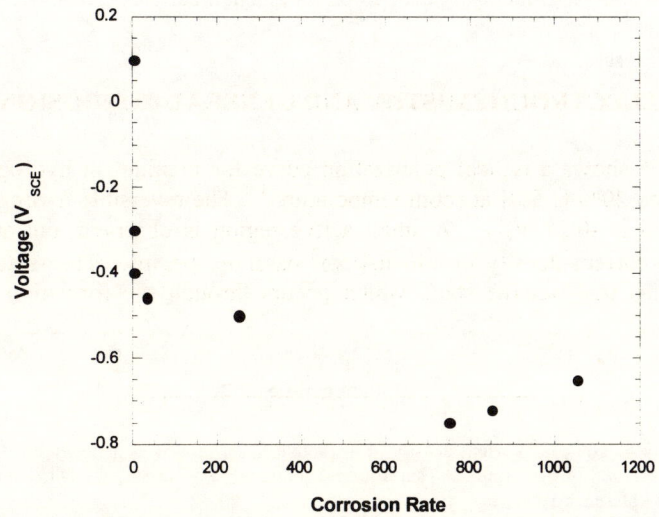

Figure 3 – The corrosion rate for titanium in 2% sulfuric acid containing 5% Na_2SO_4. Values of the corrosion rates are in mdd. Note that in the passive region the corrosion rates are nil. (Compare with Figure 2.) Data taken from Reference 19.

titanium oxide, up to very high voltages. Figure 3 shows the corrosion rate, measured as weight loss, plotted as a function of corrosion potential. It is clear that once the passive oxide is formed, the corrosion rate drops significantly. Figures 4 and 5 show the effect of temperature and concentration of the acid, respectively, on the peak current density in the active state. The results plotted in these figures show that as the concentration of acid or temperature increase, the peak anodic current in the active region increases. Although the change in the potential corresponding to this active peak did not significantly change over the temperature range examined in these experiments, it did increase to more noble values with increasing acid concentration, as shown by the numbers in parentheses on the plot in Fig. 5. For all of the temperatures and concentrations investigated in these experiments the

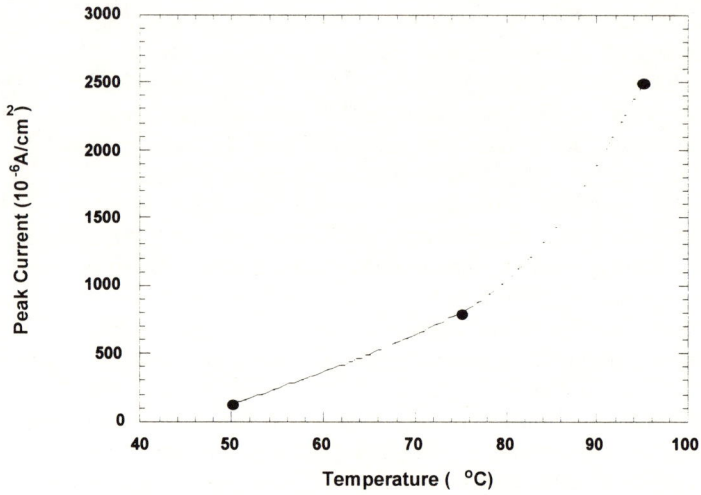

Figure 4. The peak current density in the active range plotted as a function of temperature for titanium in 5% sulfuric acid containing 5% Na_2SO_4. (Data taken from Reference 19.)

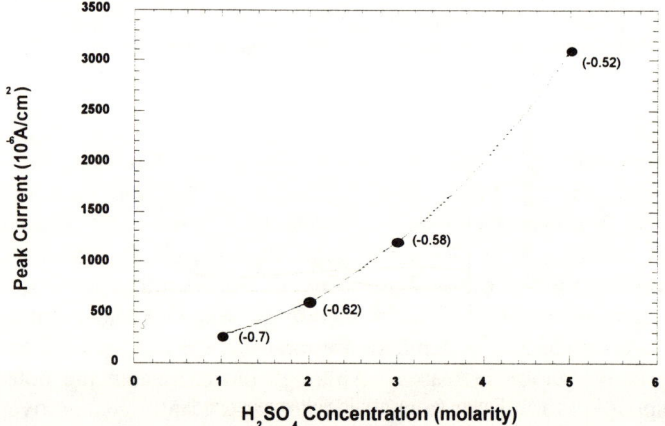

Figure 5. The peak current density in the active range plotted as a function of acid concentration. The numbers in parentheses are the electrochemical potentials (VSCE) at which the peak current density was measured. The solutions were at their boiling points and contained 5% Na_2SO_4. (Data taken from Reference 19.)

material showed an active to passive transition. Again, temperature had little effect on this transition, with all three temperatures investigated showing a critical potential near -0.3 V_{SCE}. Increasing the acid concentration from 1 to 5% changes the critical potential from -0.55 to -0.35 V_{SCE}. The passive current density also increased quite significantly, but it has been suggested that this increase may be an artifact resulting from a plating reaction from the solution.

A significant amount of research has been directed toward understanding the structure and electrochemistry of the passive oxide film. Figure 6 shows a cyclic voltammogram for a titanium electrode in 1 mol/dm^3 of sulfuric acid.[20] Two peaks are observed. One arises on the forward scan and another on the reverse scan. Note that these peaks are observed at much higher voltages than those shown on the polarization curves in Figure 2. Factors such as scan rate, solution concentration, and temperature can affect the exact magnitude and position of these peaks[20]. Oxygen evolution was associated with the occurrence of both peaks. Analysis of these curves and of the oxide films formed on the surface led to the suggestion that these peaks are associated with the cracking of the oxide film as a result of the stress caused by t he electric field at the surface. It has also been shown that

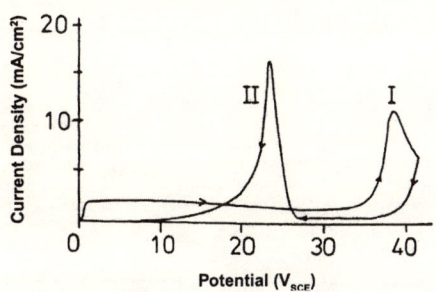

Figure 6 – A cyclic voltammogram for a titanium electrode in 1 mol dm^{-3} solution of sulfuric acid. (Reprinted from Reference 20 with permission from The Electrochemical Society, Inc.)

the initial film that is formed is amorphous and that with higher applied potentials it is converted to a crystalline oxide.[20,21]

In the previous example the oxide is always present on the surface of the metal and is usually thick. It is also of interest to consider the kinetics of the initial formation of the oxide film and the electrochemistry of the bare metal surface. Kolman and Scully[22] have examined both of these issues in chloride-containing solutions. They produced the bare surface either by scratching the surface of a piece of the metal or by fracturing a thin film of the metal. By recording the current transient as the oxide was reformed, they were able to determine the current across the clean surface as well as the kinetics of the film formation. Figure 7 shows the polarization curve for both a standard sample, which had an oxide-coated surface and is marked as steady-state in the figure, and the curve determined from the fractured thin films. The test solution was 0.6M NaCl. The first point to note is

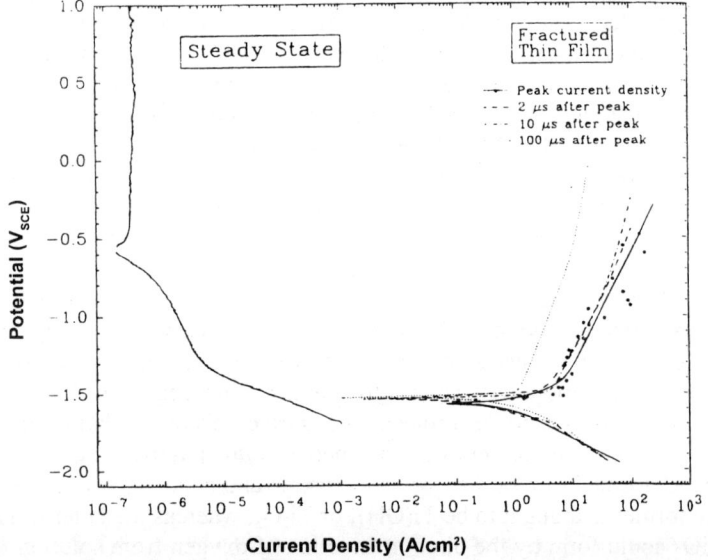

Figure 7 – A comparison of the polarization curve for oxide covered titanium (denoted as steady state) and the curves obtained from clean titanium surfaces prepared by fracturing thin films. The different lines in the data for the clean surfaces were obtained at different times after the peak current upon fracture. The solution was 0.6M NaCl.(Reproduced from Reference 22 by permission from The Electrochemical Society, Inc.)

that for the oxide-coated sample tested in neutral solution, the active region, observed in Figures 2-4 for acidic solutions, is absent. This effect is simply caused by the change in pH. However, the passive current density is similar in the two solutions. For the bare metal, the corrosion potential has decreased and the current densities measured are significantly higher. Also, the active region is now observed in the neutral solution. Raising the pH to 10 did not have a significant effect on these results. Lowering the pH to a value of one raised the corrosion potential from -1.4V_{SCE} to -0.9V_{SCE} but did not affect the current density in the active region.

Kolman and Scully[22] also investigated the decay in current density on bare surfaces at different constant potentials. In these experiments, the samples were held at a fixed potential during fracture and repassivation, so that the researchers could determine both the kinetics of passivation and the potential that was achieved. For samples tested at 0, -0.3 and -0.9 V_{SCE}, the current remained anodic throughout the test, although the current density began to decay after approximately 10^{-5} s exposure to the solution. However, for samples tested at -1.1 and -1.525 V_{SCE}, the current changed from anodic to cathodic as the potential changed; for the sample held at -1.1 V_{SCE} there was an abrupt drop in current after 4 x 10^{-3} s and for the sample held at -1.525 V_{SCE} the drop in current occurred after 10^{-4} s. The sample tested at -1.8 V_{SCE} remained cathodic throughout the test. In this case there was little change in the current density with time. In summary, these results show that the oxide film is re-formed quickly on the surface and that for a given voltage, the thickness of the oxide film can determine whether or not the sample is a cathode or anode.

Research has also been performed for samples that were oxidized in hydroxide solutions.[23,24] When titanium is placed in a strong base, the oxide film formed in air begins to dissolve immediately. However, at the same time a passive, electrochemical film begins to form. This passive film generally protects the surface from oxidation in this solution. One important difference between this passive oxide film and the air formed film or that formed in acidic and neutral solutions, is that the former is thought to be $Ti(OH)_3$ or Ti_2O_3, whereas the latter is TiO_2. TiO_2 could form by the direct interaction of oxygen from solution with the titanium. However, the formation of Ti_2O_3 requires the following series of steps. First, the formation of the oxide requires transfer of titanium and hydroxyl ions by the following reactions:

$$\text{Ti} \leftrightarrow \text{Ti}^{2+} + 2\,e^- \tag{1}$$

$$\text{Ti}^{2+} + \text{H}_2\text{O} \leftrightarrow \text{Ti}^{3+} + \text{OH}^- + 0.5\,\text{H}_2 \tag{2}$$

These reactions are then followed by

$$2\,\text{Ti}^{3+} + 6\,\text{OH}^- \leftrightarrow 2\,\text{Ti(OH)}_3 \leftrightarrow \text{Ti}_2\text{O}_3 + 3\,\text{H}_2\text{O} \tag{3}$$

In this way the oxide is formed on the surface by electrochemistry and protects the surface from significant corrosion.

In Figure 3, we compared the corrosion rate with the electrochemical potential. We now consider more information about the corrosion of titanium. First, we make some general comments about the corrosion measurements. Table 2 lists the corrosion potentials for titanium immersed in various solutions.[25] These values show that, in general, the corrosion potential of titanium increases with time. If we compare these potentials for the sulfuric acid solutions with the polarization curve shown in Figure 2, it would appear that titanium is in the passive condition in the solutions listed in Table 2 and that there should be little evidence of corrosion. This result is in agreement with the generally low corrosion rates reported in Table 3 for aerated solutions.

Given that the oxide film plays such a crucial role in the corrosion protection of titanium, we might expect that more corrosion would occur in deaerated solutions. These results are listed in Table 3. The

Table 2
Single Electrode Potentials of Freshly Abraded Titanium at Room Temperature in an Open Beaker [25]

Solution	Initial Potential (V_{SCE})	Final Potential (V_{SCE})	Total Immersion Time (h)
1% NaCl	-0.471	0.099	309.6
3% NaCl	-0.511	-0.219	188.5
1N HCl	-0.321	-0.201	71.3
7.5N HCl	-0.481	-0.812	122.3
1.1N H_2SO_4	-0.321	-0.051	23.5
2.2N H_2SO_4	-0.291	-0.049	96.3

Table 3
Corrosion of Titanium in Aerated and Deaerated Solutions [25]

Solution	Temperature (°C)	Corrosion Rate in Aerated Solution (mm per year)	Corrosion Rate in Deaerated Solution (mpy)
0.5% H_2SO_4	35	0.0 - 0.005	0.08 – 0.124
1.0% H_2SO_4	60	0	1.07
4.8% H_2SO_4	35	0.003	0.345
10% H_2SO_4	35	1.68	1.21
1% HCl	35	0.006	0
3% HCl	35	0.007	0.142
4% HCl	35	0	0.27
10% HCl	35	1.02	0.77

only exceptions in either case are for strong acid solutions in which significant corrosion was observed in both aerated and deaerated solutions. Finally, it has been found that the initial surface condition[25,26] and surface contamination[26,27] can have a significant effect on the corrosion potential of titanium. The data in Table 4 clearly show that, although the final potential is unaffected, the initial potential varies significantly with changes in the surface preparation, which occurred through furnace oxidation of the titanium. Other

Table 4
The Effect of Surface Condition on the Single Electrode Potential for Titanium Immersed in 3% NaCl [22]
(Open Beaker, Room Temperature)

Condition	Initial Potential (V_{SCE})	Final Potential (V_{SCE})	Total Immersion Time (h)
Freshly abraded	-0.511	0.291	188.5
Abraded and aged at room temperature for three hours	-0.341	0.209	191.5
Abraded and aged at 130°C for three hours	-0.251	0.209	192.5
Annealed in vacuum at 900°C for two hours.	-0.351	0.179	71.3

studies have shown that the presence of iron on the surface can also have a significant effect on the corrosion reactions that occur.[27]

We now consider corrosion data for specific solutions. Figure 8 shows the corrosion rate of titanium in sulfuric acid at room temperature plotted as a function of the acid concentration. The results show that there is a small increase in corrosion rate at concentrations near 40% but that the rate is much greater at approximately 80%. This result is generally consistent with the results shown in Figure 5, where it is clear that an increase in acid concentration will cause the material to remain active to higher potentials. Figure 9 shows the effect of temperature on corrosion of titanium in sulfuric acid solutions of different strengths. These results show that an increase in temperature has a significant effect on the corrosion rate. These changes are clearly reflections of the changes observed in the peak current density shown in Fig. 4. Figures 10 and 11 show similar plots for corrosion of titanium in nitric acid and hydrofluoric acid, respectively. The results show that increasing the temperature of nitric acid to 100 °C causes a significant increase in corrosion, although as with sulfuric acid the

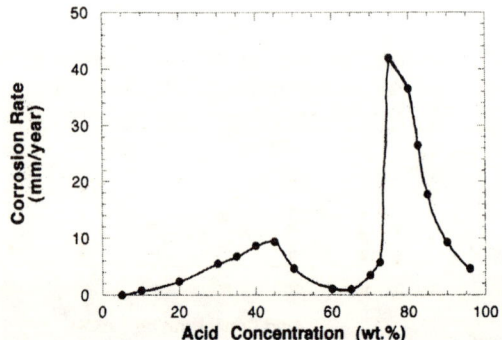

Figure 8. The effect of sulfuric acid concentration on corrosion. (Data taken from Reference 25.)

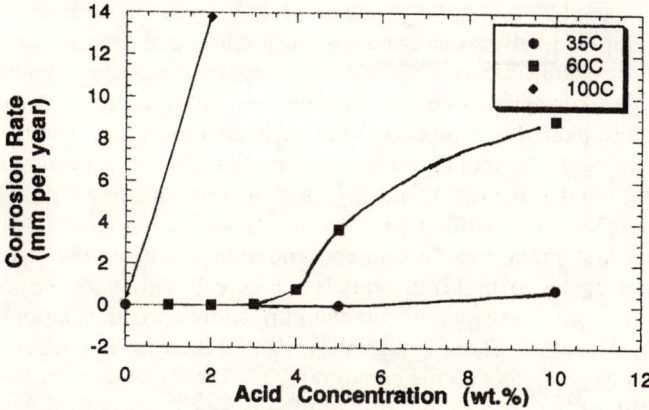

Figure 9. The corrosion rate of titanium in sulfuric acid plotted as a function of concentration. Results are given for three different temperatures.[25]

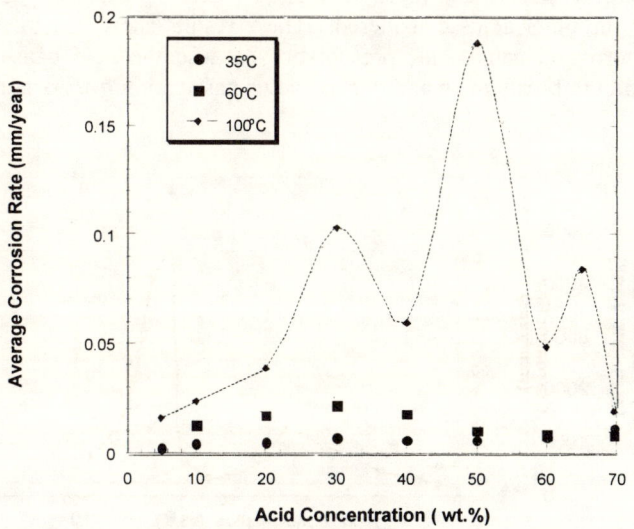

Figure 10. The corrosion rate of titanium plotted as a function of nitric acid concentration. Results are for three different solution temperatures.[25]

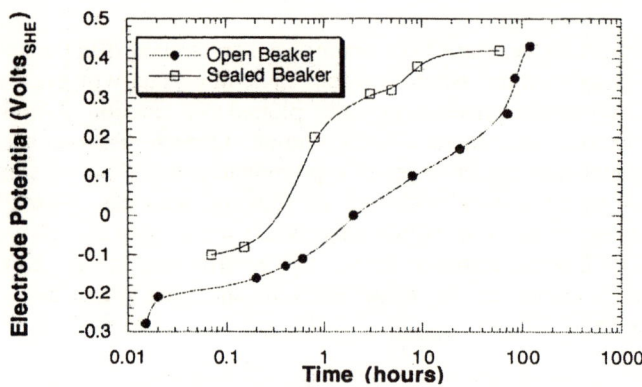

Figure 11. The corrosion of titanium in HF plotted as a function of normality. The results were obtained at room temperature.[25]

dependence on acid concentration is not linear. It has long been known that titanium corrodes rapidly in HF, and the results in Figure 11 show that the corrosion rate is more than three orders of magnitude greater than in the other acids considered. These results show that in strong acids corrosion can occur, presumably because the acid causes the oxide film to break down and thus prevents passivation from occurring.

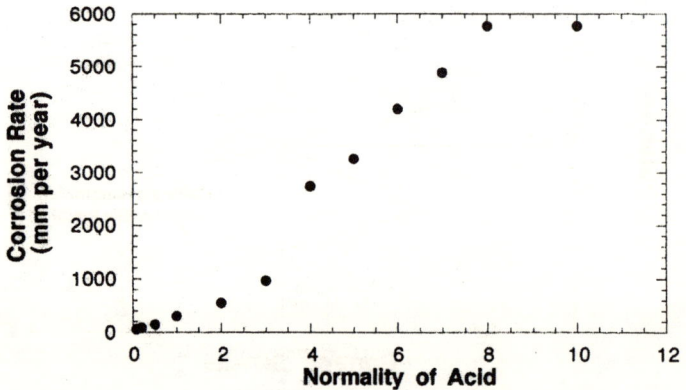

Figure 12. The corrosion potential plotted as a function of time in stagnant 3.5% NaCl solution. Data are shown for both an open and closed beaker.[25]

As mentioned above, there has recently been great interest in the use of titanium in applications where it would come in contact with seawater. Figure 12 shows the corrosion potential plotted as a function of time for titanium exposed to a stagnant NaCl solution. One can observe that the corrosion potential never exceeds approximately 0.3 V_{SCE}. A similar corrosion potential was observed in flowing seawater. Numerous corrosion tests,[25] some of which have lasted for as long as 4.5 years, have shown that the corrosion rate in this environment is nil. Reports of titanium in service in these environments also describe good performance.[2-7, 14-18]

III. GALVANIC CORROSION

One of the main concerns in the application of titanium is whether or not galvanic corrosion can occur. Galvanic corrosion refers to a process in which two metals that are electrically connected are exposed

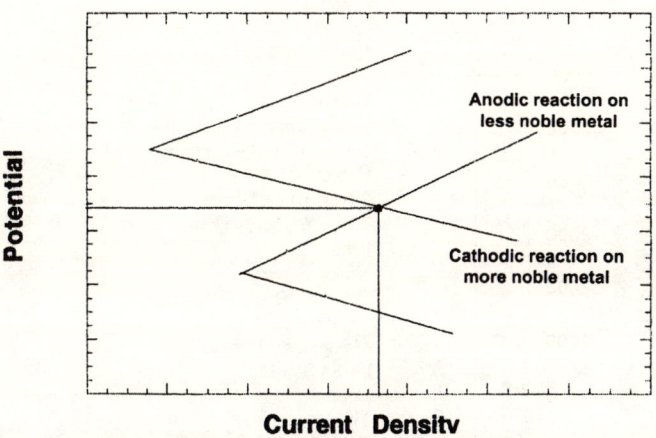

Figure 13. An example of a mixed potential diagram showing that the noble metal becomes the cathode of the couple and the less noble metal becomes the anode. The intersection of the two curves gives the potential and current density for the couple.

to the same solution. In this situation, one of the metals will become a cathode and the other a conjugate anode. The potential for the couple can be determined using mixed potential theory, as shown in Figure 13. In this figure, the polarization curves are shown separately for each metal of the galvanic pair in the solution of interest. One then determines the point at which the cathodic polarization curve for the more active material and the anodic curve for the material with the lower reversible potential intersect. This point determines the corrosion potential and the current density for the couple immersed in the solution of interest.

In practice, titanium is often galvanically coupled with another metal. Because titanium has a high reversible potential, it frequently becomes the cathode in the couple. Table 5 shows that specifically in sea water, titanium is more noble than most engineering alloys and thus would be the cathode under most circumstances.[8] As will be discussed below, if the potential is sufficiently cathodic, hydrogen will be absorbed into titanium and, consequently, hydrides can form.

Table 5
Galvanic Potential for Engineering Materials in Flowing Sea Water. (Data taken from Ref. 8)

Noble ↑	Graphite
	Platinum
	Ni-Cr-Mo Alloy C
	Titanium
	Stainless Steel – Type 316, 317
	Nickel-Copper Alloys 400
	Stainless Steel –Types 302, 304
	Nickel 200
	Silver Braze Alloys
	70-30 Copper Nickel
	Lead
	Nickel-Silver
	Stainless Steel – Type 410
	Tin Bronzes
	Manganese Bronze
	Pb-Sn Solder
	Copper
	Tin
	Naval Brass
	Low Alloy Steel
	Cast Iron
Active	Aluminum Alloys
	Zinc

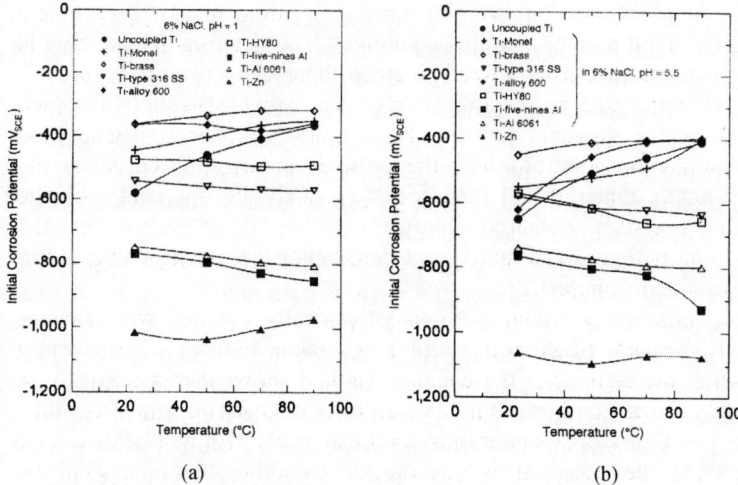

Figure 14. The corrosion potential for various galvanic couples between titanium and the indicated metal. The corrosion potential is plotted as a function of temperature. Figure 14a shows data for tests performed in 6% NaCl with a pH value of one and Figure 14b shows data for tests performed in 6%NaCl with a pH value of 5.5. (Data taken from Ref. 28)

Figure 14a shows the corrosion potential of a number of metals galvanically coupled with titanium and exposed to acidic sodium chloride.[28] One notes that as a result of this coupling the corrosion potential of titanium can vary over a significant range of values. For example, when the titanium is coupled with zinc or aluminum, the potential is very cathodic. Experiments have shown that when titanium is coupled with either of these metals in acidic solutions, titanium hydride is formed and that in the case of zinc, corrosion of the zinc is very rapid.[28] Couples with stainless steel and low alloy steels, such as HY-80, also cause the titanium to become the cathode, except at room temperature. However, the potential that is developed here is much closer to the corrosion potential for titanium and the driving force for hydride formation in titanium and corrosion of the metal is much less. Finally, in this series of metals, couples with naval brass cause titanium to become the anode in the pair. Figure 15b shows similar results for couples tested in a NaCl solution at a pH value of 5.5. Although the metals rank in the same way, the formation of hydrides was not observed in these tests.[28]

As mentioned above, zinc corrodes rapidly when coupled with titanium, and in many applications the corrosion of the anode must be considered. In any case one is also often concerned with the relative areas of the anode and cathode. For example, if titanium is coupled with a low carbon steel, the corrosion rate of the steel will be approximately 0.152 mm/year if the ratio of surface area of the steel and titanium is 10:1. If this area is reversed, the corrosion rate of the steel will be on the order of 0.5 mm/year.[8]

Another example of galvanic corrosion arises when titanium is alloyed with a more noble metal such as platinum.[29,30] In addition to being more noble than titanium, platinum has a much higher exchange current density than titanium for hydrogen-ion reduction. If a Ti-Pt alloy is placed in an acid solution, platinum acts as the cathode and titanium is dissolved, leaving a platinum-rich surface. This dissolution creates a galvanic couple between platinum and titanium. However, as the platinum becomes enriched on the surface, the exchange current titanium becomes passive in the acid solution and no further corrosion density for hydrogen reduction increases and this increase drives the potential to higher values. Eventually, the potential will reach the point that titanium becomes passive in the acid solution and no further corrosion occurs. Although platinum is an expensive metal, research has shown that only small amounts of the more noble metal are required to achieve this effect.[29,30]

IV. PITTING CORROSION

Another form of corrosion that is of concern in applications of titanium is pitting corrosion. This type of corrosion has several key characteristics. The first is that it usually occurs at very anodic potentials that can be associated with the breakdown of the oxide film. This breakdown is associated with a sudden increase in the corrosion current density, and the potential at which this increase occurs is called the pitting potential. The second characteristic is that the breakdown of the oxide film is not general but occurs at isolated points across the surface of the sample. The third is that the occurrence of pitting is usually associated with the presence of a halide ion, most commonly Cl^-, in solution.

The actual corrosion process is thought to occur in the following way. The halide ion penetrates through and locally destroys the oxide layer and exposes bare metal to the solution. The solution that forms in this area becomes very acidic, because of the high concentration of the halide ions and metal hydrolysis, and prevents repassivation of the metal. Also, because the contact with the bulk solution is through the small hole made in the oxide film, the chloride concentration remains high. Thus the metal under the oxide film is dissolved away, even though the surface of the sample, which is largely covered with oxide, appears unattacked.

Pitting corrosion of titanium has been studied by a number of researchers.[31-34] One of the most interesting points that have been observed is that Br^- is a much more effective pitting agent than either Cl^- or I^-[31-33]. The exact cause of this difference is not fully understood but presumably relates to the soluble oxo-halide salts that can form in solution. In other ways, the process shows many of the common characteristics of pitting. Figure 15a shows the pitting potential plotted as a function of oxide thickness and Figure 15b shows the pitting potential plotted as a function of temperature. The former change occurs because as the thickness decreases it becomes easier to penetrate the oxide film, and the latter occurs because the increase in temperature provides greater thermal activation of the process.

Casillas, *et al.*[34] have shown that any mechanism for pitting must include the following points: (i) The microscopic sites at which breakdown occurs must be electrochemically active. In their research they used scanning electrochemical microscopy to determine that these sites were on the order of 10-50 μm in diameter. (ii) The oxide breakdown occurs at more positive potentials with increasing oxide thickness. (iii) There is a chemical specificity in the pitting process, since Br^- is much more effective at pitting than Cl^- and I^-. The observations of Casillas, *et al.*[34] clearly show that pitting occurs at various points on the surface of titanium and that the density of these points increases with decreasing oxide thickness. Furthermore, once pitting corrosion begins, a small hole is observed in the oxide from which bubbles emerge, and around the hole a veiled region appears on the oxide. With time the entire sample can be dissolved away, leaving two oxide surfaces.

Based on these observations Casillas, *et al.*[34] propose that near the pitting potential Br^- ions become chemisorbed at electrochemically active sites on the titanium surface. These then form oxo-halide salts

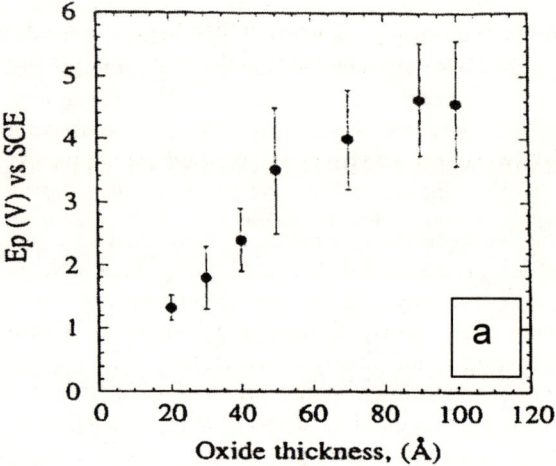

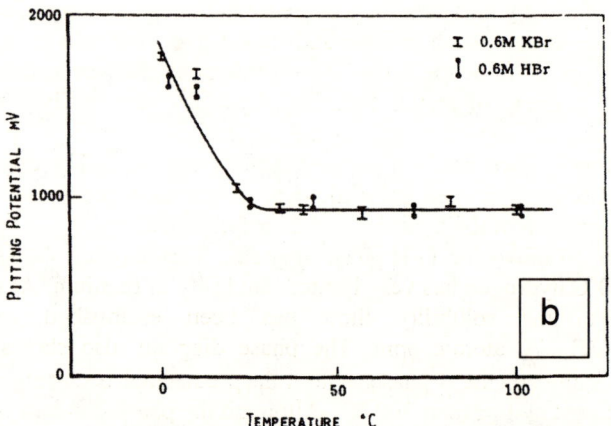

Figure 15. (a) The pitting potential of Ti Plotted as a function of oxide thickness. The solution used was 1 M KBr – 0.05 M H_8SO_4. (Data taken from Ref. 34). (b) The pitting potential of Ti in Br- containing solutions plotted as a function of temperature. (Reproduced from Ref. 33 by permission of The Electrochemical Society.)

that dissolve the film locally. Corrosion then begins to occur under the oxide surface, and this corrosion leads to dissolution of the sample.

V. HYDROGEN EMBRITTLEMENT OF TITANIUM

When atomic hydrogen enters a metal, it often causes a degradation of the mechanical properties of the material. This change can be manifested in several ways. If the sample contains a sharp crack, then the crack may grow slowly through the sample as a result of the presence of hydrogen and finally cause failure. The load required to cause this crack propagation in the presence of hydrogen is usually much less than that required to cause crack propagation in an inert environment or in air. In these types of tests the effect of hydrogen is usually measured as a time to failure or by the stress intensity factor, K_{ISCC}, required to cause the crack to grow in this environment. If the sample is a smooth bar tensile sample or one that contains a notch, the sample is often pulled slowly to failure. In this case, the effect of hydrogen is usually measured by a decrease in the elongation to failure.

In some metals, such as iron and nickel and their alloys, atomic hydrogen appears to be the cause of embrittlement. However, titanium belongs to a group of metals that form hydrides very easily. These hydrides play an important role in the embrittlement process because they are much more brittle than the titanium matrix.[35] Therefore, in order to discuss hydrogen embrittlement in titanium, we must begin with a description of the hydrides and their formation.

Figure 16 shows the Ti-H phase diagram. This diagram shows that below 100°C hydrogen has very limited solubility in titanium. At room temperature, the solubility limit has been established to be approximately 70 atomic ppm. The phase diagram also shows that hydrogen stabilizes the β-phase, with a deep eutectoid occurring at 39 atomic percent hydrogen at 300 °C. Although the phase diagram shows only the δ and ε hydrides, in practice three hydrides have been detected.[37-40] The hydride with the lowest hydrogen content is classified as γ-hydride and has a stoichiometry of $TiH_{0.1-0.3}$. This hydride appears to be ordered and has the same crystal structure as γ-ZrH, which has a face-centered-tetragonal structure. (Note that even though face-centered-tetragonal is not a Bravais lattice, the literature on these hydrides has used this description and we will follow it here.) The c/a ratio is 1.09. The δ-hydride has the stoichiometry $TiH_{1.05-2.0}$ and has the face-centered-cubic CaF_2 structure. Finally, the ε-hydride has the stoichiometry of TiH_2 and has a face-centered-tetragonal structure with c/a<1.

Particular attention has been given to the morphology of the γ and δ-hydrides, since they form at lower hydrogen concentrations and would thus be the hydrides expected to form during exposures that might be encountered during corrosion, exposure to seawater, or applications in which titanium is placed in aqueous solutions. These hydrides generally have a lath-like appearance, as shown in Figure 17, and form within the grains of the titanium. Two different habit planes (identified as hydride types I and II) have been observed and the details of the crystallography are given in Table 6. The exact causes for the occurrence of these two different orientations are not known. However, because these precipitates form at room temperature and because diffusivity is low at this temperature, the precipitation must be accomplished by a martensitic mechanism.[37,39] It should further be noted that because the hydride has a greater specific volume than titanium, the matrix around the hydride would be subjected to a large compressive stress,[41] and dislocations are nucleated around these precipitates as a result of this stress.

Because of the low solubility of hydrogen in titanium at room temperature and the degradation in mechanical properties caused by hydride formation, research has been performed to determine the effect of small quantities of interstitial elements on the solubility of hydrogen in titanium. These results all suggest that interstitial elements lower the solubility of hydrogen in titanium.[40,42] Such a decrease in solubility would be consistent with a picture of an interstitial solution in which

Table 6
Crystallography of Hydride Formation

Hydride Type	Habit Plane	Crystallographic Relationship
I	{0110}	{0110}//{110} <2110>//<110>
II	{0225}	{0113}//{110} <2110>//<110>

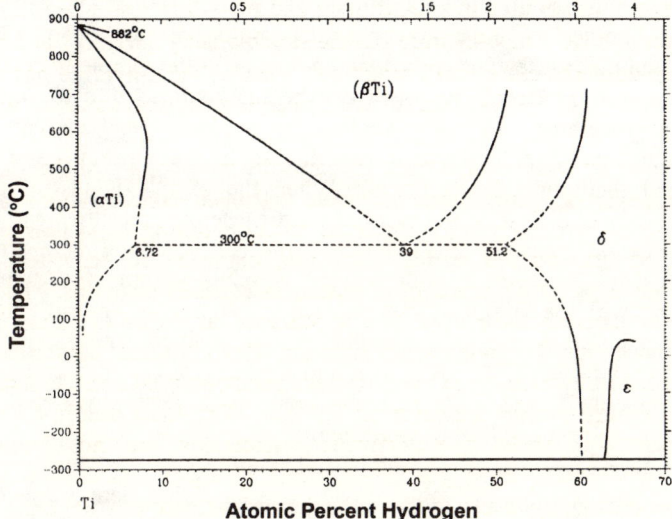

Figure 16. The Ti-H phase diagram. (Reproduced from Ref. 26 by permission of ASM International.)

other interstitial atoms occupy sites that would ordinarily be available to hydrogen. Results have also shown that impurities present in the titanium shift the α - β equilibrium line to lower hydrogen values.[42]

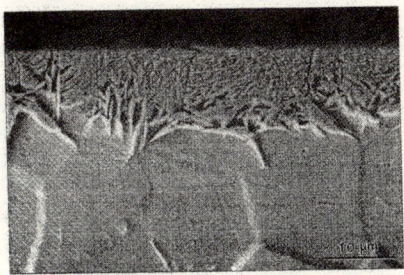

Figure 17. A layer of hydrides on grade 2 titanium. This structure is typical of γ and δ-hydrides.

Given that small amounts of hydrogen will cause hydrides to form in the sample, we must now consider sources of hydrogen. The hydrogen that causes this embrittlement can enter the sample in several ways. In some cases, hydrogen can be incorporated into the metal during processing. Melting, welding, or pickling operations can all produce atomic hydrogen that can diffuse into the sample and be retained there in solid solution. When the sample is tested, this hydrogen will diffuse to the crack tip and cause the embrittlement. In other cases, the hydrogen is external to the sample during the test and diffuses into the sample while the test is being performed. Again it interacts with the sample at the crack tip and causes the embrittlement. In the laboratory, samples are often tested in hydrogen gas, but in practice hydrogen embrittlement is most often realized in a corrosive environment in which the cathodic reaction produces hydrogen, as described in terms of the reactions below.

Figure 18a shows the cathodic polarization curve for titanium in NaCl solution with a pH value of one. It is clear that there are several places on the curve where the slope goes through an abrupt change. These can be associated with a change in the dominant cathodic reaction.[26] From the corrosion potential to the first knee in the curve, this reaction is the reduction of oxygen:

$$\frac{1}{2} O_2 + H_2O + 2 e^- \leftrightarrow 2 OH^- \qquad (4)$$

After the sharp knee in the curve, the corrosion current increases. This increase occurs because reduction of hydrogen ions,

$$2 H^+ + 2 e^- \leftrightarrow H_2 \qquad (5)$$

becomes the dominant reaction. Finally, there is another sharp knee in the curve, at which the dominant cathodic reaction changes to reduction of water,

$$H_2O + e^- \leftrightarrow OH^- + \frac{1}{2} H_2 \qquad (6)$$

Either reaction 2 or 3 generates hydrogen and thus could lead to hydrogen absorption by the metal and hydride formation. In a neutral or basic solution, the direct reduction of hydrogen ions will never be a dominant cathodic reaction. Therefore, as shown in Figure 18b, the

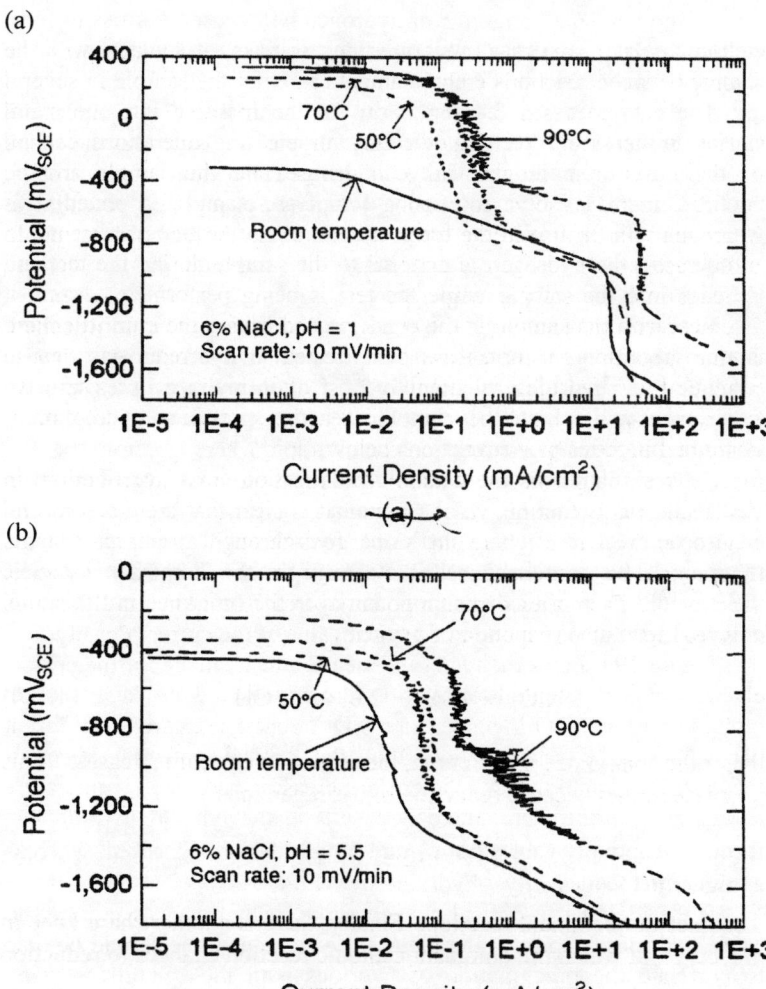

Figure 18. The cathodic portion of the polarization curves for titanium samples. Tests were run at the indicated temperatures. The curves in Figure 18a were obtained from tests performed in 3.5% NaCl with a pH value of one and the curves in Figure 18b were obtained from tests performed in 3.5% NaCl with a pH values of 5.5. (Data were taken from Ref. 28.)

cathodic polarization curve will only have one knee associated with the change between reactions 4 and 6 above.

The curves in Fig. 18 also show that an increase in temperature causes an increase in the cathodic current density. Note that the shape of the curve remains basically unchanged but that the corrosion potential increases with increasing temperature, and the potential at which there is a transition form one dominant cathodic reaction to another increases. More importantly, the cathodic current density increases with increasing temperature.

Research has shown that in a gaseous hydrogen environment there is a critical pressure, for a given temperature, that is required to cause formation of hydrides in titanium.[40,43,44] Below this pressure, an equilibrium will be established in which hydrogen is in solid solution in titanium but does not reach a concentration where precipitates will form. A similar situation exists when the source of hydrogen is an electrochemical reaction. If the amount of hydrogen produced by the electrochemical reaction is sufficient, hydrides will precipitate in the titanium. This result suggests that for a given pH and temperature, there would be a critical potential for hydride formation and that this potential would correspond to a critical value of the current density.

Figure 19a shows the results of measurements made for the critical electrochemical potential for hydride formation in a 3.5%NaCl solution plotted as a function of temperature.[45] Different curves are for different pH values. The results show that the critical potential becomes increasingly cathodic as the pH increases and the temperature decreases. Furthermore, there is a shift in the type of hydride that forms. At low pH values both γ and δ-hydrides are observed, whereas at higher pH values only γ-hydrides are reported.

If the idea that a critical concentration of hydrogen is required to cause hydrides to form in the metal is correct, then one should be able to correlate the precipitation of hydrides with the cathodic current. Figure 19b shows the current density required for hydride formation measured in a deoxygenated solution of NaCl. The results show that hydrides form at the same current density, independent of pH, which suggests that once a critical hydrogen concentration is reached in the environment next to the metal, hydrides will form.

The next issue that must be considered is how the presence of hydrogen and hydrides leads to the embrittlement of titanium. The general picture that has evolved is that hydrogen collects in regions of

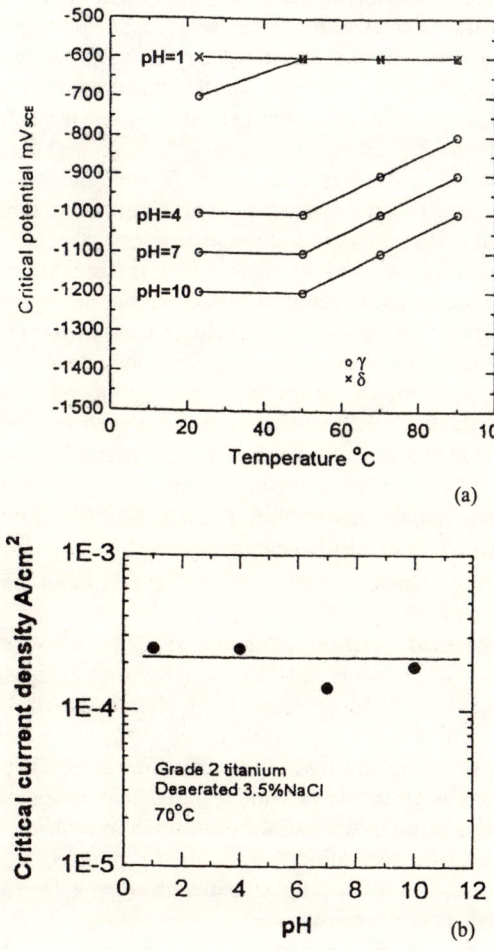

Figure 19 (a) The critical potential for hydride formation in 3.5% NaCl plotted as a function of temperature. Different curves are for different pH values. (b) The critical cathodic current density to cause hydride precipitaton.[45]

tensile stress and that because of its presence the material fails prematurely. In addition, it has been suggested that in titanium the tensile stress aids in nucleation of the hydride, since the precipitation of this phase requires an increase in volume. As a result of numerous studies it appears that hydrogen alone can impair the mechanical properties of titanium but that the presence of the hydride has a more significant effect.[41, 46-48]

This general model has been quantified by the work of Shih, *et al*.[41]. They examined hydrogen cracking of titanium in an environmental cell in a transmission electron microscope. Their results showed that at a stress intensity of less than the critical stress intensity for hydride fracture, hydrogen will first collect at the crack tip and then stress-induced hydride growth will occur in front of the crack tip. This region is favored both because of the dilatation in front of the crack tip and also because of the influx of hydrogen to that region. These hydrides will continue to grow, largely influenced by their own stress field. If the hydrides do not form directly at the crack tip, the crack will either advance to the hydride or the hydride will grow to the crack tip. The crack will remain stable until $K_{local} > K_{IC}^{hydride}$, and then it will propagate through the hydride plate until it reaches the α-solid solution. Since $K_{local} < K_{IC}^{\alpha}$, the crack will stop at that point and the process will repeat itself.

These detailed TEM studies also led to other important observations. The first is that the cracks always arose within the hydrides. Previous SEM studies[48] have shown that the fracture was transgranular, but it was not possible to tell if the cracking was through the hydrides or along the hydride-matrix interface. The second was that if the crack growth rate became too high, the crack could continue into the ⎯-solid solution without the formation of hydrides. That is, the kinetics of hydride formation could not keep up with the growing crack. When this mode of fracture occurred, the hydrogen in solid solution enhanced crack growth.

Given that hydrides and hydrogen can adversely affect mechanical properties, we must next consider penetration of hydrogen into the metal. This can occur by atomic diffusion and hydride growth. Figure 20 plots the diffusion coefficient as a function of inverse temperature, using three different values reported in the literature.[49-51] These same three values can be used to calculate the diffusion distance, x, for hydrogen for a time of one hour (x= $\sqrt{Dt}$). From this information, one finds that at room temperature diffusion distances on the order of one micron. At 100 °C, this distance has increased to almost ten microns. The importance of these results is that, even at room temperature, hydrogen can diffuse over significant distances and allow hydride formation and hydrogen embrittlement to occur.

Another important consideration is the rate of penetration of a hydride layer into the metal. In the model of Shih et al.[41], hydrides form ahead of a growing crack. However, in many applications hydrides form on the surface and slowly penetrate into the bulk as hydrogen diffuses in and exceeds the solubility limit[17]. Since this hydride is brittle,[35] this penetration can reduce the toughness of the material. Therefore, it is important to determine the rate at which this penetration occurs. Studies have shown that at potentials as cathodic as -1600V_{SCE} a hydrides can penetrate up to 150 mm in approximately 100 days.[17] At less cathodic potentials, the penetration rate decreases significantly; for example at -1400V_{SCE} this distance is approximately one-half the value obtained at -1600 V_{SCE}. In general, the growth rate is linearly proportional to the square root of time, which would be consistent with diffusion-controlled growth.[52,53] Also, the penetration rate increases rapidly with increasing temperature.

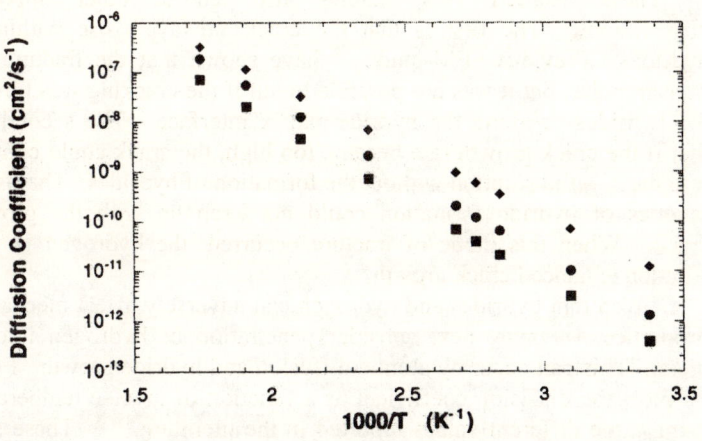

Figure 20. The diffusion coefficient of hydrogen in titanium plotted as a function of inverse temperature. [Data taken from Refs. 49 (circles), 50 (squares) and 51 diamonds).]

VI. SUMMARY

In this review, we have presented the basic electrochemistry, corrosion, and hydrogen embrittlement of titanium. The goal has been to demonstrate that, in many cases, electrochemistry determines whether or not corrosion or hydrogen embrittlement can occur. All research and engineering experience indicates that titanium is often resistant to corrosion, although its use in strong acids or in conditions where pitting can occur should be avoided. However, it can be very susceptible to hydrogen absorption, hydride formation, and hydrogen embrittlement.

REFERENCES

[1] M. Pourbaix, *Atlas of Electrochemical Equilibria in Aqueous Solutions*, M. 2nd edition, NACE, Houston, 1974.
[2] S. Sato, K. Nagata, Y. Watanabe, T. Nakamura, and T. Hamada, *Corrosion, Eng. Japan* **25** (1976) 311.
[3] K. Shimogori, H. Sato, F. Kamikubo, T. Fukuzuku, *Desalination* **22** (1977) 403.
[4] K. Kashida, S. Monisaki, M. Oshiyama, K. Suzuki, and T. Yamamoto, *Desalination* **97** (1994) 147.
[4] H. Satoh, *Desalination*, **97** (1994) 45.
[5] *Guidelines for Use of Titanium in Seawater*, The Society of Chemical Engineering, Japan, The Japan Petroleum Institute, The Japan Titanium Society, 1994.
[6] *Titanium, The Choice*, Ed. by the Titanium Development Association, Dayton, 1987.
[7] *Titanium Tubing*, TIMET Corporation, Denver, 1996.
[8] B. Little, P. Wagner, and F. Mansfeld, *Internatl. Metals Reviews* **36** (1991) 253.
[9] R.I. Jaffee, *Journal of Metals* **203**(2) (1955) 247.
[10] R. W. Schutz and D. E. Thomas, in: *Metals Handbook* Vol. 13, ASM International, Metals Park, Ohio, 1987, p. 669.
[11] R. R. Boyer, in: *Metals Handbook*, Vol. 9, ASM, Metals Park, Ohio, 1987, p. 458.
[12] A. J. Sedriks, *Materials Performance* **33**(2) (1994) 56.
[13] G. Venkataraman and A.D. Goolsby, *Corrosion 96* Paper 554.
[14] R. W. Schutz, *Offshore Technology Conference*, Paper 6909, , Houston, May, 1992, p. 319.
[15] D. A. Litvin and D. E. Smith, *Naval Eng. J.* **83**(10) (1971) 37.
[16] R.W. Schutz and J.S. Grauman, *Corrosion 89*, paper 110, NACE, Houston, 1989.
[17] S. Sato, T. Fukuzuka, K. Shimogori, and H. Tanabe, *Second International Conf. on Hydrogen in Metals*, Paris, Paper 6A1.
[18] M. Stern and A. Wissenberg, *J. Electrochemical Soc.* **106** (1959) 755.
[19] J.F. McAleer and L.M. Peter, *J. Electrochem. Soc.* **129** (1982), 1252.
[20] G. Blondeau, M. Froelicher, M. Froment, and A. Hugot-LeGoff, *J. Less Common Metals* **56** (1977) 215.
[21] D. G. Kolman and J. R. Scully, *J. Electrochem. Soc.* **143** (1996) 1847.
[22] A. Prusi, Lj. Arsov, B. Haran, and B. N. Popov, *J. Electrochem. Soc.* **149** (2002) B491.

[23] M.A.M Ibrahim, D. Pongkao, and M. Yoshimura, *J. Solid State Electrochem.* **6** (2003) 341.
[24] D. Schlain, *Bureau of Mines Bulletin 619*, US Department of the Interior, 1966.
[25] L. C. Covington, *Corrosion* **35** (1978) 278.
[26] L. C. Covington and R. W. Schutz, *ASTM-STP* **728** (1981)163.
[27] Z. F. Wang, C. L. Briant, and K. S. Kumar, *Corrosion* **55** (1999) 128.
[28] E.Mccafferty and G. K. Hubler, *J. Electrochem. Soc.* **125** (1978) 1892.
[29] N. D. Green, C. R. Bishop, and M. Stam, *J. Electrochem. Soc.* **108** (1961) 836.
[30] I. Dugdale and J. Cotton, *Corrosion Sci.* **4** (1964) 397.
[31] T. R. Beck, *J. Electrochem. Soc.* **120** (1973) 1317.
[32] T. R. Beck, *J. Electrochem. Soc.* **120** (1973)1310.
[33] N. Casillas, S. Charlebois, W. H. Sougel, and H. S. White, *J. Electrochem. Soc.* **141** (1994) 636.
[34] C. J. Beevers, M. R. Warren, and D. V. Edmonds, *J. Less Common Metals* **14** (1968) 387.
[35] *Phase Diagrams*, ASM, Metals Park, Ohio.
[36] H. Nakamura and M. Koiwa, *Acta Metall.* **32** (1984) 1799.
[37] O.T. Woo and G.J.C. Carpenter, *Scripta Metall.* **19** (1985) 931.
[38] O. T. Woo, G. C. Weatherly, C. E. Coleman, and R. W. Gilbert, *Acta Metall.* **33** (1985) 1897.
[39] William E. Mueller, in: *Metal Hydrides*, Ed. by W. M. Mueller, J.P. Blackledge, and G. G. Libowitz, Academic Press, New York, 1968, p. 336.
[40] D. S. Shih, I. M. Robertson, and H. K. Birnbaum, *Acta Metall.* **36** (1988) 111.
[41] A. D. McQuillen, *J. Inst. Metals* **79** (1951) 371.
[42] T. R. P. Gibb, Jr. and H. W. Kruschwitz, Jr., *J. Amer. Chem.Soc.* **72** (1956) 5155.
[43] C. L. Briant, Z. F. Wang, and N. Chollocoop, *Corrosion Sci.* **44** (2002) 1875.
[45] Z. F. Wang, C. L. Briant, K. S. Kumar, and N. Chollacoop, *Corrosion 99*, paper 99154, NACE, Houston, 1999.
[46] G. A. Lenning, C. M. Craighead,and R. I. Jaffee, *Trans. AIME* **200**(3) (1954) 367.
[47] M. L. Grossbeck and H. K. Birnbaum, *Acta Metall.* **25** (1977) 135.
[48] N. E. Paton and J. C. Williams, in: *Hydrogen in Metals*, Ed. by. I.M. Bernstein and A.W. Thompson, ASM, Metals Park, Ohio, 1974, p. 409
[49] I. P. Phillips, P. Poole, and L. L. Shreir, *Corrosion Science* **14** (1974) 533.
[50] T. P. Papazoglou and M. T. Hepworth, *Trans. AIME* **242** (1968) 682.
[51] J. Wasilweski and G. L. Kehl, *Metallurgia* **50** (1954) 225.
[52] L. Lunde, R. Nyborg, and H. Koonyman, Document No. IFE/KR/F-88/150, Institute for Energy Technology, Kjeller, Norway, May, 1985.
[53] I. P. Phillips, P. Poole, L. L. Shreir, *Corrosion Science* **12** (1972) 855.

3

Electrochemical Oxidation of Organics on Iridium Oxide and Synthetic Diamond Based Electrodes

György Fóti and Christos Comninellis

Unit of Electrochemical Engineering, Institute of Chemical and Biological Process Science, School of Basic Sciences, Swiss Federal Institute of Technology
CH-1015 Lausanne, Switzerland

1. INTRODUCTION

Oxidative electrochemical processes promising versatility, environmental compatibility and cost effectiveness have a continuously growing importance both in selective organic synthesis and in the electrochemical incineration (ECI) of organic pollutants in aqueous media. In the case of organic electrosynthesis selectivity is to be enhanced and in the ECI process the aim is the mineralization of the toxic and non-biocompatible pollutants with high current efficiency.[1-10] Anodic oxidation of organics may proceed by several mechanisms including direct and indirect oxidation.

In direct electrochemical oxidation, electron exchange occurs between the organic species and the electrocatalytic electrode surface. A typical example is the oxidation of organic compounds on platinum anodes at low anodic potentials. The main problem with electrocatalytic anodes of platinum group metals is the decrease of the catalytic activity during use when proceeding oxidation of organics at a fixed anodic potential, before oxygen evolution. This is mainly due to the adsorption of reaction intermediates (mainly CO) at the anode surface, commonly called poisoning effect.[11-13]

In indirect electrochemical oxidation, the organics do not exchange with the surface directly but through intermediation of some electro active species. This intermediation may be homogeneous or heterogeneous. Homogeneous mediators (e.g. Ag^{2+}/Ag^+, Ce^{4+}/Ce^{3+}, Mn^{3+}/Mn^{2+}) dissolved in the electrolyte react with the organics in a homogeneous chemical reaction then are regenerated at the electrode surface. Such processes require subsequent separation of the products from the mediator. In heterogeneous mediation the mediator is fixed at the surface of the electrode. Both its reaction with the organics and its regeneration take place at the electrode surface. An advantage over the homogeneous mediators is that no contamination of the reaction mixture with the redox mediator occurs.

A new mechanism of the indirect electrochemical oxidation of organics in aqueous media, based on intermediates of the oxygen evolution reaction, has been proposed by D.C. Johnson, et al.[14-19] The process involves anodic oxygen transfer from H_2O to organics *via* hydroxyl radicals formed by water discharge. A typical example is the oxidation of organic compounds on Bi^{5+} or Fe^{3+} doped Ti/PbO_2 electrodes.[14-19]

In our laboratory we have found that frequently electrochemical oxidation of some organics in aqueous media occurs, without any loss in electrode activity, only at high potentials with concomitant evolution of O_2.[1-5] Furthermore it has been found that the nature of electrode material influences strongly both the selectivity and the efficiency of the process.[1,6,20-22]

In order to interpret these observations a comprehensive model for anodic oxidation of organics in acid medium, including competition with oxygen evolution has been proposed.[1,6,20-22] This model permits to distinguish between two limiting cases: 'active' and 'non-active' anode. Figure 1 illustrates the reaction scheme in acid medium (e.g. $HClO_4$), where M designates an active site at the anode surface.[20-22] In all cases, the initial step is the discharge of water molecules to form adsorbed hydroxyl radicals (reaction 1):

$$M + H_2O \rightarrow M(-OH) + H^+ + e^- \qquad (1)$$

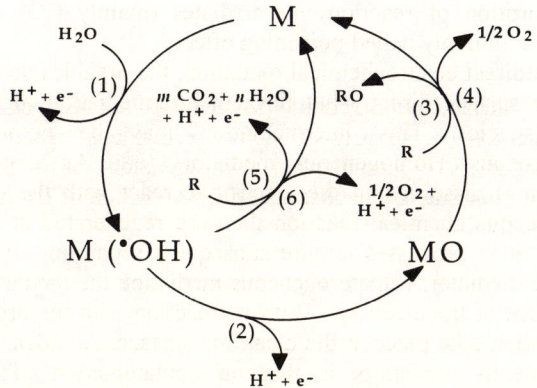

Figure 1. Scheme of the electrochemical oxidation of organic compounds on 'active' anodes (reactions 1, 2, 3, 4) and on 'non-active' anodes (reactions 1, 5, 6). (1) water discharge to hydroxyl radicals; (2) formation of the higher metal oxide; (3) partial (selective) oxidation of the organic compound, R, *via* the higher metal oxide; (4) oxygen evolution by chemical decomposition of the higher metal oxide; (5) combustion of the organic compound *via* hydroxyl radicals; (6) oxygen evolution by electrochemical oxidation of hydroxyl radicals. (Reprinted from Fóti, D. Gandini, Ch. Comninellis, A. Perret, and W. Haenni, *Electrochem. Solid-State Lett.* **2**, 228 (1999). Copyright © 1999 with permission from The Electrochemical Society, Inc.)

The electrochemical and chemical reactivity of the adsorbed hydroxyl radicals depends strongly on the nature of electrode material used. Two extreme classes of electrodes can be defined: 'active' and 'non-active' electrodes:

a) At 'active' electrodes there is a strong electrode (M) - hydroxyl radical ($^{\bullet}$OH) interaction. In this case, the adsorbed hydroxyl radicals may interact with the anode with possible transition of the oxygen from the hydroxyl radical to the anode surface, forming the so-called higher oxide (reaction 2). This may be the case when higher oxidation states on the surface electrode are available above the thermodynamic potential for oxygen evolution (1.23 V/RHE).

$$M(^{\bullet}OH) \rightarrow MO + H^+ + e^- \qquad (2)$$

The surface redox couple MO/M can act as mediator in the oxidation of organics at 'active' electrodes (reaction 3). This reaction is in competition with the side reaction of oxygen evolution due to the chemical decomposition of the higher oxide (reaction 4).

$$MO + R \rightarrow M + RO \qquad (3)$$

$$MO \rightarrow M + \tfrac{1}{2} O_2 \qquad (4)$$

The oxidation reaction *via* the surface redox couple MO/M (reaction 3) may result in the partial (selective) oxidation of organics.

b) At 'non-active' electrodes there is a weak electrode (M) - hydroxyl radical ($^{\bullet}$OH) interaction. In this case, the oxidation of organics is mediated by hydroxyl radicals (reaction 5), which may result in fully oxidized reaction products such as CO_2.

$$M(^{\bullet}OH) + R \rightarrow M + m\,CO_2 + n\,H_2O + H^+ + e^- \qquad (5)$$

In this schematic reaction equation, R is the fraction of an organic compound, which contains no heteroatom and needs one atom of oxygen to be transformed to fully oxidized elements. (Actual values of m and n depend on the elemental composition of R to be oxidized). This reaction is in competition with the side reaction of hydroxyl radical discharge (direct or indirect through formation of H_2O_2 as intermediate) to O_2 (reaction 6) without any participation of the anode surface:

$$M(^{\bullet}OH) \rightarrow M + \tfrac{1}{2} O_2 + H^+ + e^- \qquad (6)$$

The distinction between 'active' and 'non-active' behavior and the underlying mechanistic explanation are supported by several experimental observations, including measurement of the concentration of reactive intermediates in the oxygen evolution reaction, such as hydroxyl radicals produced by discharge of water. In fact, hydroxyl radicals produced at 'non-active' electrodes can be intercepted by using *p*-nitrosodimethylaniline as a selective scavenger.[6, 34]

In practice, however, most anodes will exhibit a mixed behavior since both parallel reaction paths will participate in organics oxidation and oxygen evolution reactions. As a general rule, the closer the reversible potential of the surface redox couple to the potential of oxygen evolution, the higher is the active character of the anode. We expect also that inert electrodes with weak adsorption properties are ideal non-active electrodes.

In this Chapter, dimensionally stable type anodes (DSA) based on IrO_2 (Ti/IrO_2) and synthetic boron-doped diamond (p-Si/BDD) electrodes are compared in oxidation of organics in acid media (1M H_2SO_4 or 1M $HClO_4$). Ti/IrO_2 may serve as a typical example of 'active' anode (the redox potential of the IrO_3/IrO_2 couple, 1.35V/RHE, is close to the standard potential of O_2 evolution, 1.23 V/RHE) and is thus suitable for electro synthesis, while p-Si/BDD, a typical 'non-active' electrode (BDD is known to be inert and to have weak adsorption properties), is anticipated to be an ideal anode for ECI.[22]

Kinetic models of organics oxidation competing with oxygen evolution at 'active' and 'non-active' type anodes are proposed and confirmed by preparative electrolysis using different classes of organic compounds.

II. ELECTROCHEMICAL OXIDATION OF ORGANICS IN AQUEOUS MEDIA USING ACTIVE ANODES

Here we have already defined as active anodes the electrodes, which take part in anodic reaction *via* redox catalysis. The redox system can serve as an electron transfer relay as well as a new surface for adsorption of reactants. The redox system plays only a temporary role and, in principle, it is not consumed in the overall process. F. Beck has introduced the term of "redox catalysis" in electrochemistry.[23] The same author has also proposed to use DSA type electrodes based on Cr_2O_3 (Ti/Cr_2O_3) and $CoMn_2O_4$ (Ti/$CoMn_2O_4$) for the oxidation of organic compounds.[24-26] However, a drawback of these electrodes is the low electrode service life due to the dissolution of the active coating even in absence of organics.[24-26]

Dimensionally stable anodes based on IrO_2 are very stable even at high current density and in strongly acid media.[27] Moreover, the redox potential of the surface redox catalyst IrO_3/IrO_2 (1.35 V/RHE) is close to the thermodynamic standard potential for O_2 evolution (1.23 V/RHE). This can allow predicting that this material is an ideal 'active' anode, in which oxidation of organics and oxygen evolution takes place *via* redox catalysis.

In fact, using *p*-nitrosodimethylaniline as a selective scavenger it has been found that hydroxyl radicals are practically not produced at the Ti/IrO_2 anode.[6] Furthermore, we can expect that IrO_2 based anodes are very active (low overpotential) for O_2 evolution reaction. The onset potential for O_2 evolution on Ti/IrO_2 anodes is indeed very close to the redox potential of the IrO_3/IrO_2 couple (1.35 V/RHE) as shown in Fig. 2.

In this review we present results concerning the oxidation of some model organic compounds on Ti/IrO_2 electrodes prepared by two different techniques: thermal decomposition[27] and pulse heating.[28] The

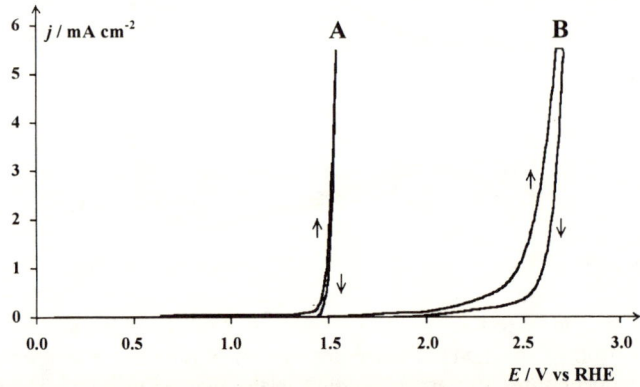

Figure 2. Cyclic voltammograms of Ti/IrO_2 (**A**) and p-Si/BDD (**B**) electrodes in 1M H_2SO_4. Scan rate: 50 mV s^{-1}; $T = 25°C$. Reprinted from I. Duo, P.-A. Michaud, W. Haenni, A. Perret, and Ch. Comninellis, *Electrochem. Solid-State Lett.* **3**, 325. Copyright © 2000 with permission from The Electrochemical Society, Inc.[40]

Electrochemical Oxidation of Organics

latter technique allows for a rapid decomposition of the precursor. Two cases will be distinguished: organics oxidation in a potential region below or above the potential of O_2 evolution. Then, a simple kinetic model will be presented which describes in a quantitative manner the partial (selective) oxidation at Ti/IrO_2 electrodes.[20, 21]

1. Oxidation of Organics on Ti/IrO$_2$ Anodes in the Potential Region before O$_2$ Evolution

The electrochemical activity of Ti/IrO_2 electrodes, prepared by the thermal decomposition technique, has been studied with various organic compounds in acid medium. As organic model compounds, simple alcohols (methanol, ethanol, 1-propanol, *i*-propanol, *tert*-butanol),

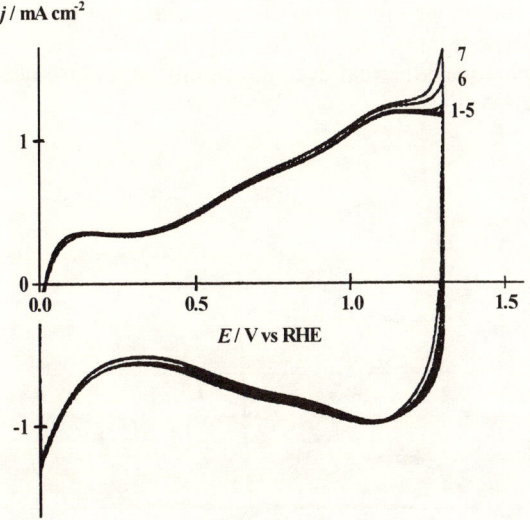

Figure 3. Cyclic voltammograms at a Ti/IrO_2 electrode, prepared by thermal decomposition technique, recorded at different *i*-propanol concentrations: **(1)** 0M; **(2)** 0.1M; **(3)** 0.25M; **(4)** 0.5M; **(5)** 1M; **(6)** 2.5M; **(7)** 5M. Scan rate: 50 mV s^{-1}; electrolyte: 1M H_2SO_4; $T = 25°C$. Reprinted from M.A. Rodrigo, P.-A. Michaud, I. Duo, M. Panizza, G. Cerisola, and Ch. Comninellis, *J. Electrochem. Soc.* **148**, D60 Copyright © 2001 with permission from The Electrochemical Society, Inc.[38]

phenol and carboxylic acids (formic acid, oxalic acid, maleic acid) have been chosen.[20] This choice is due either to their potential use for synthesis in combustion fuel cells or to their role as intermediate products in the electrochemical degradation of aromatic pollutants. Series of voltammograms were determined as a function of the concentration of the organic compound in the range of zero to 5M, depending on the solubility in the supporting electrolyte. The cyclic voltammograms (CV) were recorded with a scan rate of 50 mV s^{-1} in 1M H_2SO_4 at 25°C.

The CV of i-propanol on Ti/IrO_2 electrodes is illustrated in Figure 3. The voltammograms recorded below oxygen evolution display no significant change in presence of i-propanol with respect to the voltammogram of the supporting electrolyte. The only difference is a slight decrease in the onset potential of oxygen evolution, indicating an effect of the organic compound on the overpotential of oxygen evolution.[20]

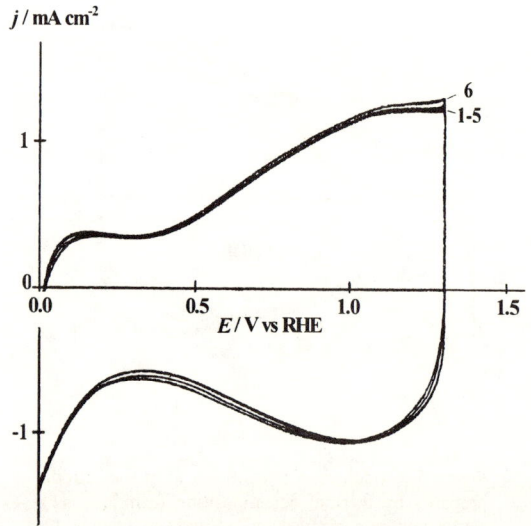

Figure 4. Cyclic voltammograms at a Ti/IrO_2 electrode, prepared by thermal decomposition technique, recorded at different tert-butanol concentrations: (**1**) 0M; (**2**) 0.1M; (**3**) 0.25M; (**4**) 0.5M; (**5**) 1M; (**6**) 2.5M. Scan rate: 50 mV s^{-1}; electrolyte: 1M H_2SO_4; T = 25°C.

In presence of methanol, ethanol, 1-propanol and formic acid, the Ti/IrO$_2$ electrodes show the same behavior as with i-propanol yielding pictures identical to Fig. 3. In the case of *tert*-butanol, even the starting potential of oxygen evolution does not seem to be affected by increasing concentration of the organics, as illustrated by Fig. 4. One can conclude that in the region of water stability the Ti/IrO$_2$ electrode shows no significant electrocatalytic activity in oxidation of any of the tested alcohols and carboxylic acids.

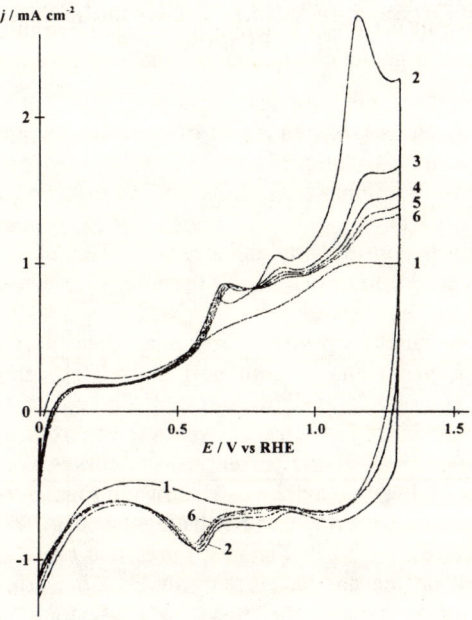

Figure 5. Oxidation of phenol at a Ti/IrO$_2$ electrode, prepared by thermal decomposition technique. (**1**): cyclic voltammogram recorded with no phenol. (**2-6**): evolution of cyclic voltammograms at 0.1M phenol concentration with increasing number of polarization cycles: (**2**) 1st; (**3**) 4th; (**4**) 7th; (**5**) 10th; (**6**) 13th. Scan rate: 50 mV s^{-1}; electrolyte: 1M H$_2$SO$_4$; $T = 25°C$.

On the other hand, in the oxidation of phenol the Ti/IrO_2 electrode displays a significant but rapidly changing catalytic activity, as seen in Fig. 5. This system was examined at one concentration (0.1M) under continuous potential cycling. Several oxidation and reduction peaks can be observed in the first scan, among them a high oxidation peak around 1.10 V/RHE which disappears after some ten cycles, whereas a pair of anodic and cathodic peaks around 0.60 V/RHE remains significant after stabilization of the system. The rapid decrease in the catalytic activity is probably due to the deposition of polymeric product at the surface of the electrode.[20]

2. Oxidation of Organics on Ti/IrO_2 Anodes in the Potential Region of O_2 Evolution

Cyclic voltammetric measurements showed that in absence of electrolytically generated oxygen the Ti/IrO_2 electrodes prepared by the thermal decomposition technique have no electrocatalytic activity in the oxidation of various aliphatic alcohols and carboxylic acids in acid medium. However, in the potential region of oxygen evolution the electrode activity is considerably enhanced.[20-22] This observation gives reason for the study of the oxidation of organics in the potential region of O_2 evolution.

Low carbon number primary, secondary and tertiary alcohols (1-propanol, *i*-propanol and *tert*-butanol) as well as carboxylic acids (formic, oxalic and maleic acid) were tested. The concentration of the organic compounds in 1M H_2SO_4 was varied from zero to 5M depending on solubility. Anodic current density curves measured with a linear scan rate of 20 mV s^{-1} at 25°C are shown in Figs. 6-9.

Several types of behavior can be distinguished: *(i) tert*-butanol is completely inactive as seen in Fig. 6, its presence up to 2.5M has no significant effect on the anodic current curves; *(ii) i*-propanol clearly diminishes the overpotential, the curves are essentially parallel and shifted to lower potentials with increasing concentration (see Fig. 7); *(iii)* maleic acid exhibits an opposite behavior as shown in Fig. 8, giving rise to a significant increase in the anodic overpotential; *(iv)* finally, a more complex behavior was observed with 1-propanol, formic acid and

Electrochemical Oxidation of Organics

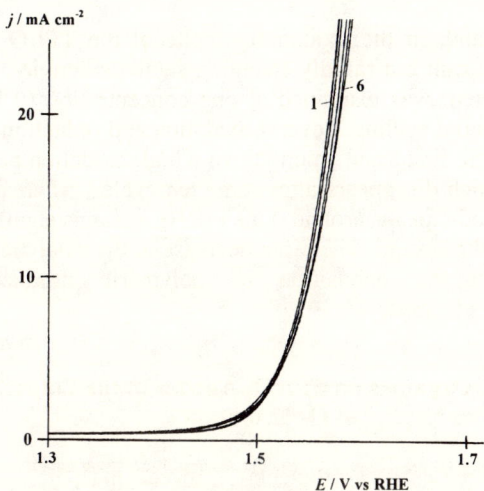

Figure 6. Effect of *tert*-butanol on the polarization curves for water discharge at a Ti/IrO$_2$ electrode prepared by the thermal decomposition technique. The *tert*-butanol concentrations are as in Fig. 4. Scan rate: 20 mV s^{-1}; electrolyte: 1M H$_2$SO$_4$; $T = 25°C$.

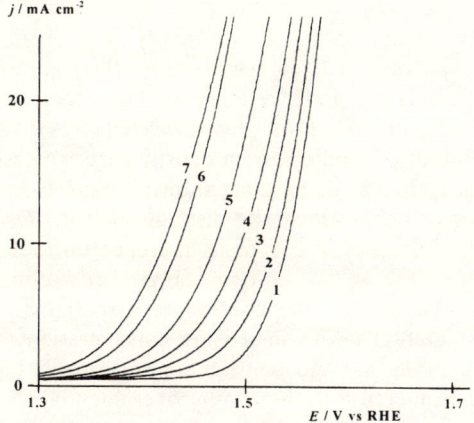

Figure 7. Effect of *i*-propanol on the polarization curves for water discharge at a Ti/IrO$_2$ electrode prepared by thermal decomposition technique. *i*-propanol concentrations as in Fig. 3. Scan rate: 20 mV s^{-1}; electrolyte: 1M H$_2$SO$_4$; $T = 25°C$.

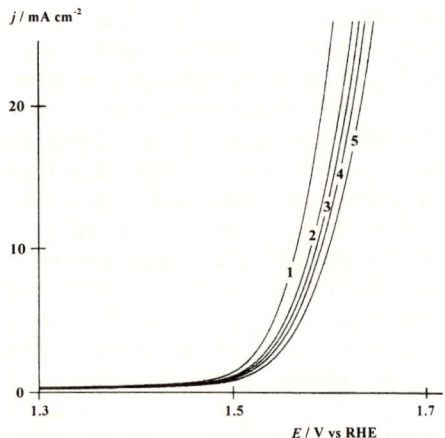

Figure 8. Effect of maleic acid on the polarization curves for water discharge at a Ti/IrO$_2$ electrode prepared by the thermal decomposition technique. Maleic acid concentrations: (**1**) 0M; (**2**) 0.1M; (**3**) 0.25M; (**4**) 0.5M; (**5**) 1M. Scan rate: 20 mV s^{-1}; electrolyte: 1M H$_2$SO$_4$; T = 25°C.

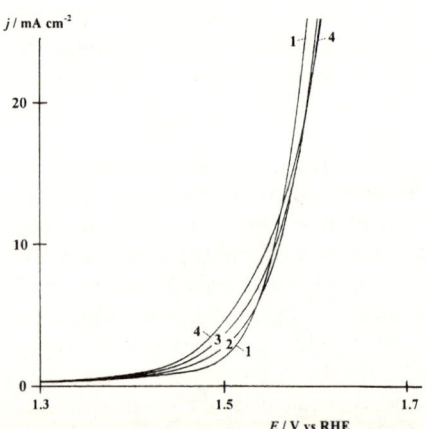

Figure 9. Effect of oxalic acid on the polarization curves for water discharge at a Ti/IrO$_2$ electrode prepared by thermal decomposition technique. Oxalic acid concentrations: (**1**) 0M; (**2**) 0.1M; (**3**) 0.25M; (**4**) 0.5M. Scan rate: 20 mV s^{-1}; electrolyte: 1M H$_2$SO$_4$; T = 25°C.

oxalic acid, where the slope and even the shape of the anodic current curves changes in presence of the organic compound (see the example of oxalic acid in Fig. 9). At lower potentials, increasing concentration enhances the rate of water discharge whereas, at higher potentials, an inhibition becomes predominant. In the case of these latter compounds, a decrease in anodic current was also observed and some ten to twenty minutes were necessary to arrive at a steady state. The curves shown in Fig. 9 were recorded after stabilization of the system. This time dependent inhibition is probably due to adsorption phenomena.

The observed decrease in the anodic overpotential, either in a simple manner as in the case of i-propanol or in a more complex manner as in the case of 1-propanol, formic acid and oxalic acid, might indicate an important change in the electrochemical activity of the Ti/IrO_2 electrode. New reaction paths in the oxidation of the organic compounds are suggested to open, requiring water decomposition intermediates, which are only available in conditions of oxygen evolution. Cyclic voltammetric measurements exceeding the anodic overpotential of oxygen evolution can make this hypothesis evident.

The effect of simultaneous oxygen evolution to the oxidation of organics is further illustrated using Ti/IrO_2 electrodes prepared by pulse heating.[28] When oxygen evolution is excluded, these electrodes exhibit the same inactive voltammetric behavior in presence of organics as shown with electrodes prepared by the thermal decomposition upper potential limit and the concentration of i-propanol. Experiments were made with a scan rate of 50 mV s^{-1} at 25°C using 1M H_2SO_4 as supporting electrolyte.

Typical voltammograms, recorded between 0 and 1.50 V/RHE of Ti/IrO_2 electrodes prepared by the pulse heating technique, are shown in Fig. 10. They exhibit a pair of anodic and cathodic peaks around 0.90 V/RHE, characteristic of iridium electrodes and attributed to the formation of IrO_2 by electrochemical oxidation of metallic iridium present in the coating.[29-32] In the cathodic scan, an intensive reduction peak at about 1.30 V/RHE is also observed. Repetition of the same voltammetric cycles results in a significant increase in current explained by continuous electrochemical oxidation of the electrode surface. For further experiments, the electrodes have been conditioned by continuous cycling in the range of 0 to 1.50 V/RHE requiring typically several hundreds of cycles until the voltammetric signal levels off.

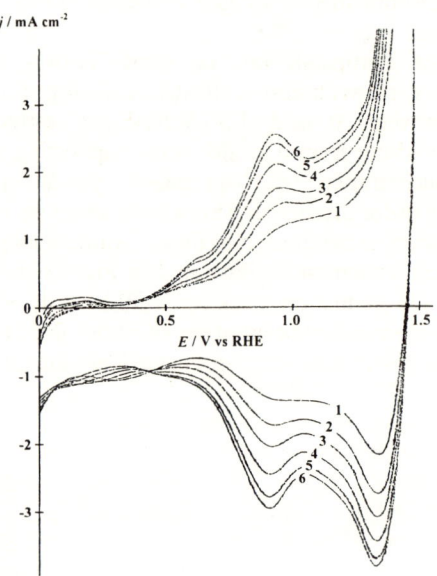

Figure 10. Cyclic voltammograms of a Ti/IrO$_2$ electrode prepared by pulse heating technique. Shape modification with increasing number of polarization cycles: (1) 1st; (2) 10th; (3) 25th; (4) 60th; (5) 100th; (6) 150th. Scan rate: 50 mV s^{-1}; electrolyte: 1M H$_2$SO$_4$; T = 25°C.

Figure 11 shows the effect of the upper potential limit. The voltammograms have been recorded in 1M H$_2$SO$_4$ containing no organics, increasing the inversion potential from 1.36 to 1.56 V/RHE by an increment of 0.02 V. It is seen that the increase in the upper potential limit does not influence the overpotential of oxygen evolution but generates a reduction peak at 1.30 V/RHE. This peak, becoming more and more important with higher inversion potentials, is attributed to the electrochemical reduction of higher oxidized surface sites (IrO$_3$) formed during the anodic scan.[32] The corresponding oxidation peak is masked by the high anodic current of oxygen evolution. In fact, the two phenomena are closely related: on one hand, formation of the higher surface oxide requires adsorbed water decomposition products and, on the other hand, simultaneous oxygen evolution needs surface sites of higher oxide type.

The CV of *i*-propanol on the same Ti/IrO_2 electrodes with simultaneous oxygen evolution is illustrated in Fig. 12. Measurements were made between 0 and 1.50 V/RHE at different *i*-propanol concentrations varying from 0 to 2M in the supporting acid electrolyte. Increasing concentration of *i*-propanol yields a decrease in overpotential of water discharge. This effect, small on the current scale of a voltammogram, is better seen in Fig. 7. Another effect of increasing *i*-propanol concentration, well illustrated in Fig. 12, is the progressive diminution of the reduction peak at 1.30 V/RHE. This apparently disappears at *i*-propanol concentration of about 2M. Both effects are explained with oxidation of *i*-propanol at the higher oxidized

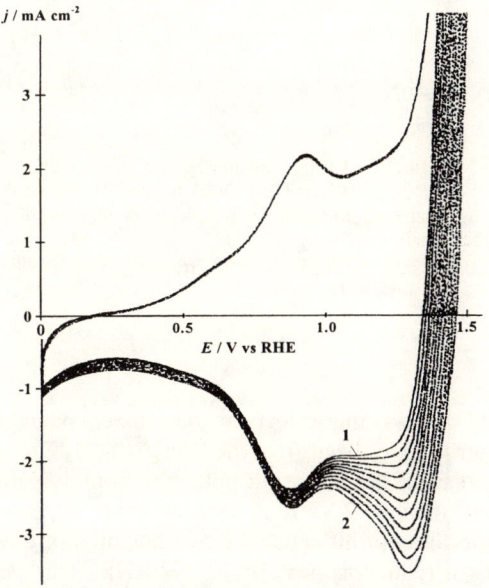

Figure 11. Effect of the upper inversion potential, varied from (**1**) 1.36 to (**2**) 1.56 V/RHE by an increment of 20 mV, on cyclic voltammograms of a Ti/IrO_2 electrode prepared by pulse heating technique. Scan rate: 50 mV s^{-1}; electrolyte: 1M H_2SO_4; $T = 25°C$.

surface sites. The higher the concentration of the organics, the lower the stationary concentration of the surface intermediate, resulting in lowering the potential of the IrO_3/IrO_2 redox couple and in diminishing the current required for reduction of IrO_3 to IrO_2. Comparison of Figs. 3 and 12 clearly shows that electro-oxidation of *i*-propanol on Ti/IrO_2 anodes occurs only under conditions of simultaneous oxygen evolution.[20]

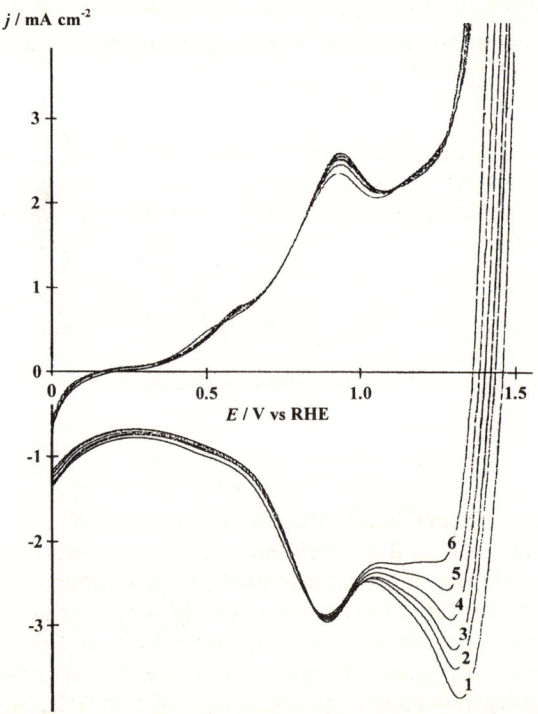

Figure 12. Oxidation of *i*-propanol at a Ti/IrO_2 electrode prepared by pulse heating technique. Cyclic voltammograms recorded at different *i*-propanol concentrations: (**1**) 0M; (**2**) 0.1M; (**3**) 0.2M; (**4**) 0.5M; (**5**) 1M; (**6**) 2M. Scan rate: 50 mV s^{-1}; electrolyte: 1M H_2SO_4; $T = 25°C$.

3. Kinetic Model of Organic Oxidation on Ti/IrO$_2$ Anodes

In this model describing oxidation of organics (R) on Ti/IrO$_2$ anode, it is assumed that both organics oxidation and oxygen evolution take place exclusively *via* intermediation by the same surface active sites of type IrO$_3$, *i.e.* the electrode does not exhibit any 'non-active' character.[20,21]

This assumption simplifies the presented general mechanism (Fig. 1) to give a reaction scheme involving three different reactions. The first reaction, consisting of two consecutive steps, is an electrochemical oxidation leading to the formation of the active species (IrO$_3$) by discharge of water. All other reactions of the intermediate IrO$_2$ (•OH) being excluded, the overall equation of this reaction can be formulated as (reaction 7):

$$IrO_2 + H_2O \xrightarrow{r_0} IrO_3 + 2 H^+ + 2 e^- \qquad (7)$$

The active IrO$_3$ species is then consumed by two competing reactions: partial (selective) oxidation of the organic (reaction 8)

$$IrO_3 + R \xrightarrow{r_1} IrO_2 + RO \qquad (8)$$

and the side-reaction of oxygen evolution (reaction 9)

$$IrO_3 \xrightarrow{r_2} IrO_2 + \tfrac{1}{2} O_2 \qquad (9)$$

both leading to regeneration of the electrode active sites.

A kinetic model of this simplified reaction scheme is developed by making the following further assumptions: (*i*) adsorption of all organic compounds (both R and RO) at the electrode surface is negligible; (*ii*) the reaction of organics oxidation (reaction 8) is of first order with respect both to the organics (R) and to the active species (IrO$_3$); (*iii*) the reaction of oxygen evolution (reaction 9) is of first order with respect to IrO$_3$.[20,21]

Under galvanostatic conditions, the rate of formation of the active species (IrO$_3$), r_0 (mol m^{-2} s^{-1}), at the electrode surface (reaction 7) is given by the equation

$$r_o = \frac{j}{2F} \tag{10}$$

where j is the applied current density (A m^{-2}), F is Faraday's constant (C mol^{-1}) and 2 is for the number of electrons involved. The rate of organics oxidation (reaction 8), r_1 (mol m^{-2} s^{-1}), can be written as

$$r_1 = k_1 \cdot \theta \cdot \Gamma_s \cdot C_R^E \tag{11}$$

where k_1 is the rate constant of the organics oxidation reaction (m^3 mol^{-1} s^{-1}), θ the relative surface coverage of IrO$_3$, Γ_s the saturation surface concentration of this species (mol m^{-2}) and C_R^E the concentration of organic R near the electrode surface (mol m^{-3}). The side-reaction of oxygen evolution (reaction 9) follows the kinetics represented by the equation

$$r_2 = k_2 \cdot \theta \cdot \Gamma_s \tag{12}$$

where r_2 is the reaction rate (mol m^{-2} s^{-1}) and k_2 is the rate constant (s^{-1}) of oxygen evolution.

Using the steady-state assumption $r_o = r_1 + r_2$, the surface coverage of IrO$_3$ can be formulated as

$$\theta = \frac{j}{2F \cdot \Gamma_s \cdot (k_1 \cdot C_R^E + k_2)} \tag{13}$$

This expression shows that the surface coverage of the active species depends on the electrode material (characterized by the parameters Γ_s and k_2). At a given anode, θ increases linearly with the applied current density (j) and decreases with increasing reactivity and concentration of the organics (k_1 and C_R^E). Obviously, the relative surface coverage may only rise up to 1, meaning that in a given system (electrode material, nature and composition of electrolyte) the current is limited.

The current efficiency, η, is defined as the fraction of current used for organics oxidation and is given by the following ratio of reaction rates:

$$\eta = \frac{r_1}{r_o} \tag{14}$$

It can also be expressed as a function of rate constants and organic concentration and written in a convenient linear form as[20,21]

$$\frac{1}{\eta} = 1 + \frac{k_2}{k_1} \cdot \frac{1}{C_R^E} \tag{15}$$

For a given electrode material, the ratio k_2/k_1 expresses the reactivity of the organics to be oxidized: the lower the ratio, the stronger the competition of organics oxidation against oxygen evolution. It is also seen, that increasing concentration of the organic favors its oxidation, resulting in a higher current efficiency.

Equations (10) and (12), formulated as a function of the local concentration of organics at the electrode surface, C_R^E, are not directly applicable since this variable is *a priori* unknown. Since adsorption of organic species has been neglected in this model, the mass transfer in the liquid phase determines the relation between local and bulk concentrations. At steady state, the mass transfer rate is equal to the rate of organics oxidation:

$$k_m \left(C_R^S - C_R^E \right) = k_1 \cdot \theta \cdot \Gamma_s \cdot C_R^E \tag{16}$$

where k_m is the mass transfer coefficient (m s^{-1}) and C_R^S is the concentration of the organic in the bulk solution (mol m^{-3}). The local concentration is then related to the bulk concentration by the equation

$$C_R^E = \frac{C_R^S}{1 + \phi} \tag{17}$$

where the parameter ϕ is defined as:

$$\phi = \frac{k_1}{k_m} \cdot \theta \cdot \Gamma_s \tag{18}$$

From Eqs. (15) and (17), one obtains the following expression of current efficiency as a function of the bulk organic concentration.[20,21]

$$\frac{1}{\eta} = 1 + \frac{k_2}{k_1}(1+\phi)\frac{1}{C_R^S} \qquad (19)$$

When using Eq. (19) one should remember that, in the general case, ϕ is also dependent on the organic concentration.

Obviously, the model considerably simplifies when neglecting concentration polarization. In fact, if the rate of oxidation (r_1) is much slower than the diffusion of organics towards the anode, ϕ tends to zero and the organic concentration at the electrode equals the bulk concentration: $C_R^E = C_R^S$. Thus, when plotting the inverse of current efficiency against the inverse of bulk concentration, a straight line with a slope of k_2/k_1 is predicted. This provides a suitable experimental method to determine the ratio of the rate constants k_2/k_1 by measuring the current efficiency for organics oxidation at different bulk concentrations. Current efficiencies can be determined either by analyzing the oxidation products or by measuring the oxygen evolution rate.

The effect of the mass transfer on current efficiency has also been treated assuming a concentration independent ϕ.[21] This description is limited to low surface coverage of the active species, involving a strong limitation of the applicable current density.

In order to get an experimental verification of the model in a simple case, oxidation of *i*-propanol in 1M H_2SO_4 at 20 to 60°C was investigated on Ti/IrO$_2$ electrodes prepared by the thermal decomposition technique.[29] Under the chosen experimental conditions (high organic concentration, low conversion) no important diffusion limitation is expected ($\phi \approx 0$). The applied current density ranged from 30 to 60 mA cm^{-2}. For almost all investigated conditions the oxidation of *i*-propanol at Ti/IrO$_2$ anodes with electrogenerated IrO$_3$ is found to be selective yielding acetone as the oxidation product (reaction 20):

$$(CH_3)_2CHOH + IrO_3 \rightarrow (CH_3)_2CO + IrO_2 + H_2O \qquad (20)$$

No combustion of the organic was observed, as foreseen when using 'active' type electrodes. A typical example of *i*-propanol oxidation to acetone is given in Fig. 13. Note that even at i-propanol conversions higher than 0.9, the selectivity of acetone production is better than 80%.[22] Fig. 14 shows the inverse of current efficiency as a function of the inverse of bulk i-propanol concentration at different current densities. The plot is linear with an intersect close to unity, as predicted by the kinetic model. It is also seen that, up to 50 mA cm^{-2}, the current efficiency is fairly independent of the current density, as it should be when diffusion limitation is negligible. It was also found that the slope of the $1/\eta$ vs. $1/C_R^S$ plot, which equals the ratio k_2/k_1, decreases with increasing temperature. This implies that the selective oxidation of i-propanol to acetone by the intermediation of IrO$_3$ active sites has higher activation energy than that of oxygen evolution at the same sites. All these results are in good agreement with the proposed kinetic model and confirm its applicability for the case of negligible diffusion limitation.

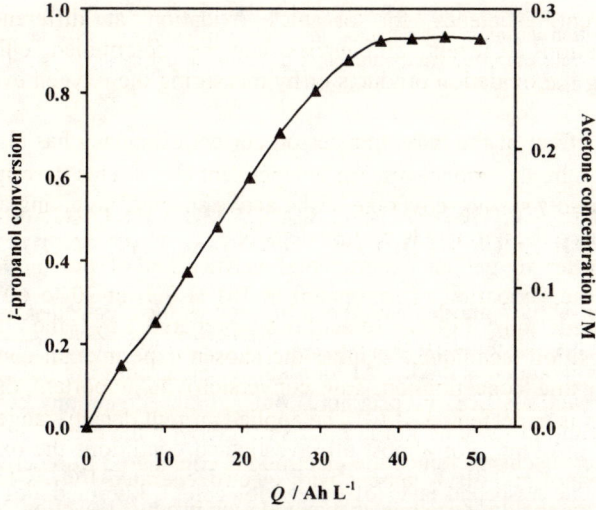

Figure 13. Preparative electrolysis of i-propanol (initial concentration 0.3M) at a Ti/IrO$_2$ electrode ($A = 40$ cm^2) prepared by the thermal decomposition technique. The conversion of i-propanol (defined as its fraction consumed in the electrochemical oxidation) and the concentration of produced acetone as a function of the specific charge. Electrolyte: 1M H$_2$SO$_4$; volume: 150 mL; $j = 30$ mA cm^{-2}; $T = 20°$C.

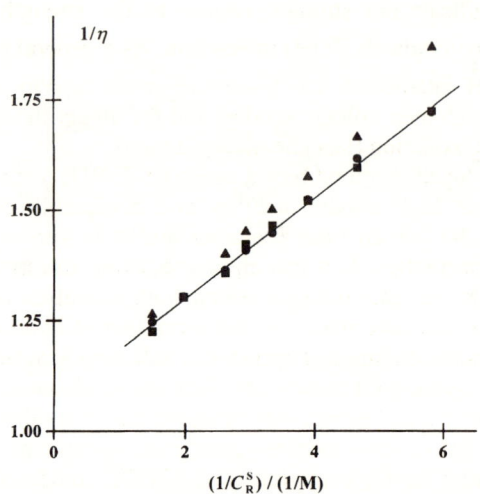

Figure 14. Oxidation of *i*-propanol at a Ti/IrO$_2$ electrode prepared by thermal decomposition technique. The inverse of current efficiency, $1/\eta$, as a function of the inverse of bulk *i*-propanol concentration at different current densities, ■: 30 mA cm^{-2}; ●: 50 mA cm^{-2}; ▲: 60 mA cm^{-2}. Electrolyte: 1M H$_2$SO$_4$; $T = 45°C$.

III. ELECTROCHEMICAL OXIDATION OF ORGANICS IN AQUEOUS MEDIA USING NON-ACTIVE ANODES

The 'non-active' anodes have been defined as the electrodes, which do not provide any catalytic active site for adsorption of reactants and/or products in aqueous media. At 'non-active' electrodes the only possible anode reactions are, in principle, outer sphere reactions (when the reactant and product do not interact strongly with the electrode surface) and water discharge (since the electrode is considered to be covered by at least one adsorbed layer of water molecules). Intermediates such as hydroxyl radicals produced by water discharge at 'non-active' anodes (reaction 1 in Fig. 1) are considered to be involved in the oxidation of organic compounds in aqueous media. This can result in the electrochemical incineration (ECI) of the organic compounds (reaction 5 in Fig. 1).[20,21] The electrochemical activity (overpotential for oxygen evolution) and chemical reactivity (rate of organics oxidation with electrogenerated hydroxyl radicals) of adsorbed hydroxyl radicals are strongly related to the strength of the anode (M) - hydroxyl radicals (•OH) interaction. As a general rule, the weaker the M - •OH interaction, the lower the anode activity toward oxygen evolution (high over voltage anodes) and the higher the anode reactivity for organics oxidation (fast chemical reaction).

Boron doped diamond based anode (p-Si/BDD) is a new electrode material with high anodic stability and acceptable conductivity. In addition, BDD has an inert character and it is known to have weak adsorption properties. This can allow predicting that this material is an ideal 'non-active' electrode on which both oxidation of organics and oxygen evolution take place *via* the formation of hydroxyl radicals. In fact using *p*-nitrosodimethylaniline as a selective scavenger it has been found that hydroxyl radicals are formed at diamond anodes [35]. Furthermore, we can expect that diamond based anodes have very low activity for the O_2 evolution reaction (high overpotential). In fact, the onset potential for O_2 evolution on p-Si/BDD anodes (2.2 V/RHE) is about 1V higher than the standard redox potential of O_2 evolution

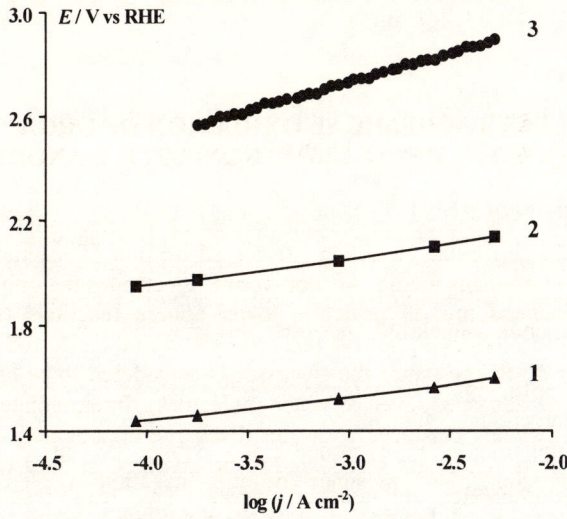

Figure 15. Tafel plots in 0.5M H_2SO_4. (**1**) Ti/IrO$_2$, (**2**) Ti/SnO$_2$-Sb$_2$O$_5$ (**3**) p-Si/BDD. T = 25°C.

(1.23 V/RHE) as was shown in Fig. 2 (curve B). However, the Tafel slope (0.23V/decade) obtained on BDD anode (see Fig. 15) is about twice as high as on other 'non-active' (SnO_2-Sb_2O_5) and 'active' (IrO_2) anodes. This is probably due to the semimetal/semiconductor nature of the diamond film.

In this work we present the results concerning the oxidation of some model organic compounds on BDD based electrodes. Two cases will be distinguished: by excluding and by including simultaneous oxygen evolution. Then, a simple kinetic model will be presented which describes in a quantitative manner the electrochemical incineration (ECI) at BDD based anodes.[36-38] Boron-doped diamond films were synthesized by the hot filament chemical vapor deposition technique (HF-CVD) on a conducting p-Si substrate (1-3 mΩ cm, Siltronix). The temperature of the filament was about 2500°C and that of the p-Si substrate was kept at 830°C. The reactive gas used was methane in an excess of dihydrogen (1% CH_4 in H_2). The doping gas was trimethylboron with a concentration of 1-3 ppm. The gas mixture was supplied to the reaction chamber at a flow rate of 5 L min^{-1} to give a growth rate of 0.24 μm h^{-1} for the diamond layer. The thickness of the obtained diamond film was about 1 μm and the resistivity 10-30 mΩ cm. This HF-CVD process produced columnar, randomly textured, polycrystalline films.[36-38]

1. Oxidation of Organics on p-Si/BDD Anode in the Potential Region before O_2 Evolution

The reactivity of p-Si/BDD electrodes was studied with various organic compounds in acid medium. As organic model compounds, simple alcohols (methanol, ethanol, *i*-propanol, *tert*-butanol), carboxylic acids (formic acid, oxalic acid, maleic acid), aromatic compounds (phenol, 4-chlorophenol, 2-naphthol, benzoic acid), and heterocyclic compounds (3-methylpyridine) were chosen. This choice is due either to their role as model pollutants or as intermediate products in the electrochemical incineration (ECI).

The voltammograms recorded below oxygen evolution display no significant change in presence of the investigated alcohols and carboxylic acids with respect to the voltammogram of the supporting electrolyte. The only difference is a slight decrease in starting potential of oxygen evolution, indicating an effect of the organic compound on the overpotential of water discharge. These results prove that the p-Si/BDD electrode has no electrocatalytic activity for the oxidation of the alcohols and the carboxylic acids investigated. The situation is, however, different for the investigated aromatic and heterocyclic compounds. In this case, the aromatic organic compounds are oxidized at potentials before oxygen evolution resulting in electrode deactivation. A typical example is 4-chlorophenol (4-CP) oxidation in acid solution. Fig. 16 shows cyclic voltammograms (first cycle) obtained with a scan rate of 50 mV s^{-1} for different 4-CP concentrations in 1M H_2SO_4. An anodic current peak corresponding to the oxidation of 4-CP is observed at about 1.7 V/RHE. The current peak increases almost linearly with the 4-CP concentration and the peak potential shifts to more positive potentials as the concentration of 4-CP increases. However, in the case of continuous potential cycling (with stirring the solution before each recording) it has been found that the anodic current peak decreases as the number of cycles increases until a steady state is reached after about 5 cycles (Fig. 17). This decrease in electrode activity appears to be due to deposition of polymeric adhesive products

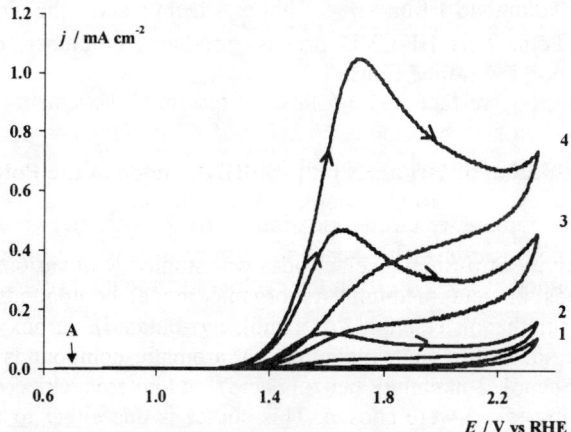

Figure 16. Cyclic voltammograms (first cycles; scan rate: 50 mV s^{-1}) of p-Si/BDD in 1M H_2SO_4 + 4-chlorophenol (4-CP) at different 4-CP initial concentrations: (**1**) 0mM; (**2**) 0.5mM; (**3**) 2mM; (**4**) 5mM. **A**: anodic start of the cyclic voltammograms. T = 25°C. (Reprinted from M.A. Rodrigo, P.-A. Michaud, I. Duo, M. Panizza, G. Cerisola, and Ch. Comninellis, *J. Electrochem. Soc.* **148**, D60 Copyright © 2001 with permission from The Electrochemical Society, Inc.)

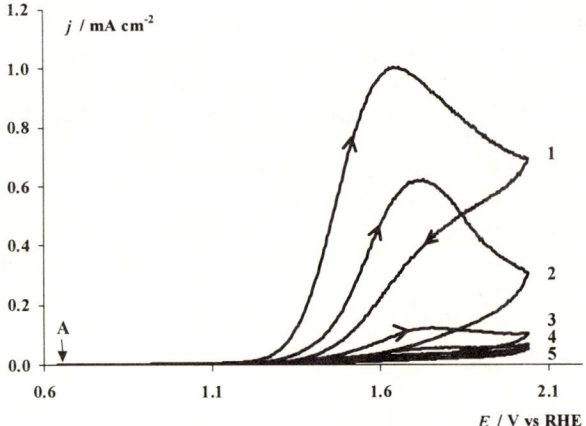

Figure 17. Consecutive cyclic voltammograms (1st to 5th cycle; scan rate: 50 mV s^{-1}) of p-Si/BDD in 1M H$_2$SO$_4$ + 5mM 4-chlorophenol. A: anodic start of the cyclic voltammograms. $T = 25°C$. Reprinted with permission from Scientific Publishing Division of MY K. K., Tokyo.

on the electrode surface. The following reaction mechanism has been proposed to explain the anodic oxidation of 4-CP in the potential region of the supporting electrolyte stability:[38]

1) Two-step one-electron oxidation of 4-CP to phenoxy radical and phenoxy cation. This reaction corresponds to the anodic peak obtained at 1.7 V/RHE in the cyclovoltammogram (Fig. 17).

2) Formation of polymeric material due to the coupling reaction involving phenoxy radicals or phenoxy cations.[38]

Similar results have been obtained with other aromatic (phenol, 2-naphthol [37,36]) and heterocyclic (4-methylpyridine[39]) compounds.

2. Oxidation of Organics on p-Si/BDD Anode in the Potential Region of O_2 Evolution

Cyclic voltammetric measurements have shown that in the potential region of water stability p-Si/BDD electrodes prepared by the hot filament chemical vapor deposition technique (HF-CVD) have no electrocatalytic activity in the oxidation of aliphatic alcohols and carboxylic acids in acid medium. Moreover, the electrode is deactivated during oxidation of aromatic compounds.

At high anodic potential close to the potential of water decomposition, however, the behavior is quite different: the activity of p-Si/BDD anodes is considerably enhanced and there is no evidence of loss of electrode activity [33,36-41]. This observation gives reason for the study of the oxidation of organics in the potential region of O_2 evolution.

The activity of BDD electrodes was investigated at high anode potential close to the potential of O_2 evolution using various aliphatic (alcohols and carboxylic acids), aromatic (phenols, naphthols, benzoic

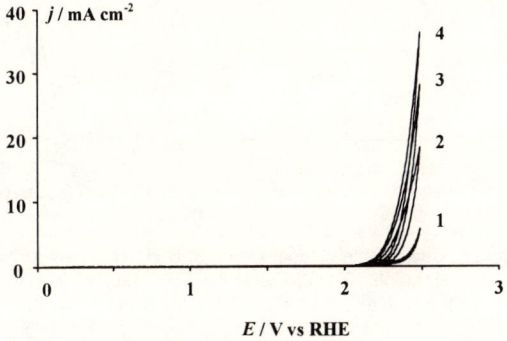

Figure 18. Cyclovoltammograms of p-Si/BDD at different *i*-propanol concentrations: (**1**) 0M; (**2**) 0.2M; (**3**) 0.5M; (**4**) 1M. Electrolyte: 1M H_2SO_4; scan rate: 50 mV s^{-1}; $T = 25°C$. (Reprinted from J. Iniesta, P.-A. Michaud, M. Panizza, G. Cerisola, A. Aldaz, and Ch. Comninellis, *Electrochim. Acta* **46**, 3573. Copyright © 2001 with permission from Elsevier Science.)

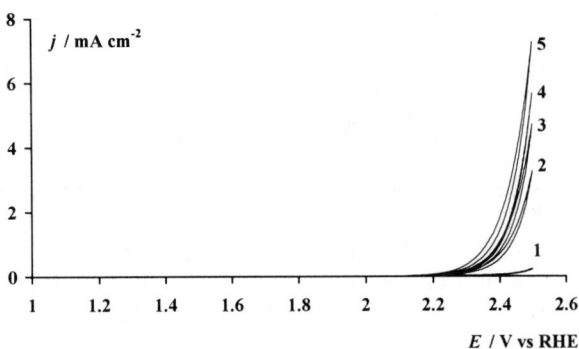

Figure 19. Cyclovoltammograms of p-Si/BDD at different formic acid concentrations: (**1**) 0M; (**2**) 0.05M; (**3**) 0.1M; (**4**) 0.2M; (**5**) 0.5M. Electrolyte: 1M H_2SO_4; scan rate: 50 mV s^{-1}; $T = 25°C$.

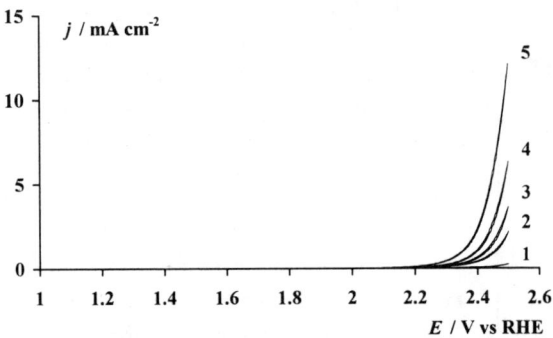

Figure 20. Cyclovoltammograms of p-Si/BDD at different oxalic acid concentrations: (**1**) 0M; (**2**) 0.05M; (**3**) 0.1M; (**4**) 0.2M; (**5**) 0.5M. Electrolyte: 1M H_2SO_4; scan rate: 50 mV s^{-1}; $T = 25°C$.

acid), and heterocyclic (methylpyridine) organic compounds in acidic medium at 25°C by cyclic voltammetry. The concentration of the organic compounds in 1M H_2SO_4 or in 1M $HClO_4$ was varied from zero to max 0.5M depending on solubility. For almost all investigated aliphatic compounds, the current density at a given potential in the region of water decomposition increases with organic concentration. Typical cyclovoltammograms for i-propanol, and formic and oxalic acid up to the potential of water decomposition are given in Figs. 18-20. The observed decrease in the anodic overpotential might indicate an important change in the electrochemical activity of the p-Si/BDD electrode. New reaction paths in oxidation of the organic compounds are likely to open, requiring water decomposition intermediates (hydroxyl radicals), which are only available in conditions of oxygen evolution.

In the case of aromatic organic compounds, it has been found that the polymeric film formed in the region of water stability (see above) can be destroyed by treating the electrode at high anodic potential (E > 2.3 V/RHE) in the region of O_2 evolution. This treatment can restore the electrode activity. In fact, the applied potential is located in the region of water discharge and at BDD it involves the production of reactive intermediates (hydroxyl radicals), which oxidize the polymeric film on the anode surface. Figure 21 shows a typical example of electrode activation in the case of phenol oxidation (2.5 mM phenol in 1 M $HClO_4$). In the first voltammetric scan (Fig. 21, curve a) an anodic current peak corresponding to the oxidation of phenol is observed at about 1.67 V/RHE. As the number of cycles increases, the anodic current peak decreases, and it nearly vanishes after about 5 cycles (curve b). Curves (c) and (d) of Fig. 21 show the voltammetric responses obtained during the reactivation at the fixed anode potential of 2.84 V/RHE for 10 and 20 s. When the polarization time exceeds 40 s, the phenol oxidation peak comes back to its initial position (Fig. 21, curve e), indicating the complete reactivation of the electrode surface.[37] The trend of the normalized current density peaks (j_{peak}/j_{peak}^o, where j_{peak}^o is the current density peak during the first scan) as a function of polarization time is illustrated in the inset of Fig. 21.

Similar results have been obtained for the oxidation of other aromatic compounds (2-naphthol, 4-chlorophenol, 3-methylpyridine).[36,38,39]

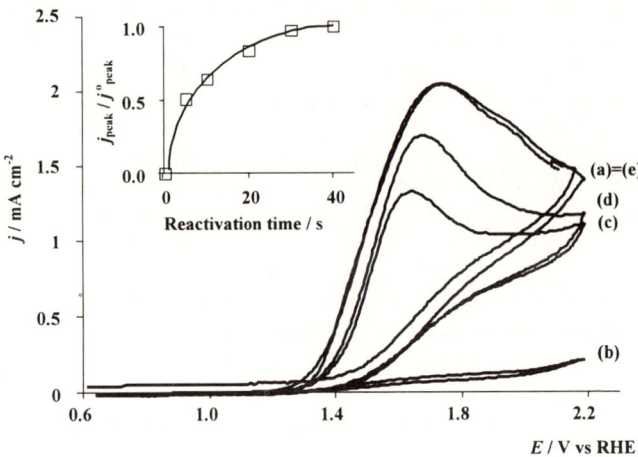

Figure 21. Cyclovoltammograms of p-Si/BDD in 2.5 mM phenol solution in 1M HClO$_4$. (a) 1st cycle; (b) 5th cycle; and after reactivation at +2.84 V/RHE for (c) 10 s; (d) 20 s; (e) 40 s. Scan rate: 100 mV s^{-1}; T = 25°C. Inset: effect of reactivation time on the current density peak, j$_{peak}$, normalized to its value during the first scan, j^o_{peak}. (Reprinted from reference M. Panizza, P.-A. Michaud, G. Cerisola, and Ch. Comninellis, *J. Electroanal. Chem.* **507**, 206. Copyright © 2001 with permission from Elsevier Science)

3. Modeling of Organics Incineration on p-Si/BDD Anode

In this model describing incineration of organics (R) on p-Si/BDD anode, it is assumed that both organics oxidation and oxygen evolution take place exclusively *via* intermediation of hydroxyl radicals, *i.e.* the electrode does not exhibit any 'active' character. This assumption simplifies the presented general mechanism (Fig. 1) to give a reaction scheme involving three different reactions.

The first reaction is the electrochemical discharge of water leading to the formation of hydroxyl radicals (Eq. 21)

$$BDD + H_2O \rightarrow BDD(^{\bullet}OH) + H^+ + e^- \tag{21}$$

These hydroxyl radicals are then consumed by two competing reactions, *i.e.* incineration of organics (reaction 22)

$$BDD\,(^{\bullet}OH) + R \rightarrow BDD + mCO_2 + nH_2O + H^+ + e^- \quad (22)$$

and oxygen evolution (reaction 23)

$$BDD\,(^{\bullet}OH) \rightarrow BDD + \tfrac{1}{2}O_2 + H^+ + e^- \quad (23)$$

It is worth to notice that dioxygen (formed at the anode according to reaction 23) participates very probably also in the combustion of organics according to the following reaction scheme:

1) Formation of organic radicals by a hydrogen abstraction mechanism:

$$RH + BDD\,(^{\bullet}OH) \rightarrow R^{\bullet} + H_2O + BDD \quad (24)$$

2) Reaction of the organic radical with dioxygen

$$R^{\bullet} + O_2 \rightarrow ROO^{\bullet} \quad (25)$$

3) Further abstraction of a hydrogen atom with the formation of an organic hydroperoxide (ROOH):

$$ROO^{\bullet} + R'H \rightarrow ROOH + R'^{\bullet} \quad (26)$$

Since the organic hydroperoxides are relatively unstable, decomposition of such intermediates often leads to molecular breakdown and formation of subsequent intermediates with lower carbon numbers. These scission reactions continue rapidly until formation of carbon dioxide and water.

A kinetic model of this reaction scheme is developed by making the following further assumptions: *(i)* adsorption of the organic compounds at the electrode surface is negligible; *(ii)* the global rate of the electrochemical incineration of organics (involving hydroxyl radicals and formation of hydroperoxide) is a fast reaction and is controlled by mass transport of organics to the anode surface.

Under these conditions, the limiting current density for the electrochemical incineration of a given organic compound can be given by Eq. (27):

$$j_{lim} = n \cdot F \cdot k_m \cdot C \tag{27}$$

where j_{lim} is the limiting current density for organics incineration (A m^{-2}), n is the number of electrons involved in the organics incineration reaction, F is Faraday's constant (C mol^{-1}), k_m is the mass transport coefficient (m s^{-1}), and C is the concentration of organics in the solution (mol m^{-3}).

For the electrochemical incineration of a generic organic compound, it is possible to calculate the number of exchanged electrons, from the following electrochemical reaction:

$$C_xH_yO_z + (2x-z)H_2O \rightarrow \tag{28}$$
$$xCO_2 + (4x+y-2z)H^+ + (4x+y-2z)e^-$$

Replacing the value of $n = (4x + y - 2z)$ in Eq. (27) we obtain:

$$j_{lim} = (4x + y - 2z) \cdot F \cdot k_m \cdot C \tag{29}$$

From the equation of the chemical incineration of the organic compound (Eq. 30):

$$C_xH_yO_z + \left(\frac{4x+y-2z}{4}\right)O_2 \rightarrow xCO_2 + \frac{y}{2}H_2O \tag{30}$$

it is possible to obtain the relation between the organics concentration (C in mol $C_xH_yO_z$ m^{-3}) and the chemical oxygen demand (COD in mol O_2 m^{-3}):

$$C = \frac{4}{(4x+y-2z)}COD \tag{31}$$

From Eqs. (29) and (31) and at a given time t during electrolysis, we can relate the limiting current density for the electrochemical incineration of organics and the chemical oxygen demand of the electrolyte (Eq. 32):

$$j_{\lim}(t) = 4F \cdot k_m \cdot COD(t) \tag{32}$$

At the beginning of electrolysis, at time $t = 0$, the initial limiting current density ($j_{\lim}^o$) is given by:

$$j_{\lim}^o = 4F \cdot k_m \cdot COD^o \tag{33}$$

where COD^o is the initial chemical oxygen demand.

Working under galvanostatic conditions, it is possible to identify two different operating regimes: at $j < j_{\lim}$ the electrolysis is controlled by the applied current, while at $j > j_{\lim}$ it is controlled by the mass transport.

(i) Electrolysis Under Current Control ($j < j_{lim}$)

In this operating regime, the current efficiency is 100%, and the rate of COD removal (mol O_2 m^{-2} s^{-1}) is constant and can be expressed via Eq. (34):

$$r = j_{\lim}^o / 4F \tag{34}$$

where

$$\alpha = j / j_{\lim}^o \quad \text{with } 0 < \alpha < 1 \tag{35}$$

Using Eq. (33), the constant rate of COD removal (Eq. 34) can be expressed as:

$$r = \alpha \cdot k_m \cdot COD^o \tag{36}$$

It is now necessary to consider the mass-balances over the electrochemical cell and the reservoir in order to describe the temporal evolution of COD in the batch recirculated reactor system shown in Fig. 22.

Since the volume of the electrochemical reactor (V_E) is much smaller than the reservoir volume (V_R), the mass-balance on COD for the electrochemical cell yields the following relation:

$$\dot{V} \cdot COD_{out} = \dot{V} \cdot COD_{in} - \alpha \cdot A \cdot k_m \cdot COD^0 \qquad (37)$$

where $\dot{V}$ is the flow-rate (m³ s⁻¹) through the electrochemical cell, COD_{in} and COD_{out} are the chemical oxygen demands (mol O_2 m⁻³) at the inlet and at the outlet of the electrochemical cell, respectively, and A the anode area.

For the well-mixed reservoir in Fig. 22, the mass balance on COD can be expressed as:

$$\dot{V} \cdot (COD_{out} - COD_{in}) = V_R \frac{d\,COD_{in}}{d\,t} \qquad (38)$$

where again outlet and inlet are with respect to the electrochemical cell, as defined for Eq. (37). Obviously, in the given batch recirculated reactor system, the outlet of the cell means the inlet of the reservoir, and *vice versa*. Combining Eqs. (37) and (38), and replacing COD_{in} by the temporal evolution of COD, one obtains:

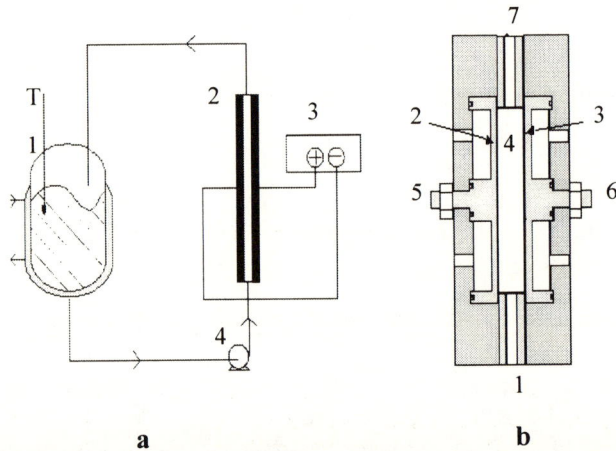

Figure 22. Electrochemical cell for bulk oxidation of organics on p-Si/BDD electrode. (**a**) Set-up used: (1) thermoregulated reservoir; (2) electrochemical cell; (3) power supply; (4) pump. (**b**) Electrochemical cell: (1) inlet; (2) anode; (3) cathode; (4) electrolysis compartment; (5) and (6) electric contacts; (7) outlet.

$$\frac{d\,COD}{dt} = -\frac{\alpha \cdot A \cdot k_m \cdot COD^o}{V_R} \tag{39}$$

Integrating this equation subject to the initial condition $COD = COD^o$ at $t = 0$ gives the temporal evolution of $COD(t)$ in this operating regime ($j < j_{\lim}$):

$$COD(t) = COD^o\left(1 - \frac{\alpha \cdot A \cdot k_m}{V_R}\,t\right) \tag{40}$$

This behavior persists until a critical time (t_{cr}) of electrolysis when the excess of the limiting current density over the applied current density vanishes at:

$$j = j_{\lim} \quad : \quad t = t_{cr} \tag{41}$$

The value of COD at this moment, called critical chemical oxygen demand, COD_{cr}, is given as:

$$COD_{cr} = \alpha \cdot COD^o \tag{42}$$

Substituting Eq. (42) in Eq. (40), the critical time can be formulated as:

$$t_{cr} = \frac{1-\alpha}{\alpha}\,\frac{V_R}{A \cdot k_m} \tag{43}$$

or given in terms of critical specific charge, Q_{cr} (Ah dm^{-3}):

$$Q_{cr} = \frac{(1-\alpha) \cdot j_{\lim}^o}{k_m \cdot 3.6 \times 10^6} \tag{44}$$

At the critical time, the process passes from current-controlled regime to mass-transport-controlled regime.

(ii) Electrolysis Under Mass Transport Control ($j > j_{\lim}$)

When the applied current exceeds the limiting current, secondary reactions (such as oxygen evolution) commence, resulting in a decrease in the instantaneous current efficiency, *ICE*, defined as

$$ICE = \frac{j_{\lim}}{j} = \frac{COD(t)}{\alpha \cdot COD^{\circ}} \qquad (45)$$

This regime is realized either

(i) at $j < j_{\lim}^{\circ}$ (*i.e.* $\alpha < 1$) when the electrolysis is continued over the critical time (with $COD < \alpha \cdot COD^{\circ}$ at $t > t_{cr}$), or

(ii) at $j > j_{\lim}^{\circ}$ (*i.e.* $\alpha > 1$) meaning application of a higher current density than its initial limiting value (hence, $COD < \alpha \cdot COD^{\circ}$ at any finite time).

In these cases, the *COD* mass balances on the electrochemical cell and on the reservoir can be expressed as:

$$\frac{d\,COD}{dt} = -\frac{A \cdot k_m \cdot COD}{V_R} \qquad (46)$$

Integration of this equation from the initial conditions (i) $COD = \alpha \cdot COD^{\circ}$ at $t = t_{cr}$, and (ii) $COD = COD^{\circ}$ at $t = 0$ leads to:

(i) $COD(t) = \alpha \cdot COD^{\circ} \exp\left(-\frac{A \cdot k_m}{V_R} t + \frac{1-\alpha}{\alpha}\right)$ at $t > t_{cr}$ (47a)

(ii) $COD(t) = COD^{\circ} \exp\left(-\frac{A \cdot k_m}{V_R} t\right)$ (47b)

From Eqs. (45) and (47) the instantaneous current efficiency, *ICE*, is now given by:

(i) $ICE = \exp\left(-\dfrac{A \cdot k_m}{V_R} t + \dfrac{1-\alpha}{\alpha}\right)$ at $t > t_{cr}$ \hfill (48a)

(ii) $ICE = \dfrac{1}{\alpha} \exp\left(-\dfrac{A \cdot k_m}{V_R} t\right)$ \hfill (48b)

(iii) Experimental Verification

In order to verify the validity of this model, the anodic oxidation of various aromatic compounds in acidic solution has been performed under different current densities, temperatures, and initial organics concentrations. Figures 23-25 show the results obtained using 2-naphthol.[36] It is seen that COD decreases down to zero for all the conditions investigated. This indicates that under these conditions the final oxidation product of 2-naphthol is CO_2.

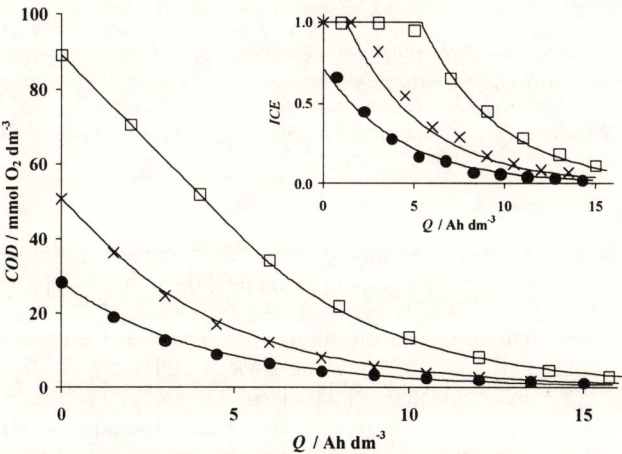

Figure 23. Influence of the initial 2-naphthol concentration, □: 9 mM; ×: 5 mM; ●: 2 mM, on the trends of COD and ICE (inset) during electrolysis, using a p-Si/BDD anode. Electrolyte: 1M H_2SO_4; $j = 30$ mA cm^{-2}; $T = 30°C$. The solid lines represent model prediction. (Reprinted from reference M. Panizza, P.-A. Michaud, G. Cerisola, and Ch. Comninellis, *J. Electroanal. Chem.* **507**, 206. Copyright © 2001 with permission from Elsevier Science)

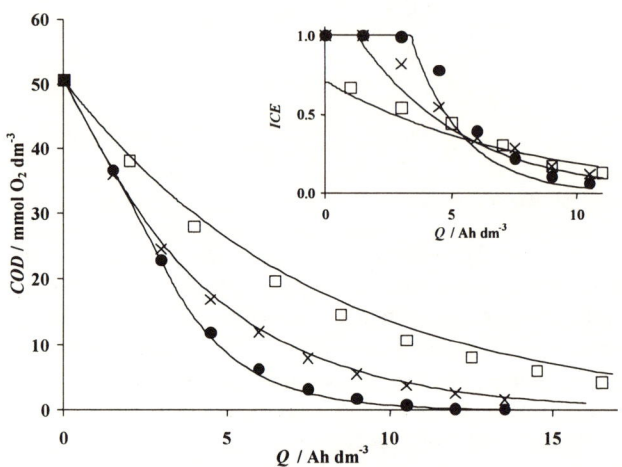

Figure 24. Influence of the applied current density, ●: 15 mA cm^{-2}; ×: 30 mA cm^{-2}; □: 60 mA cm^{-2}, on the trends of COD and ICE (inset) during electrolysis, in 1M H_2SO_4 + 5mM 2-naphthol, using a p-Si/BDD anode. $T = 30°C$. The solid lines represent model prediction. (Reprinted from reference M. Panizza, P.-A. Michaud, G. Cerisola, and Ch. Comninellis, *J. Electroanal. Chem.* **507**, 206. Copyright © 2001 with permission from Elsevier Science)

Figure 23 shows the effect of the initial concentration of 2-naphthol on the trends of COD and ICE during electrolysis under galvanostatic conditions (30 mA cm^{-2}) at 30°C. At the beginning of the electrolysis with high 2-naphthol concentration (9 mM), the COD decreased linearly with charge and ICE remained about 100%, which means that under these conditions the oxidation of 2-naphthol is under current control ($j < j_{lim}$). On the contrary, at low concentration (2 mM), COD decreased exponentially and ICE was below 100%, because the oxidation was under mass-transport control ($j > j_{lim}$). In the same figure, the theoretical values of COD and current efficiency trends calculated from the model are reported. As can be seen, the model can satisfactorily predict the experimental data for all investigated 2-naphthol concentrations.

Figure 24 shows the influence of anodic current density on the trends of COD and ICE during the oxidation of 5 mM 2-naphthol at

30°C. Increasing current density resulted in a decrease in current efficiency of 2-naphthol oxidation due to the side reactions of peroxodisulfuric acid formation (Eq. 49) and oxygen evolution.

$$2H_2SO_4 \rightarrow H_2S_2O_8 + 2H^+ + 2e^- \qquad (49)$$

Figure 25 shows the influence of temperature on the variation of *COD* and *ICE* during galvanostatic electrolysis (30 mA cm^{-2}). At the beginning of the electrolyses, no significant differences were found between the different temperatures investigated, while after passing about 4.5 Ah dm^{-3}, the *COD* removal rate at 40°C and 60 °C was higher than that predicted by the model. This behavior can be explained by the production of peroxodisulfuric acid in a secondary reaction, according to Eq. (49), which can act as a mediator for organics oxidation at high temperature. In fact, the rate of oxidation of organic compounds with peroxodisulfuric acid increases with temperature.[29,30] Oxidation of

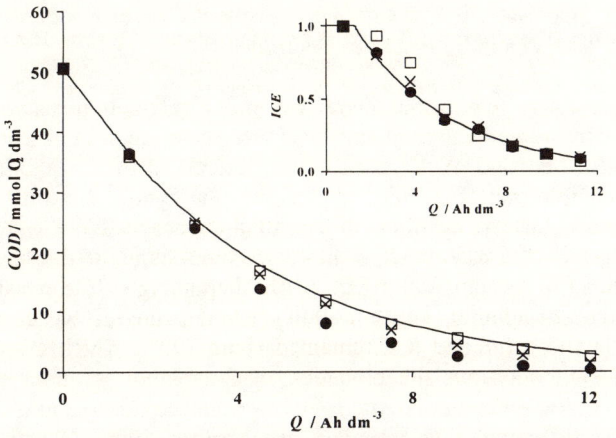

Figure 25. Influence of the temperature, □: 30°C; ✕: 40°C; ●: 60°C, on the trends of *COD* and *ICE* (inset) during electrolysis, in 1M H$_2$SO$_4$ + 5mM 2-naphthol, using a p-Si/BDD anode. $j = 30$ mA cm^{-2}. The solid lines represent model prediction. (Reprinted from reference M. Panizza, P.-A. Michaud, G. Cerisola, and Ch. Comninellis, *J. Electroanal. Chem.* **507**, 206. Copyright © 2001 with permission from Elsevier Science)

organics in the bulk of the electrolyte with electrogenerated oxidants has not been considered in the present model.

The validity of the model has been also verified for the oxidation of phenol [37,42], 4-chlorophenol,[38] 3-methylpyridine,[39] and polyacrylates.[43]

IV. CONCLUSION

Electrochemical oxidation of organics in aqueous media frequently occurs only at high potentials with concomitant evolution of O_2, and without any loss in electrode activity. As a general rule, the nature of the electrode material strongly influences both the selectivity and the efficiency of the process.

In order to explain this behavior, a simplified mechanism for the anodic oxidation of organic compounds with simultaneous oxygen evolution has been put forward (Fig. 1). The initial step in the proposed mechanism is the discharge of water to form adsorbed hydroxyl radicals (reaction 1 in Fig. 1). The electrochemical and chemical reactivity of the adsorbed hydroxyl radicals depends strongly on the nature of electrode material used. Two extreme classes of electrodes can be defined: 'active' and 'non-active' electrodes :

a) At 'active' electrodes there is a strong interaction between the hydroxyl radicals and the electrode material. In this case, the adsorbed hydroxyl radicals may interact with the anode with possible transfer of the oxygen from the hydroxyl radical to the anode surface, forming a higher oxide (reaction 2 in Fig. 1). This may be the case when higher oxidation states of the electrode material are available above the thermodynamic potential for oxygen evolution (1.23 V/RHE). Iridium dioxide based anodes (Ti/IrO_2), forming the higher oxide IrO_3 at 1.35 V/RHE, can be considered as typical active electrodes.

A kinetic model has been proposed considering that at IrO_2 based 'active' electrodes the surface redox couple IrO_3/IrO_2 can act as mediator in the oxidation of organics (reaction 8). This reaction is in competition with the side reaction of oxygen evolution due to the chemical decomposition of the higher oxide (reaction 9). The model predicts a simple relationship between the current efficiency and the concentration of the organic compound (see Eq. 19). It provides an experimental method to determine the ratio of the rate constants of the competing reactions. The prediction of the model was confirmed by preparative electrolysis experiments.

b) At 'non-active' electrodes there is a weak interaction between the hydroxyl radicals and the electrode surface. In this case, the oxidation of organics is mediated by hydroxyl radicals, which may result in fully oxidized reaction products such as CO_2 (reaction 5 in Fig. 1). This reaction is in competition with a side reaction, the discharge of hydroxyl radicals to form O_2 (reaction 6 in Fig. 1), without any participation of the anode surface.

Boron-doped diamond anodes (BDD), having an inert character and weak adsorption properties, can be considered as typical 'non-active' electrodes. A simple model has been proposed considering BDD based 'non-active' electrodes. In this case, the electrogenerated hydroxyl radicals can act as mediator

Table 1
Equations Describing *COD* and *ICE* Evolution During Oxidation at BDD Electrode. COD^0: Initial Chemical Oxygen Demand (mol O_2 m^{-3}); V_R: Reservoir Volume (m^3); k_m: Mass Transfer Coefficient (m s^{-1}); A: Electrode Area (m^2); $\alpha = j / j^o_{lim}$, $t_{cr} = [(1-\alpha)/\alpha] \cdot (V_R/A \cdot k_m)$: Critical Time (s) Defined in Eq. (41).

	Instantaneous Current Efficiency, ICE	Chemical Oxygen Demand, COD(t) (mol O_2 m^{-3})
$j < j^o_{lim}$		
$t < t_{cr}$ current control	1	$COD^o \left(1 - \dfrac{\alpha \cdot A \cdot k_m}{V_R} t \right)$
$t > t_{cr}$ mass transport control	$\exp\left(-\dfrac{A \cdot k_m}{V_R} t + \dfrac{1-\alpha}{\alpha}\right)$	$\alpha \cdot COD^o \exp\left(-\dfrac{A \cdot k_m}{V_R} t + \dfrac{1-\alpha}{\alpha}\right)$
$j > j^o_{lim}$		
mass transport control	$\dfrac{1}{\alpha} \exp\left(-\dfrac{A \cdot k_m}{V_R} t\right)$	$COD^o \exp\left(-\dfrac{A \cdot k_m}{V_R} t\right)$

in the incineration of organics (reaction 22). This reaction is in competition with the side reaction of oxygen evolution due to the anodic discharge of hydroxyl radicals (reaction 23).

The model can allow prediction of the temporal evolution of the chemical oxygen demand (*COD*) and instantaneous current efficiency (*ICE*) during oxidation of the organic compound (see equations in Table 1). The prediction of the model was confirmed by preparative electrolysis of different classes of organic compounds.

LIST OF SYMBOLS

A	electrode surface area
C	concentration of organics in the solution
C_R^E	concentration of organic R near the electrode surface
C_R^S	concentration of organic R in the bulk solution
COD	chemical oxygen demand
COD^o	initial chemical oxygen demand
COD_{cr}	critical chemical oxygen demand
COD_{in}	chemical oxygen demand at the inlet of the electrochemical cell
COD_{out}	chemical oxygen demand at the outlet of the electrochemical cell
E	potential
F	Faraday constant
ICE	instantaneous current efficiency
j	current density
j_{lim}	limiting current density for incineration of organics
j_{lim}^o	initial limiting current density
j_{peak}	current density peak
j_{peak}^o	current density peak in the first voltammetric scan
k_m	mass transfer coefficient
k_1	rate constant of oxidation of organics
k_2	rate constant of oxygen evolution
n	electron stoichiometry
Q	specific charge
Q_{cr}	critical specific charge
r	rate of *COD* removal
r_o	rate of formation of surface active species (*e.g.* IrO_3)
r_1	rate of oxidation of organics
r_2	rate of oxygen evolution
t	time
t_{cr}	critical time
T	absolute temperature

V_E	volume of the flow-through electrochemical reactor
V_R	volume of the reservoir
$\dot{V}$	volume flow-rate through the electrochemical cell
α	parameter relating the current density to its initial limiting value (see Eq. 35)
Γ_s	saturation surface concentration of active species
η	current efficiency of organics oxidation
θ	relative surface coverage of active species
ϕ	parameter relating local and bulk concentrations of the organic (see Eq. 18)

ACKNOWLEDGEMENT

Financial support from the *Fonds National Suisse de la Recherche Scientifique* is gratefully acknowledged.

REFERENCES

[1] Ch. Comninellis, and A. De Battisti, *J. Chim. Phys.* **93** (1996) 673.
[2] Ch. Comninellis, and E. Plattner, *Chimia* **42** (1988) 250.
[3] Ch. Comninellis, and C. Pulgarin, *J. Appl. Electrochem.* **21** (1991)703.
[4] Ch. Comninellis, and C. Pulgarin, *J. Appl. Electrochem.* **23** (1993)108.
[5] Ch. Comninellis, and A. Nerini, *J. Appl. Electrochem.* **25** (1995) 23.
[6] Ch. Comninellis, *Electrochim. Acta* **39** (1994) 1857.
[7] S. Stucki, R.Kötz, B. Carcer, and W. Suter, *J. Appl. Electrochem* **21** (1991) 99.
[8] C. Pulgarin, N. Adler, P. Péringer, and Ch. Comninellis, *Wat. Res.* **28** (1994) 887.
[9] Ch. Comninellis, E. Plattner, C. Seignez, C. Pulgarin, and P. Péringer, *Swiss Chem.* **14** (1992) 25.
[10] Ch. Comninellis, *Gaz–Eaux–Eaux usées* **11** (1992) 792.
[11] C. Lamy, *Electrochim. Acta* **29** (1984) 1581.
[12] V. Solis, T. Iwasita, A. Pavese, and W. Vielstich, *J. Electroanal. Chem.* **255** (1988) 155.
[13] M. Manzanares, A. Pavese, and V. Solis, *J. Electroanal. Chem.* **310** (1991) 159.
[14] J. Feng, and D.C. Johnson, *J. Electrochem. Soc.* **137** (1990) 507.
[15] H. Chang, and D. C. Johnson, *J. Electrochem. Soc.* **137** (1990) 2452.
[16] H. Chang, and D .C. Johnson, *J. Electrochem. Soc.* **137** (1990) 3108.
[17] J. E. Vitt, and D. C. Johnson, *J. Electrochem. Soc.* **139** (1992) 774.
[18] J. Feng, D.C. Johnson, S.N. Lowery, and J. Carey, *J. Electrochem. Soc.* **141** (1994) 2708.
[19] S. E. Treimer, J. Feng, M. D Scholten, D. C. Johnson, and A. J. Davenport, *J. Electrochem. Soc.* **148** (2001) E459.
[20] G. Fóti, D. Gandini, and Ch. Comninellis, in *Current Topics in Electrochemistry*, Vol. 5, Research Trends, Trivandrum (1997), p. 71.
[21] O. Simond, V. Schaller, and Ch. Comninellis, *Electrochim. Acta* **42** (1997) 2009.

[22] G. Fóti, D. Gandini, Ch. Comninellis, A. Perret, and W. Haenni, *Electrochem. Solid-State Lett.* **2** (1999) 228.
[23] F. Beck, *Ber. Bunsenges. Phys. Chem.* **77** (1973) 353.
[24] F. Beck, and H. Schultz, *Electrochim. Acta* **29** (1984) 1569.
[25] F. Beck, B. Wermeckes, and E. Zimmer, *Dechema Monogr.* **112** (1988) 257.
[26] E. Lodowicks, and F. Beck, *Chem. Eng. Techn.* **17** (1994) 338.
[27] F. Cardarelli, P. Taxil, A. Savall, Ch. Comninellis, G. Manoli, and O. Leclerc, *J. Appl. Electrochem.* **28** (1998) 245.
[28] C. Mousty, G. Fóti, Ch. Comninellis, and V. Reid, *Electrochim. Acta* **45** (1999) 451.
[29] J.M. Otten, and W. Visscher, *Electroanal. Chem. Interfac. Electrochem.* **55** (1974) 1.
[30] D. A.J. Rand, and R. Woods, *Electroanal. Chem. Interfac. Electrochem.* **55** (1974) 375.
[31] S. Trasatti, and G. Lodi, in *Studies in Physical and Theoretical Chemistry*, Vol. 11, *Electrodes of Conductive Metallic Oxides*, Part A, S. Trasatti, ed., Elsevier, Amsterdam (1980), p. 152.
[32] R. Kötz, in *Advances in Electrochemical Science and Engineering*, Vol. 1, H. Gerischer, and C. Tobias, eds., VCH, Weinheim, (1990), p. 109.
[33] Gherardini, P.-A. Michaud, M. Panizza, Ch. Comninellis, and N. Vatistas, *J. Electrochem. Soc.* **148** (2001) D78.
[34] D. Wabner, and C. Grambow, *J. Electroanal. Chem.* **195** (1985) 95.
[35] M. Tanaka, Y. Nishiki, S. Nakamatsu, and K. Suga, *Denki Kagaku* **66** (1998) 856.
[36] M. Panizza, P.-A. Michaud, G. Cerisola, and Ch. Comninellis, *J. Electroanal. Chem.* **507** (2001) 206.
[37] J. Iniesta, P.-A. Michaud, M. Panizza, G. Cerisola, A. Aldaz, and Ch. Comninellis, *Electrochim. Acta* **46** (2001) 3573.
[38] M.A. Rodrigo, P.-A. Michaud, I. Duo, M. Panizza, G. Cerisola, and Ch. Comninellis, *J. Electrochem. Soc.* **148** (2001) D60.
[39] J. Iniesta, P.-A. Michaud, M. Panizza, and Ch. Comninellis, *Electrochem. Comm.* **3** (2001) 346.
[40] I. Duo, P.-A. Michaud, W. Haenni, A. Perret, and Ch. Comninellis, *Electrochem. Solid-State Lett.* **3** (2000) 325.
[41] D. Gandini, E. Mahé, P.-A. Michaud, W. Haenni, A. Perret, and Ch. Comninellis, *J. Appl. Electrochem.* **30** (2000) 1345.
[42] M. Panizza, P.-A. Michaud, G. Cerisola, and Ch. Comninellis, *Electrochem. Comm.* **3** (2001) 336.
[43] R. Bellagamba, P.-A. Michaud, Ch. Comninellis, and N. Vatistas, *Electrochem. Comm.* **4** (2002) 171.

4

Fuel Cells

K. Hemmes

Faculty Technology Policy and Management Delft University of Technology, Delft The Netherlands

I. INTRODUCTION

This chapter is not an introduction to fuel cells. General introductions to fuel cells describing history and presenting types, principles, and problems can be found in many articles and books on fuel cells.[1,2,3] It is assumed here that the reader is familiar with this literature. The goal of this chapter is to present a comprehensive and general theory to describe the typical characteristics of fuel cells irrespective of their type and application and to outline their possibilities. Fuel cell theory is largely based on thermodynamics. In addition, modeling of fuel cells by analytical mathematics, as opposed to numerical mathematics, is very useful for understanding the principles and characteristics of a fuel cell and for quantifying them. Model predictions are compared with experimental results on a 110 cm^2 MCFC single cell. It will be shown that fuel cells need not only be seen as devices that generate electricity as their main purpose, but they can also be applied to improve the chemical process into which they can be incorporated by making appropriate use of their specific characteristics.

The rapid development of the PEMFC has led to small fuel cells replacing secondary batteries. Further miniaturization is possible by integration of micro fuel cells on a silicon chip. The geometry of a fuel cell may be simple flat plates or tubes, but also radial symmetries are developed. The most common fuel cell is one with two inlets and two outlets where both the reactants and products are in the gas phase. But many other configurations and reactants and products are possible. So we need a fuel cell theory describing the general features and laws of fuel cells regardless of their type, size, geometry, operating temperature and application. This theory can be used to model present and future types of fuel cells and can provide us with a better understanding of the principles, limitations and possibilities of fuel cells.

It is well known that fuel cells are not limited by the Carnot efficiency in the conversion of chemical energy into Work (electricity) but it is less well known that by electrochemical conversion in some cases, for example electrochemical gasification of carbon, heat can be converted into power without suffering from the Carnot limitation. The principle reason for this is that a fuel cell can approach the reversible limit thereby minimizing entropy production and thus conserving exergy. So, 200 years after Volta and 100 years after Becquerel, electrochemists are now challenged to provide the world with the most efficient and clean energy systems possible. The development of such an energy conversion system that uses the fuel exergetic quality of (solid) fuel much more efficiently is one of the greatest challenges to electrochemists.

II. FUEL CELLS AND CARNOT ENGINES

1. Fuel Cells versus Carnot Engines

It has been said many times that fuel cells are not limited by the Carnot efficiency, and that, therefore, conversion of internal energy into power in principle takes place with a higher efficiency. In general this is not true. Present fuel cells are also limited in efficiency by thermodynamic laws; however, the dependency on temperature is just the opposite as that for a Carnot Engine (CE). A CE is defined here as a hypothetical apparatus that is converting heat into power in an isolated process at the highest theoretically possible efficiency (i.e., at the Carnot efficiency).

Fuel Cells

The Carnot efficiency is given by the following well-known formula that can be found in any textbook on thermodynamics:

$$\eta_c = 1 - \frac{T_0}{T} \tag{1}$$

For a better understanding of the differences between the fuel cell efficiency and the Carnot efficiency η_c of a CE, we will next give a derivation of the Carnot factor, as it is sometimes called, following, for example Ref.[4] The derivation is based on the first and second law of thermodynamics stating respectively that energy is conserved and that in any isolated process the entropy change is positive (or zero for a reversible process). The conversion of heat into work (e.g., electric power) can be schematically pictured as in Fig. 1.

According to the first law of thermodynamics, we have:

$$Q_1 = Q_2 + W \tag{2}$$

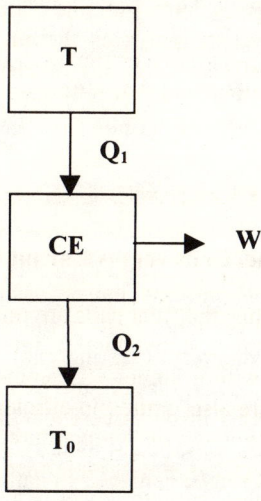

Figure 1. Conversion of heat into power by a Carnot Engine ($T > T_0$).

And according to the second law, we have:

$$\frac{Q_2}{T_0} - \frac{Q_1}{T} \geq 0 \tag{3}$$

The efficiency of the CE is defined as:

$$\eta_c = \frac{W}{Q_1} \tag{4}$$

So combining the above three equations we find:

$$\eta_c = \frac{W}{Q_1} = \frac{Q_1 - Q_2}{Q_1} = 1 - \frac{Q_2}{Q_1} \leq 1 - \frac{T_0}{T} \tag{5}$$

The conversion in a CE is by definition reversible, so for this case the equality sign holds and Eq. (5) presents the well-known Carnot factor, i.e. the maximum efficiency with which heat can be converted into Work (or power) in an isolated cyclic process.

In a reversible operating fuel cell only part of the heat reaction of ΔH is converted into power; i.e., Gibbs Free Energy; ΔG:

$$V_{eq} = -nF\Delta G \tag{6}$$

The difference between ΔG and ΔH is the term $T\Delta S$. This term accounts for the inevitable entropy changes associated with the overall reaction. If ΔS is negative, then $T\Delta S$ is the amount of heat that is dissipated even in a reversible operating fuel cell. This must be so because according to the second law in the overall process the entropy change is positive (or zero in the reversible limit). So a decrease in entropy due to the reaction must be compensated by an equal amount of entropy production. This can be achieved in nature by the conversion of part of the chemical energy into an amount of heat Q. The entropy production involved in this heat dissipation is equal to Q/T; hence, this amount of heat Q must equal TΔS. Therefore, the maximum (electric) efficiency of a fuel cell is given by:

$$\eta_{fc} = \frac{\Delta G}{\Delta H} = \frac{\Delta H - T\Delta S}{\Delta H} = 1 - T\frac{\Delta S}{\Delta H} = 1 - \frac{T}{T_{\Delta G=0}} \qquad (7)$$

In the first approximation, i.e. assuming that ΔS and ΔH do not depend on temperature, we can define a characteristic temperature at which the fuel cell delivers only heat. It is the temperature where ΔG = 0; called the spontaneous combustion temperature:[3]

$$T_{\Delta G=0} \equiv \frac{\Delta H}{\Delta S} \qquad (8)$$

For the overall reaction of hydrogen oxidation:

$$H_2 + \frac{1}{2}O_2 \rightarrow H_2O \qquad (9)$$

ΔS is about -50 J/mol.K, and ΔH is about -250 kJ/mole, hence $T_{\Delta G=0}$ ~ 5000 K.

It is worth noting here that the entropy change of a reaction involving gaseous species is largely determined by the change in the number of gas molecules in the forward overall reaction. For example, the entropy change for reaction (1) is negative since the number of gas molecules is reduced in the ratio 3:2, and since one mole of H_2 and half a mole of O_2 react to one mole of H_2O (steam). The number of quantum states of the new (statistical mechanical) system is reduced. Hence, the total entropy decreases. Or in other words, we are creating more order in the system. This would be a violation of the second law. However, part (TΔS) of the available energy (ΔH) cannot be converted into power, but is dissipated as heat that compensates exactly for the decrease in entropy as explained above.

We see from Eq. (7) that the maximum fuel cell efficiency of a hydrogen-oxygen fuel cell is a linear decreasing function of temperature, whereas the Carnot efficiency is a monotone increasing function approaching 100% efficiency at very high temperatures. In Figure 2, both efficiencies are plotted as a function of temperature. The fuel cell is assumed to operate reversibly on hydrogen and oxygen at atmospheric pressure.

We see that there is a break-even point for a certain temperature, T_{BE}, where both efficiencies are equal:

$$1 - \frac{T_0}{T_{BE}} = 1 - \frac{T_{BE}}{T_{\Delta G=0}} \tag{10}$$

or

$$T_{BE} = \left(T_0 T_{\Delta G=0}\right)^{\frac{1}{2}} \tag{11}$$

Using T_0 = 298 K and $T_{\Delta G=0}$ = 5000 K, T_{BE} is found to be about 1200 K which falls within the range of operating temperatures of a SOFC.

For temperatures lower than T_{BE} a fuel cell in principle can indeed obtain a higher efficiency than a Carnot engine; however, for temperatures higher than T_{BE} the opposite holds. CE's and FC's are complementary devices, and it is interesting to study the combination of a FC and a CE.

2. Fuel Cells Combined with Carnot Engines

A well-known saying is: "If you can't beat them join them." This certainly holds for fuel cells. The efficiency of a reversible system in which the dissipated heat from a fuel cell is used in a CE will be higher than that of a FC alone, as shown in Fig. 2.

(i) *FC- CE Systems Having an Overall FC Reaction with $\Delta S < 0$ (Reversible)*

In most fuel cells the overall cell reaction is the reaction of hydrogen with oxygen to water (steam). For this reaction the change in entropy ΔS is negative. Heat is dissipated even if the cell operates reversibly.

The reversible fuel cell generates power with an efficiency of η_{fc}. A CE converts the waste heat, being proportional to $(1- \eta_{fc})$, with a

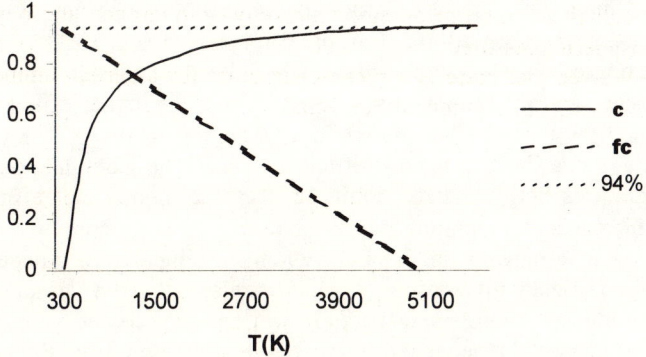

Figure 2. Reversible efficiencies of a H2/O2 fuel cell (FC) a Carnot Engine (CE) and combined FC-CE system as a function of operating temperature.

Carnot efficiency of η_c. Hence, the total system efficiency η_s of such a combination of a fuel cell and CE as shown in Fig. 2 is given by:

$$\eta_s = \eta_{fc} + (1 - \eta_{fc})\eta_c \tag{12}$$

Substituting Eqs. (5) and (7), it is found that the system efficiency is constant and independent of the operating temperature and is simply given by:

$$\begin{aligned}\eta_s &= 1 - \frac{T}{T_{\Delta G=0}} + \left[1 - \left(1 - \frac{T}{T_{\Delta G=0}}\right)\right]\left(1 - \frac{T_0}{T}\right) \\ &= 1 - \frac{T_0}{T} \\ &= 1 - T_0 \frac{\Delta S}{\Delta H}\end{aligned} \tag{13}$$

Since the spontaneous combustion temperature $T_{\Delta G=0}$ is much higher than the temperature T_0 of the environment, the efficiency of this system is very high. Again using T_0 = 298 K and $T_{\Delta G=0}$ = 5000 K, η_S is found to be 0.94. This is equal to the efficiency of a fuel cell at room temperature, where a Carnot engine cannot deliver any power, as well as it is equal to the Carnot efficiency at $T_{\Delta G = 0}$ at which temperature a H_2/O_2 fuel cell under otherwise standard conditions only produces heat. For temperatures lower than T_0 a higher efficiency is obtained by a fuel cell alone while at T higher than $T_{\Delta G = 0}$ a Carnot Engine alone has a higher efficiency than the combination. For temperatures between these limits, a decrease in fuel cell efficiency with increasing temperature due to the term $T\Delta S$ is compensated by the increase in efficiency of the CE yielding a temperature independent system efficiency given by Eq. (13). So a fuel cell and a Carnot engine are completely complementary and form an ideal couple with a very high efficiency independent of temperature in the reversible limit. This is the principle reason for the high efficiency of high temperature fuel cell systems. So-called bottoming cycles are used to recover the waste heat. Often a steam cycle is applied in the system but also combinations of SOFC's and MCFC's with gas turbines are proposed in systems that are calculated to have very high efficiencies.[5][6]

(ii) FC- CE Systems Having an Overall FC Reaction with $\Delta S < 0$ (Irreversible)

In practice however, losses, i.e. irreversibilities occur. In this section we will analyze first in a simple approach the influence of irreversibilities on the system efficiency. It is found that if we take irreversibilities to account an optimum temperature for the FC-CE system can be calculated.

If polarization losses in the fuel cell are taken into account when the CE is assumed to operate reversibly, we can expect that the optimum temperature for the system will be found at high temperature where the CE produces most of the power and at high efficiency. Vice versa, if the FC is considered ideal and losses occur in the CE, the optimum temperature will be found at low T. If both are non-ideal, as in all practical situations, we expect an optimum at intermediate temperatures.

In a first approximation, the losses of a CE can be modelled by a loss factor q_1:

$$\eta_c = q_1\left(1 - \frac{T}{T_0}\right) \tag{14}$$

This factor accounts for losses inside the CE but also for losses that occur during transport of heat from the FC to the CE. Practically, we use a value for q_1 between 0.8 and 0.9. Combined with a reversible fuel cell we find an overall system efficiency decreasing linearly with T:

$$\begin{aligned}\eta_s &= \eta_{fc} + (1-\eta_{fc})\eta_c \\ &= 1 - q_1\frac{T_0}{T_{\Delta G=0}} - (1-q_1)\frac{T}{T_{\Delta G=0}}\end{aligned} \tag{15}$$

Polarization losses in fuel cells have been studied in great detail and depend on a number of parameters. However, for the purpose of these general calculations, we first assume that the efficiency is a fraction p_1 lower than the efficiency of a reversible fuel cell:

$$\eta_{fc} = 1 - p_1 - \frac{T}{T_{\Delta G=0}} \tag{16}$$

Combined with a CE we find a system efficiency of:

$$\eta_s = 1 - p_1\frac{T_0}{T} - \frac{T_0}{T_{\Delta G=0}} \tag{17}$$

As expected, the system efficiency is higher for higher temperatures since in this case the CE is more efficient than the FC. However, if both operate irreversibly, then a combination of Eqs. (14) and (16) with Eq. (12) yields

$$\eta_s = 1 - p_1 - \frac{T}{T_{\Delta G=0}} + q_1\left(p_1 + \frac{T}{T_{\Delta G=0}}\right)\left(1 - \frac{T_0}{T}\right) \tag{18}$$

Here a true maximum can be found at a temperature of:

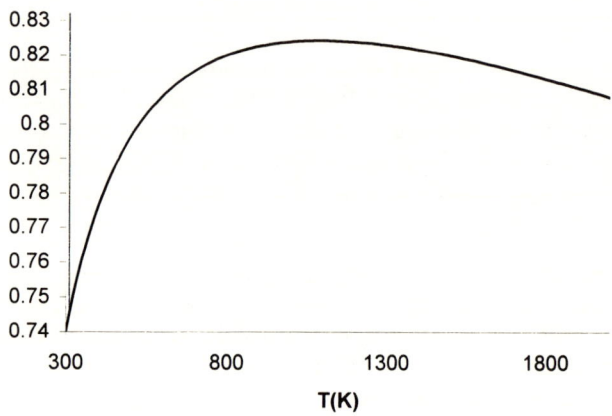

Figure 3. Example calculation of the efficiency of an irreversible FC-CE system as a function of temperature according to Eq. (20) using $p_1=0.2$, $q_1=0.8$, $T_0 = 298$K and $T_{\Delta G=0}= 5000$ K.

$$T_{max} = \left[T_0 T_{\Delta G=0} \frac{p_1 q_1}{1-q_1} \right]^{\frac{1}{2}} \qquad (19)$$

The maximum efficiency is given by:

$$\eta_s^{max} = 1 - p_1(1-q_1) - q_1 \frac{T_0}{T_{\Delta G=0}} - 2 p_1 q_1 T_0 \left(\frac{1-q_1}{p_1 T_0 T_{\Delta G=0}} \right)^{\frac{1}{2}} \qquad (20)$$

In Fig. 3, the system efficiency is shown as a function of temperature for $q_1=0.8$ and $p_1=0.2$ (and $T_0 = 300$ K, $T_{\Delta G=0} = 5000$ K). In this example T_{max} is found to be 1095 K and $\eta_s^{max} = 0.824$. This

Table 1
Temperature T_{max} (in K) for Maximum η_s as Defined in Eq. (19) for Several q_1 and p_1.

q_1	p_1			
	0.1	0.2	0.3	0.4
0.5	387	548	671	775
0.6	474	671	822	949
0.7	592	837	1025	1183
0.8	775	1095	1342	1549
0.9	1162	1643	2012	2324

maximum efficiency is just a little higher than the 80% maximumefficiencies of the FC and CE alone that were used in this example. This maximum is very flat though in principle there is an optimum temperature at which to operate an irreversible FC-CE system, and we have the freedom to deviate from this optimum temperature for other, more practical, reasons without having to sacrifice much in efficiency. It is interesting to see though that the optimum temperature lies in the range where MCFC's and SOFC's operate (900 to 1300 K).

This is illustrated in Fig. 3 and in Table 1. In this table T_{max} is given as a function of q_1 and p_1. Only for very high FC efficiency and low CE efficiency, T_{max} is lower than the operating range of present high-T fuel cells.

Of course, this is just a very rough order of magnitude calculation, which can be refined with better (temperature dependent) relations for the efficiencies. Also, in these calculations it is implicitly assumed that there is one fuel cell and one CE which we can vary the operating temperature as desired between say 300 K and 5000 K, which of course is just hypothetical. The general conclusions, however, will not change. S. F. Au, et al. give an example of a more detailed study that confirms the above conclusions for an MCFC system in which the temperature dependence of the polarization of the MCFC is taken into account by using experimental data. [7]

(iii) FC- CE Systems Hhaving an Overall FC Reaction with $\Delta S = 0$ (Reversible)

If ΔS is zero for the overall reaction, a reversible FC reaches 100% efficiency irrespective of temperature. Application of a CE in a bottoming cycle is not necessary and not even possible since the term $T\Delta S$ is zero and no heat is dissipated.

For example, the oxidation of methane to water and carbon dioxide ΔS approaches zero:

$$CH_4 + 2O_2 \rightarrow CO_2 + 2H_2O \tag{21}$$

We may wonder if it is of importance whether or not the actual electrochemical reaction involves direct methane oxidation at the anode:

$$CH_4 + 4O^{2-} \rightarrow CO_2 + 2H_2O + 8e^- \tag{22}$$

Or if the reaction proceeds via the methane reforming reaction followed by the electrochemical oxidation of hydrogen:

$$CH_4 + 2H_2O \rightarrow CO_2 + 4H_2 \tag{23}$$

$$4H_2 + 2O_2 \rightarrow 4H_2O \tag{24}$$

In principle, it does make a difference since the same overall reaction proceeds via different reaction steps and in each step losses may occur. In this case, the direct electrochemical oxidation of methane as well as the indirect path via the reforming reaction can take place in parallel and so the fastest path will be most dominant. Direct oxidation of methane in SOFC's with Cu-cermets as anodes is reported by S. Park, et al.[8] However, in general in MCFC's and SOFC's it is assumed that the reforming reaction is predominant. The heat $T\square S$ dissipated in the electrochemical reaction of hydrogen can be used directly (i.e., on a molecular level without macroscopic heat transfer) for the endothermic reforming reaction. However, this raises the question as to where exactly the heat $T\Delta S$ is dissipated. This will be discussed in the paragraph titled Heat Effects (Section VII.5.).

(iv) FC- CE Systems Having an Overall FC Reaction with $\Delta S = 0$ (Irreversible)

Due to polarization losses heat is dissipated and the application of a CE becomes meaningful. For the CE and, hence, for the FC-CE system, a higher temperature increases efficiency:

$$\Delta S = 0$$
$$\eta_{fc} = 1 - p_1 \quad (25)$$
$$\eta_s = 1 - p_1 + p_1\left(1 - \frac{T_0}{T}\right) = 1 - p_1 \frac{T_0}{T}$$

Since fuel cell losses of p_1 normally occur with higher temperature, it is even more advantageous to operate at higher temperature. If we are to have polarization losses in a fuel cell, as we always will, it is desirable to have the heat that is dissipated due to these irreversible processes at a high temperature. This is the principle reason that fuel cell power plants are designed using high temperature fuel cells such as the MCFC and the SOFC. This is in sharp contrast with the present research trends to lower the operating temperature of both the MCFC and SOFC. Materials' issues, cost, and endurance related problems are the main reasons for this. The thermodynamic advantage of operating at higher temperatures is a strong driving force for research to solve the materials problems and the reward for success in this area is great.

In addition, for a system with non-ideal CE, we can draw the same conclusion. Again, using an efficiency reduction of q_1 on the Carnot efficiency, we find similar temperature dependence for the system efficiency:

$$\eta_s' = 1 - p_1 + p_1 q_1 \left(1 - \frac{T_0}{T}\right)$$
$$= 1 - p_1(1 - q_1) - p_1 q_1 \frac{T_0}{T} \quad (26)$$

So for FC-CE systems on natural gas with internal reforming it is advantageous to operate at the highest possible temperature.

(v) *FC- CE Systems Having an Overall FC Reaction with* $\Delta S > 0$ *(Reversible)*

For some reactions ΔS is positive. For example, in the conversion of methanol in the DMFC the overall reaction has a positive entropy change due to the increase in the number of moles of gas molecules. In addition, in the conversion of carbon into CO the entropy change is positive because out of each mole of oxygen two moles of CO are formed:

$$C + \frac{1}{2}O_2 \to CO \qquad (27)$$

Should it be possible to construct a fuel cell that can be supplied with the solid fuel (C), then this Direct Carbon Fuel Fell or DCFC would - following the definition of fuel cell efficiency - have an efficiency greater than 100%. This does not violate the first law of thermodynamics (energy conservation); it just means that the fuel cell extracts heat from the environment and converts that into power.

$$\eta_{fc} = 1 + T\frac{\Delta S}{|\Delta H|}$$

so $\qquad (28)$

$\eta_{fc} > 1 (\Delta S > 0)$

Therefore, heat can be converted into work (electricity) with 100% efficiency in a process that is not limited by the Carnot factor. This does not violate the second law of thermodynamics because the conversion does not take place in an isolated process, but it is coupled to the fuel cell reaction involving an increase in entropy $\Delta S > 0$). So in an overall reversible process, the reaction (with $\Delta S > 0$) must be complemented by a process that destroys entropy; i.e., the conversion of heat into power. This is completely analogous to why the production of reversible heat $T\Delta S$ occurs in an H_2/O_2 fuel cell.

(vi) FC- CE Systems Having an Overall FC Reaction with $\Delta S > 0$ (Irreversible)

Because overall $\Delta S > 0$, heat is converted into power in this fuel cell as explained above. When polarization losses in the fuel cell are taken into account, heat is dissipated in the fuel cell and an operating temperature T_1 exists where all the dissipated heat from the polarization losses is converted into power by the fuel cell itself. Then the fuel cell efficiency is 100%. We can calculate the fuel cell efficiency and this particular operating temperature T_1, respectively, as:

$$\eta_{fc} = 1 + T\left|\frac{\Delta S}{\Delta H}\right| - p_1$$
$$\eta_{fc} = 1 \rightarrow T_1 = p_1\left|\frac{\Delta H}{\Delta S}\right| \quad (29)$$

For the carbon gasification reaction Eq. (27) ΔH = -110 kJ/mol and $\Delta S \sim 0.090$ kJ/mol.K.[9] Hence, $\Delta H/\Delta S$ = 1200 K. Depending on the polarization losses in the fuel cell the operating temperature T_1 where the fuel cell neither consumes nor produces, net heat varies according to Eq. (29). Large polarization losses are obtained by drawing sufficiently high current from the DCFC so that $p_1 = 1$, and the fuel cell efficiency equals 1 at $T = T_1 = \Delta H/\Delta S$ = 1200K. At this point the cell voltage equals the voltage equivalent of the reaction enthalpy for the carbon gasification reaction: $V_{cell} = \Delta H/2F$ = 550 mV. However, if we operate at lower current densities, polarization losses decrease so $p_1 < 1$ and $T_1 < 1200$ K. A complication now arises because at lower temperatures the conversion to CO_2 becomes favored thermodynamically relative to the conversion into CO. Hence, the entropy change of the reaction also changes and becomes approximately zero for the conversion into CO_2. If we want the DCFC to produce mainly CO, a high operating temperature ($T > 1200$ K) should be maintained at all times, and unless high polarization losses and low cell voltage are accepted heat should also be supplied continuously to the DCFC.

If we assume that polarization losses decrease as a function of temperature, for example, inversely proportional to temperature, the full cell efficiency reads:

$$\eta_{fc} = 1 + T\left|\frac{\Delta S}{\Delta H}\right| - p_2 \frac{T_0}{T} \qquad (30)$$

Then the operating temperature T_1 at 100% efficiency can be calculated as:

$$\eta_{fc} = 1 \rightarrow T_1 = \left(\frac{p_2 T_0 |\Delta H|}{\Delta S}\right)^{\frac{1}{2}} \qquad (31)$$

The exact temperature dependence is not so very important here. It is just emphasized that there is a temperature at which the fuel cell operates at very high efficiency and which does not need additional cooling facilities. Moreover, the process is self-regulating. When somehow the temperature increases, the term $T\Delta S$ increases and more heat is extracted from the fuel cell, hence cooling it. Vice versa, if the temperature decreases, more heat is dissipated than extracted and the cell temperature rises again. This self-regulating process is enhanced by a temperature dependent polarization loss according to the example above.

However, processes with $\Delta S < 0$ are scarce and often involve a solid or liquid as fuel. For example, the Direct Methanol Fuel Cell (DMFC) exhibits a positive entropy change in the overall reaction. Since the operating temperature of a DMFC is low, the amount of heat converted into power is limited. The Direct Carbon Fuel Cell will be treated in more detail in Section X.

III. NERNST LAW

1. Derivation of Nernst Law

In most fuel cell introductions only Eq. (32) below and/or the Nernst law are given to calculate the open cell voltage OCV. Here it is shown that, of course, both expressions are the same and that the logarithmic terms in the Nernst equation refers to the extra entropy change of the reaction with respect to the standard state. Hence, the ΔS of a reaction depends on the partial pressures of the reactant and product gases. Moreover, in principle, we can obtain negative as well as positive ΔS

values for almost all fuel cell reactions depending on the partial pressure or concentrations of the reactants and products.

The following derivation of the Nernst law given by F.R.A.M Standaert[10] is particularly useful for gaining insight in the operating principles of fuel cells. Standaert starts his argument with the fact that if all the species are in their standard state, the cell potential is given by:

$$E_0 = -nF\Delta G_0 \tag{32}$$

This is because if one mole of fuel is converted reversibly ΔG_0 can be converted into work (electric power), while $T\Delta S$ is converted into heat. Per mole of fuel n moles of electrons, hence, a charge of nF Coulomb is transferred from anode to cathode in an external circuit through an electric field E_0. Conservation of energy leads to:

$$W_0 = nFE_0 = -\Delta G_0 \tag{33}$$

The minus sign appears since by convention for a spontaneously occurring reaction we define the cell voltage positive and ΔG_0 is negative. The heat dissipated in this reversible process is $q_0 = T_0 S_0$.

If the reactants and products are not in their standard state, we first have to bring them to the standard state (at least hypothetically) to derive the cell potential in this case. Standaert has defined five hypothetical steps for a simple H_2/O_2 fuel cell in which H_2O is released at the anode. The cathode is supplied with pure O_2 in the standard state ($pO_2 = 1$ atm.)(See Fig. 4).

In the first step, hydrogen with a partial pressure of $pH_2 < P_0$ in the anode gas stream is separated by means of an ideal membrane; hence, preserving its partial pressure (no pressure drop across the membrane). Next by an isothermal compression step the partial pressure is increased to the standard pressure P_0 (1 atm). To do this W_{comp} (compression work) is needed and heat q_{comp} is released. Since, for an ideal gas, the enthalpy is independent of pressure energy conservation, which yields:

$$q_{comp} = W_{comp} \tag{34}$$

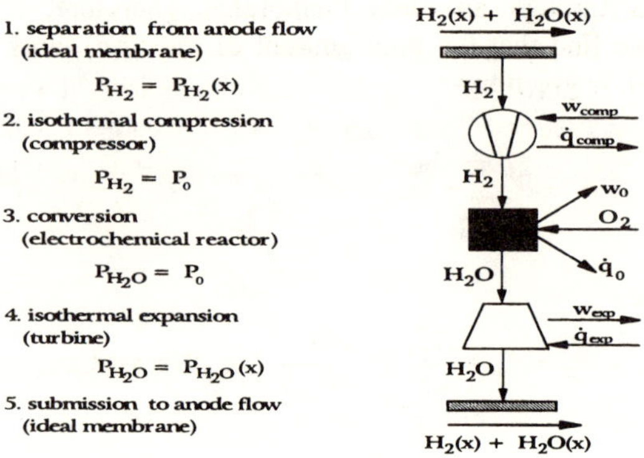

Figure 4. Derivation of the Nernst law. (From PhD thesis F. Standaert.[10])

The work of compression can be calculated as

$$W_{comp} = \int_{pH_2}^{p_0} -V dp \tag{35}$$

For an ideal gas, $V = RT_0/p$, hence:

$$W_{comp} = -\int_{pH_2}^{p_0} \frac{RT_0}{p} dp = -RT_0 \ln \frac{p_0}{pH_2} \tag{36}$$

The electrical work generated in the standard state (step 3) was already given by Eq. (33). In the fourth step, the produced water is reversibly expended to the partial pressure of water in the anode gas stream. Analogous to the second step we derive:

$$W_{exp} = -\int_{p_0}^{pH_2O} V dp = -RT_0 \ln \frac{pH_2O}{p_0} = q_{exp} \tag{37}$$

Fuel Cells

In the final step, the product water (gas phase) is brought to the anode gas through an ideal membrane. Therefore, for the whole process, we find that the total amount of (electric) work produced by the fuel cell is given by:

$$W_0 + W_{comp} + W_{exp} = nFE_0 + RT_0 \ln \frac{pH_2}{pH_2O} = nFV_{eq} \quad (38)$$

Hence,

$$V_{eq} = E_0 + \frac{RT_0}{nF} \ln \frac{pH_2}{pH_2O} \quad \text{(i.e., Nernst Law)} \quad (39)$$

The total amount of heat produced by the reversibly operating fuel cell is given by

$$q_0 + q_{comp} + q_{exp} = T_0 \Delta S_0 + RT_0 \ln \frac{pH_2}{pH_2O} \quad (40)$$

Where ΔS_0 is the entropy change of the overall reaction under standard conditions,

$$\Delta S_0 + R \ln \frac{pH_2}{pH_2O} \quad (41)$$

is the change in entropy when the reactants and products are not in their standard state. So for arbitrary temperature and arbitrary partial pressures of fuel and oxidant, we find the H_2/O_2 fuel cell of our example is:

$$\Delta G_0(T) = \Delta H_0(T) - T\Delta S_0(T) - RT \ln \frac{pH_2}{pH_2O} - \frac{1}{2} RT \ln pO_2 \quad (42)$$

where the last term is introduced to account for the hypothetical compression of oxygen from pO_2 to p_0 (1 atm) by analogous reasoning. The factor 1/2 appears because for every mole of H_2 we only have to compress half a mole of oxygen. Hence, by dividing by 2F, we find the Nernst law for the H_2-O_2 fuel cell reaction:

$$V_{eq} = E_0(T) + \frac{RT}{2F} \ln \frac{(pH_2)_a (pO_2)_c^{\frac{1}{2}}}{(pH_2O)_a} \tag{43}$$

If water is released at the cathode side (e.g. for a PAFC and PEFC) then $(pH_2O)_a$ should be replaced by the partial pressure of water at the cathode side.

For an MCFC, also, CO_2 participates in the reaction. And, by an analogous derivation, we find that:

$$V_{eq} = E_0(T) + \frac{RT}{2F} \ln \frac{(pH_2)_a (pO_2)_c^{\frac{1}{2}} (pCO_2)_c}{(pH_2O)_a (pCO_2)_a} \tag{44}$$

We note that $(pCO_2)_a$ and $(pCO_2)_c$ need not be the same. In fact, usually they are not. When calculating the partial pressures one should also consider the presence of CO that may be present from the start or formed in the water gas shift reaction. Also, the possibility of carbon deposition according to the Boudouard equilibrium cannot always be excluded. Much can be found in the literature on these reactions and their equilibria and this will not be elaborated further here.[11] Looking at this Nernst equation a bit further we can derive a few, sometimes very paradoxical, conclusions.

2. MCFC with Diluted Fuel Gas

If we dilute the anode gas of an MCFC by some inert gas all partial pressures of the gasses in the anode stream will decrease by a factor of say β compared to the original situation ($0 < \beta < 1$). We find that the Nernst potential related to the new gas composition is:

$$V'_{eq} = E_0(T) + \frac{RT}{2F} \ln \frac{\beta(pH_2)_a (pO_2)_c^{\frac{1}{2}} (pCO_2)_c}{\beta(pH_2O)_a \beta(pCO_2)_a} \tag{45}$$

or:

$$V_{eq}' = E_0(T) + \frac{RT}{2F} \ln \beta^{-1} \qquad (46)$$

Hence for $\beta = 1/2$ (50% dilution with an inert gas), we find that the cell potential will be $RT/2F \ln 2$ volts higher (!). At $T = 650$ °C the normal operating temperature of an MCFC is 40mV x ln(2) or about 60mV higher!! This increase in cell voltage is due to a smaller change in entropy, i.e. the absolute value of ΔS has become smaller. Of course, the energy content of the gas is lower and larger amounts of gas have to flow through the MCFC. This can be done to cool the cell instead of, or in addition to, cooling the MCFC by a large cathode flow. It is probably not advantageous to dilute the fuel gas on purpose. However, some fuel gases are diluted by an inert gas (Nitrogen) due to the way they are produced, for instance coal and biomass gas produced by gasifiers. The nitrogen from the air in air blown gasifiers dilutes the fuel gas produced significantly. This gives an MCFC system a significant advantage over other systems such as SOFC in this important application area. (See section XI.2.)

3. Pure Hydrogen as a Fuel

By lowering pH_2O and pCO_2 in the MCFC anode feed gas, a richer fuel is obtained which leads to a higher cell voltage. When we strictly apply the Nernst law for pure hydrogen supplied to the anode of an MCFC (or SOFC), the cell voltage should become infinite. Of course, in practice we never have a 100% H_2 gas available. Secondly, other electrochemical reactions will occur preventing the potential from becoming extremely high. Yet applying pure H_2 from a bottle to an MCFC laboratory scale cell can be helpful in detecting possible leakages between anode and cathode. A small leakage of gas from cathode to anode will drastically lower the cell potential relative to a good cell showing no leakage.

Starting with an arbitrary H_2/H_2O anode gas mixture for an SOFC, the gas composition as a function of utilization 'u' is given by:

$$pH2(u) = pH2(0)(1-u)$$
$$pH2O(u) = pH2O(0) + u\,pH2(0)$$
$$\overline{\quad pH2(u) + p(H2O)u = pH2(0) + pH2O(0) \quad} \quad (47)$$

And for an MCFC:

$$pH2(u) = pH2(0)(1-u)/(1+u)$$
$$pCO2(u) = [pCO2(0) + u\,pH2(0)]/(1+u)$$
$$pH2O(u) = [pH2O(0) + u\,pH2(0)]/(1+u)$$
$$\overline{\quad pH2(u) + pCO2(u) + p(H2O)u = pH2(0) + pCO2(0) + pH2O(0) \quad} \quad (48)$$

As a check, it is noted that independent of u the sum of the terms on the left hand side equals the sum of the partial pressures at u=0, hence the total pressure remains constant as it should. Note that since:

$$H_2 + CO_3^{2-} \Leftrightarrow H_2O + CO_2 + 2e^- \quad (49)$$

one mole of H_2 has expanded into two moles of gas at the anode of an MCFC. If we feed the cell with pure hydrogen of 1 atm:

$$pH2(0) = 1 \qquad pH2(u) = (1-u)/(1+u) \quad (50)$$

$$pH2O(0) = 0 \quad \text{and} \quad pH2O(u) = u/(1+u) \quad (51)$$

$$pCO2(0) = 0 \qquad pCO2(u) = u/(1+u) \quad (52)$$

assuming $pO_2 = pCO_2 = 1$ atm on the cathode side. The Nernst cell voltage as a function of utilization u is now given:

$$V_{eq}(u) = E_0 + \Delta V_a \quad (53)$$

with:

$$\Delta V_a = \frac{RT}{2F} \ln \frac{(1-u)(1+u)^2}{(1+u) \, u^2} = \frac{RT}{2F} \ln \frac{1-u^2}{u^2} \tag{54}$$

This last equation yields the deviation of the (local) cell voltage from the situation where the gases at the anode would all be in their standard state (i.e. all having a partial pressure of 1 atm). In the following, we continue to assume the oxidant gases to be in their standard state and recall that changes in the Nernst potential are due to changes in the entropy terms. These entropy changes can be found explicitly by using Eq. (54) as the term in the brackets below.

$$\Delta G = \Delta H_o - T \left\{ \Delta S_o + R \ln \frac{(1-u^2)}{u^2} \right\} \tag{55}$$

Defining the term in brackets as ΔS' we note that ΔS' > 0 for $u \leq u^*$ with u^* given as the solution of:

$$\Delta S_0 + R \ln \frac{(1-u^2)}{u^2} = 0 \tag{56}$$

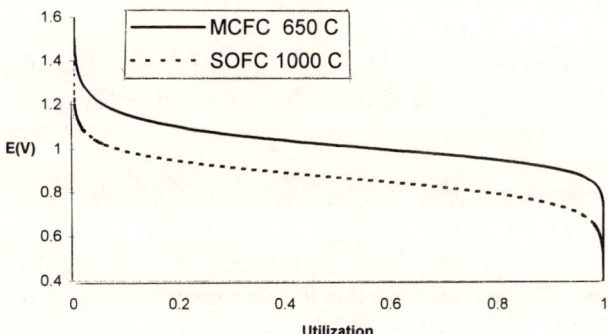

Figure 5. Calculated reversible cell voltage between anode and cathode of an MCFC and SOFC as a function of hydrogen utilization using pure hydrogen as input fuel gas.

Thus, with ΔS_o = -50 J/mole.K, we find u^* = 0.044. At the inlet of the MCFC, it is fed with pure hydrogen ΔS'> 0 and as long as the utilization has not reached 4% the heat is extracted from the cell and converted reversibly into electric power (see Section II.2.). We can calculate this amount of heat as:

$$Q = \frac{RT}{u^*} \int_{0^+}^{u^*} \ln\left(\frac{1-u^2}{u^2}\right) du \qquad (57)$$

Note that 0^+ is written as the under bound of the integral since strictly speaking for pure hydrogen ($u = 0$) the Nernst potential would be infinite and the integral as well. Of course, never in practice will this be obtained. First, hydrogen is never 100% pure; second, feeding an MCFC with pure hydrogen will probably lead to carbon deposition from the CO_2 reaction product released at the anode side. These calculations are just performed to explore the limits and to get a better understanding of the fuel cell operating under different conditions. For instance relative to operating the MCFC at constant gas composition for which $T \Delta S$ would be zero, the increase in cell voltage due to the converted heat Q is obtained by dividing it by 2F so we have:

$$\Delta V = \frac{Q}{2F} (Volt) \qquad (58)$$

From $u = 0.001$ to $u = u^* = 0.04$ the average Nernst potential is 74 mV higher than $\Delta H/2F$. Standaert defines this cell potential at $u = u^*$ where $\Delta S = 0$ by:

$$\chi \equiv \frac{\Delta H}{2F} \qquad (59)$$

Standaert calls this potential the "turning point potential," because at this potential a transition occurs from a region ($u < u^*$) where reversibly heat is converted into electric power to a region ($u > u^*$) where reversibly heat is produced. The formula shows that the turning point potential is the voltage equivalent of the enthalpy change of the reaction or, in other words, the enthalpy change expressed in units of

Fuel Cells

Volt. He notes that since ΔH is not completely independent of T, neither is χ. However, the temperature dependence is only very weak and in the range 0-1000 °C it can be approximated by:

$$\chi(T) = 1255 + 0.039(T - 273)\, mV \qquad (60)$$

so at 1000 °C χ = 1294 mV. Also, ΔS hardly depends on T so in this chapter we will further ignore these temperature dependences.

As long as the logarithmic term in Eq. (55) is positive there is a positive effect on cell voltage compared to E_0; the situation with all the gasses being in the standard state, i.e. for:

$$\frac{(1-u^2)}{u^2} \geq 1 \quad \Rightarrow u^2 \leq 0.5 \quad or \quad u \leq u^{**} \approx 0.71 \qquad (61)$$

Nevertheless, heat ($T\Delta S'$) will still be dissipated in the cell, although less than in the situation where all the gases are in the standard state:

$$T\Delta S\ <T\Delta S_o \qquad (62)$$

However, for $u > u^{**}$, the entropy term will become more negative and more heat is dissipated while the cell potential drops. In fact, it decreases rapidly. This is a situation where the very last amount of hydrogen must be removed from the fuel gas to be converted, and this is a process opposing the second law of thermodynamics. Although it is possible to convert the last few % of hydrogen from the fuel, one has to pay for it by a lower cell voltage and larger heat dissipation. The cell voltage will be zero when $T\Delta S'$ equals ΔH_o. With $T\Delta S_o/2F$ = -258 mV, $\Delta H/2F$ = 1294 mV and $RT/2F$ = 40 mV the logarithmic term would then have to be about -27 hence the argument of the logarithm equals $\exp(-27)$ = 1.8×10^{-12}. This is achieved for a utilization approaching 1 up to 10^{-12}. At say 99% utilization still a high Nernst potential exists, and, therefore, almost full utilization can be achieved in fuel cells. Looking at it from another perspective, one can use very dilute fuel gases in fuel cells and still obtain high cell voltages. Only at extremely low hydrogen concentration will the cell voltage be lowered significantly, and the voltage may become zero or even negative.

IV. ANALYTICAL FUEL CELL MODELING

1. Analytical versus Numerical Modeling

In order to optimize a specific type of fuel cell, models are needed to simulate and predict the performance as a function of parameters that can be optimized. Although the overall purpose is to optimize a fuel cell system, modeling efforts differ in their aggregation level as well as in their approach to reach sub goals. Different aggregation levels can be distinguished as depicted in Fig. 6. For example, in this chapter fuel cell systems are optimized through obtaining a better understanding of fuel cell principles by analytical modeling on the level of fuel cell systems and the single cell, while the stack is considered to consist of N identical cells in series. Others focus on the most essential part of the fuel cell: the electrodes where the actual conversion reactions take

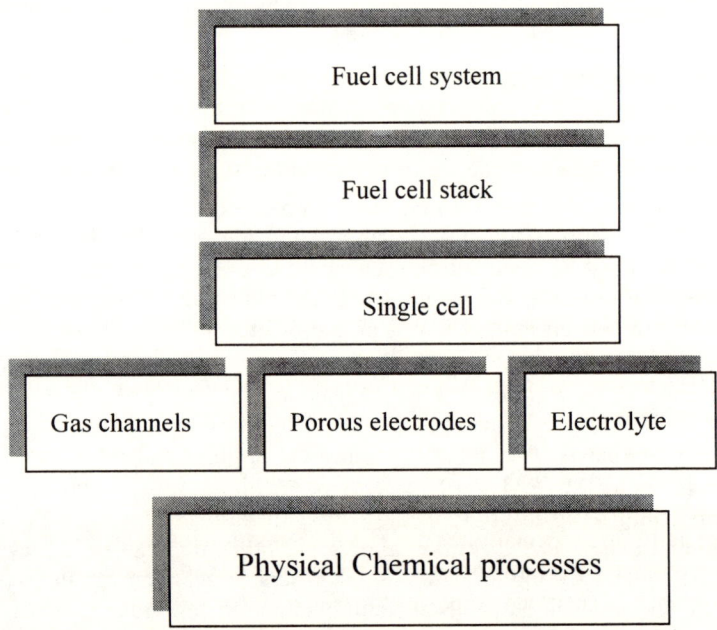

Figure 6. Aggregation levels in fuel cell modeling.

place, through the development of porous electrode models. Modeling of porous electrodes is fuel cell specific. Different (electro-) chemical reactions occur. Morphologies of the electrodes are different, transport processes differ, etc. The porous electrode models are particularly useful for determining the relative contribution of the different losses and determine R&D strategies to minimize the most important losses inside the electrodes. This is a very specialized field of high importance, but it falls outside the scope of this chapter. Sometimes models on different levels are combined into one larger model. Bessette and Wepfer[12] give a fine example of such an approach for an SOFC. They call it multi-level modeling. Evidently, the solution has to be found by numerical methods. Bistolfi, et al.[13] note that in general different knowledge at different scales is required in R&D. In other words, the modeling approach, tools and degree of detail should relate to the particular research problem one wants to solve. They illustrate this by their modeling work on a SOFC.

First, an analytical approach will be followed that will provide us with a better understanding of fuel cell principles just as the fuel cell systems in section II were modelled analytically on a simple conceptual level where we learned for example that high efficiencies can be obtained by FC-CE combinations in the medium to high temperature range. As will be shown below, it is not always necessary to solve and describe the detailed processes and current densities at every position in the cell to be able to predict the cell performance that can be measured externally. However, in some types of fuel cells it is essential to know internal details to be able to optimize performance. For example in low temperature fuel cells ($T < 100$ °C), special attention is often needed for a proper water management. Water is not just a product gas, as in the high T-fuel cells, but is predominantly in liquid form and may dilute the electrolyte, as in an AFC, or provide sufficient conductivity of the polymer electrolyte membrane in a PEMFC. On the other hand, too much water may flood the electrodes or gas channels, thus decreasing the effective reaction surfaces in the porous electrodes or preventing access of the reactant gases to the electrodes. So for low temperature fuel cells, and especially for the PEMFC, water management is the key to optimized performance. This requires detailed modeling of the physical and chemical processes for which numerical methods are often necessary. In section X the literature on numerical fuel cell models for various types of fuel cells will be reviewed.

A large difference between the modeling approach followed here and most other approaches in the literature is the lumping of polarization sources into one parameter; the local (quasi-ohmic) resistance $r(x)$. As explained below, one can take this resistance constant $r(x) = r$, or include more detailed dependencies $r(u(x))$ or $r(i(x))$. In the latter case Eq. (83), which will be derived later can still be used, but will be more difficult to evaluate. The determination of the detailed dependencies of r on certain parameters often requires models for the porous electrodes in which polarization losses predominantly occur. Losses in the electrolyte are ohmic and can be added simply. The example of the MCFC shows that it is not always necessary to know and model all the details to be able to describe the performance of the fuel cell. Moreover, it is emphasized here that it is not always necessary to solve the problem. Solving the problem means that one determines exactly the current density profile $i(x)$ inside the cell, for example. As will be shown, a non-uniform current density does influence the performance to a certain extent but the exact shape of this non-uniformity has very little influence on the overall performance (see section VI.1.).

Many fuel cell companies and research institutes have developed their own numerical fuel cell model in order to be able to optimize the performance of a specific type of fuel cell and to perform a detailed engineering study of the construction of the fuel cell stack. Although this is a very powerful approach and accurate results may be obtained, its drawback is that calculations are performed just for one specific case and results cannot be transferred to other cases, let alone other types of fuel cells. Tendencies and interrelations between parameters and performance can be obtained by running many simulations, but one is never sure if the results just hold for the specific choice of parameters one has used, or are more general. Often many parameters need to be put in, either obtained from experimental data or derived approximately using theoretical arguments. For every change, the numerical model has to run on a computer again.

The analytical approach to fuel cell modelling outlined in this chapter is very helpful for a better understanding of fuel cells. Using this analytical mathematical approach, the developed model is not limited to one particular type of fuel cell or fuel cell geometry, and the influence of a parameter is directly visible in the derived model equations. However, often an analytical solution of a problem can only be obtained if certain approximations are made. This also holds for

Fuel Cells. Nevertheless, with just a few assumptions, very simple but accurate analytical solutions can be obtained as will be shown in section VIII where these models are compared with experimental results on a MCFC. Accuracy can be further improved by quantifying the influence of the approximations. In general, these correction terms are found to be in the order of only 10 to 20 mV relative to an OCV of about 1000 mV, hence, only a few percent. Machielse[14] was the first to derive an analytical expression for the performance of an MCFC in what he called a simple fuel cell model. He was inspired by the fact that experiments on 10 by 10 cm^2 MCFC's showed an almost perfect linear behavior of cell voltage as a function of the two main adjustable parameters, i.e., current density and gas utilization if the other was held constant. In his footsteps an alternative derivation using hypothetical sub-cells was obtained.[15] For didactic purposes and because it is so similar to many numerical models, the latter derivation will first be given below. Next, the more general and elegant derivations by Standaert[10,16] are highlighted because his work provides a rigorous mathematical base for the solutions which clearly reveals the nature and order of magnitude of the approximations that can be made to find analytical solutions.

It will be shown that an integral Eq. (83) can be derived for the cell voltage, which can be solved approximately yielding a bilinear relation also found experimentally e.g. for the MCFC. Correction terms are introduced to describe the performance of a fuel cell even more accurately as will be shown in the comparison of experimental results for a MCFC. Perhaps, more importantly, a good prediction of experimental results is obtained and one can reach a better understanding of the results from the analysis. We will be able to answer questions such as:

1. What exactly is Nernst loss?
2. How can we calculate it?
3. How can we avoid or minimize Nernst loss?
4. And how can we not minimize it?
5. What is the effect of networking and how large is this effect?

These and others questions will be answered in the next sections.

2. Analytical FC Model by Division in Hypothetical Sub-Cells

For didactical reasons, we will first derive an expression for the current as a function of cell voltage by dividing the cell into a number of imaginary sub cells analogous to the approach followed in numerical 2D-models (see Fig. 7). These sub cells are not to be mistaken with the individual cells that build a fuel cell stack. The sub cells are imaginary parts of one cell over the full width of a cell while assuming the fuel flow in the length direction of the cell. Assuming all cells in a stack to be equal, we can interpret the sub cell equally as well as the imaginary part of a fuel cell stack. By finally taking the limit for an infinite number of sub cells, we obtain an analytical expression for the cell voltage as will be shown below. By analogous reasoning, the analytical modelling using differential equations can be understood much better, as well as the sequential nature of how a fuel cell operates.

We assume that all polarization losses (except the Nernst loss) can be modelled as a quasi-ohmic specific resistance, which may be assumed constant throughout the fuel cell. Often the cathode gas is used to cool the cell and is supplied at high flow rates. So its composition is essentially constant over the cell, yielding a low utilization. Therefore, for the moment, here only utilization of the fuel gas will be considered.

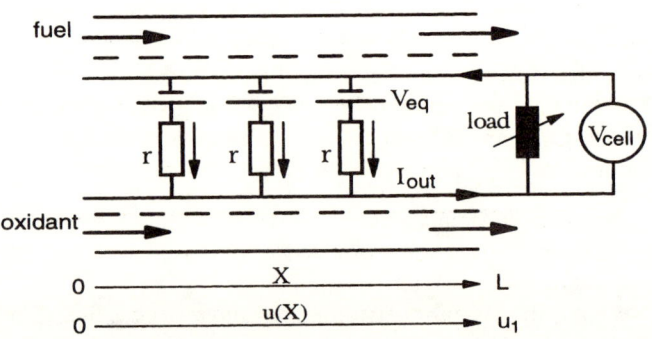

Figure 7. Schematic electric representation of a fuel cell divided in hypothetical sub cells.

Fuel Cells

A model will be derived for a hydrogen/oxygen fuel cell in which water is released at the cathode. The gas enters the first sub cell located at the inlet of the fuel cell and the OCV or $V_{eq}(1)$ of this first sub cell is given by the Nernst law:

$$V_{eq}(1) = E_0(T) + \frac{RT}{2F} \ln \frac{(pH_2)_a (pO_2)_c^{\frac{1}{2}}}{(pH_2O)_c} \tag{63}$$

The first sub cell delivers current and part $u(1)$ of the hydrogen is utilized and converted into water. For simplicity, we assume that $pH_2(u) = (1 - u)pH_2(0)$. In general, we have for the reversible cell voltage of the ith-sub cell:

$$V_{eq}(i) = E_0(T) + \frac{RT}{2F} \ln \frac{(pH_2)_a \{1-u(i-1)\}(pO_2)_c^{\frac{1}{2}}}{(pH_2O)_c} = V_{eq}(1) + \frac{RT}{2F} \ln\{1-u(i-1)\} \tag{64}$$

with $u(i)$ the utilization of the hydrogen after the ith sub cell. We can now calculate the current supplied by each sub cell, but first we will introduce dimensionless parameters. The flow of hydrogen represents an input of 2 moles of electrons that can be converted into a current for each mole of hydrogen. Hence, a number of electrons per second enter the cell. We define this current as the equivalent input current I_{in}. In other words, I_{in} is the current that can be drawn from the cell if all the hydrogen that flows into the cell is fully converted.

The total gas utilization of the fuel cell is given by the utilization after the last, i.e., the Nth sub cell, and is equal to the ratio of the total current produced (I_{tot}) relative to the equivalent input current (I_{in}):

$$u_{tot} = u(N) = \frac{I_{tot}}{I_{in}} \tag{65}$$

By adjusting the flow or the (imaginary) width of the cell we choose to model, we can always set I_{in} = 1 unit of current and obtain a dimensionless current. Note that $u_{tot} = I_{tot}$ and $u(1) = I(1)$ with these dimensionless currents.

With N the number of sub cells and R_c the total (quasi ohmic) internal resistance of the whole fuel cell, the resistance describing the polarization losses in one cell is given by $R_c.N$. As said, we assume the resistance to be constant over the whole fuel cell. We also assume the current collectors to be good conductors so that the voltage across each sub cell equals the cell voltage of the whole fuel cell, which we set equal to V_{cell}. Hence, the current delivered by the ith sub cell is given by:

$$I(i) = \frac{V_{eq}(i) - V_{cell}}{R_c N} \tag{66}$$

The utilization after the ith sub cell has increased by $I(i)$ relative to that of the previous sub cell:

$$u(i) = u(i-1) + I(i) \tag{67}$$

Combining Eqs. (64), (66), and (67) we find the following recursive relation for $u(i)$ and note that $u(0) = 0$:

$$\begin{aligned} u(i) &= u(i-1) + \frac{V_{eq}(1) - V_{cell} + \frac{RT}{2F}\ln(1 - u(i-1))}{R_c N} \\ &= u(i-1) + u(1) + \frac{RT}{R_c N.2F}\ln(1 - u(i-1)) \end{aligned} \tag{68}$$

Note that we can now subsequently calculate $u(1)$ through $u(N)$ from Eq. (68) and, hence, using Eq. (67) also $I(1) \ldots I(N)$ can be calculated for a certain OCV, V_{cell} and R_c.

In Figure 8, $u(i)$ and $I(i)$ are plotted for the parameter values OCV = 1 V and various V_{cell} and R_c as indicated. We see that the current density is decreasing near the outlet, while the utilization is an increasing function starting approximately linearly then levelling off near the outlet. So as a first order approximation, we can write:

$$u(i) \approx 1 - \exp\left(-\frac{\gamma i}{N}\right) \tag{69}$$

Fuel Cells

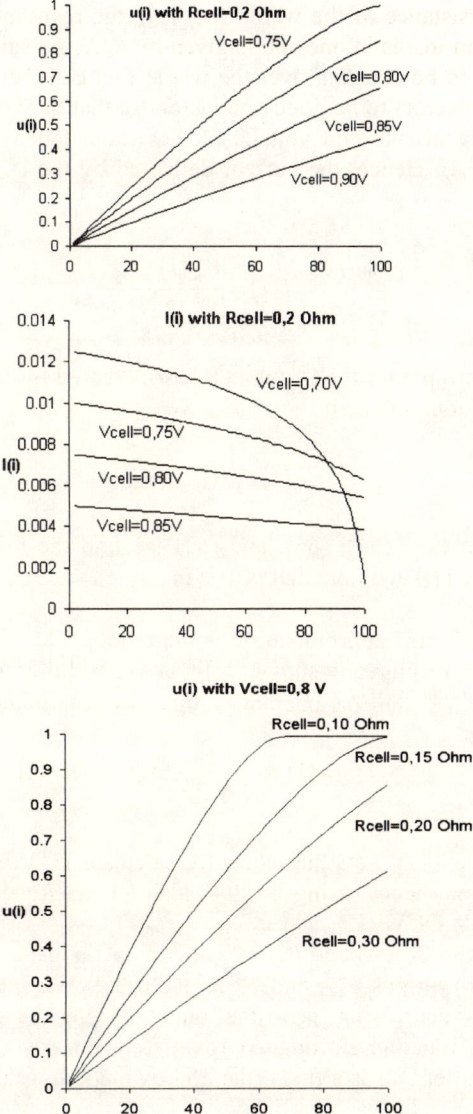

Figure 8 Calculated utilization and current distribution in a fuel cell divided in 100 sub cells numbered from fuel inlet to outlet.

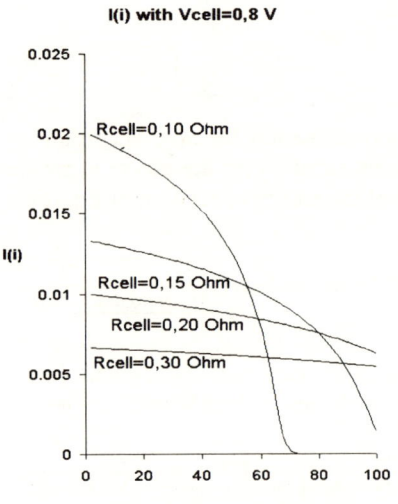

Figure 8. Continuation

In Figure 9, this approximation is plotted together with $u(i)$ calculated for $R_c = 0.3$ as already shown in Figure 8. Because the total utilization equals $u(N)$, the parameter γ roughly equals the total fuel cell utilization.

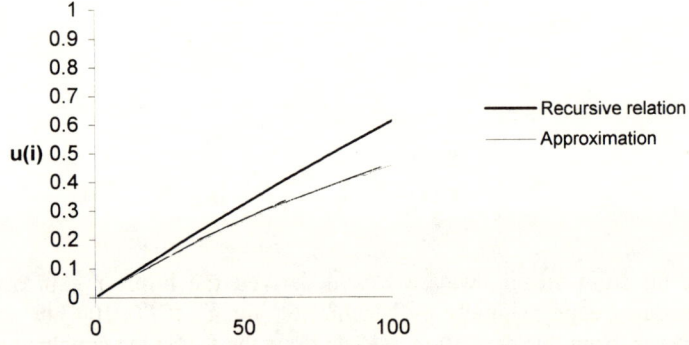

Figure 9. Calculation and approximation of $u(i)$ using Eqs. (68) and (69) respectively with $V_{cell} = 0{,}8$ V and $R_{cell} = 0.3\,\Omega$.

Fuel Cells

$$u_{tot} = u(N) = 1 - \exp(-\gamma)$$
$$\text{or} \tag{70}$$
$$\gamma = -\ln(1 - u_{tot}) \approx u_{tot}$$

Using the first approximation for $u(i)$, according to Eq. (68), we can find a better one by substituting the above expression for u_{tot} into the logarithmic term of the recursive relation of Eq. (68):

$$u(i) = u(i-1) + u(1) - \frac{RT}{R_c N.2F} \frac{\gamma(i-1)}{N} \tag{71}$$

This recursive relation can now be solved (e.g. with the help of computer algebra packages such as Maple) to yield:

$$u(i) = N.u(1) - \frac{RT}{2F} \frac{\gamma}{R_c.N^2} \sum_{k=2}^{i} (k-1) \tag{72}$$

$$u(i) = \frac{V(1) - V_{cell}}{R} - \frac{RT}{2F} \frac{\gamma}{R_c.N^2} \left\{ \frac{1}{2}(i+1)^2 - \frac{3}{2}i - \frac{1}{2} \right\} \tag{73}$$

By substituting $i = N$, we can calculate $u(N) = I_{tot}$ and take the limit for N to infinity.

$$u(N) = I_{tot} = \frac{V(1) - V_{cell}}{R_c} - \frac{1}{2}\left(\frac{RT}{2F}\right) \frac{\gamma}{R_c} \tag{74}$$

Rearranging and using $\gamma \approx u_{tot}$ we find:

$$V_{cell} \approx V(1) - I_{tot} R_c - \frac{1}{2}\left(\frac{RT}{2F}\right) u_{tot} \tag{75}$$

In a far from elegant way we have derived the bilinear expression determined experimentally by Machielse[14] for a MCFC 100 cm^2 cell. However, from this derivation, we can draw the following conclusions.

In the first approximation, the cell voltage is a bilinear function of cell current and gas utilization. We can distinguish two polarization terms; one representing the (quasi) ohmic polarization losses we assumed in the model, and a loss term representing the so-called Nernst loss caused by utilization of the (fuel) gas. This latter term is proportional to temperature; hence, Nernst loss is more severe in high T fuel cells. In Section (V), this important loss term is discussed in more detail.

Furthermore, note that we did not need to specify the shape of the fuel cell. Gasses may either flow in the largest direction of a rectangular cell or across the width of the cell. In the first order this makes no difference. The calculation we performed is not an iterative procedure despite its similarities with a numerical simulation. Calculation of the current in each sub cell is straightforward. With a given inlet gas composition, the reversible cell potential of the first sub cell is determined by the Nernst law and its current is determined by its quasi-ohmic resistance and the imposed V_{cell}. From the current the output gas composition is calculated and subsequently treated as input composition for the second sub cell etc. Note that the current produced in a particular sub cell does not depend on the current produced in the sub cells down stream. Therefore, when a certain cell voltage is imposed on the cell (potentiostatic control), the first part of the cell takes from the fuel stream whatever it needs to obtain the current as shown in Eq. (66). Processes down stream near the outlet side do not influence what happens at the inlet side of the cell. The consequences for fuel cell operation are further discussed in Section VII.

In more detailed numerical simulations, one generally does have to use iterative procedures until convergence is achieved since sometimes one wants to model more precisely the local polarization losses as a function of gas composition and temperature. However, as shown by the experimental results of Machielse,[14] the behavior of a MCFC is essentially bilinear in I_{tot} and u_{tot}, and this behavior can be understood as shown by the previous and following derivation.

3. Analytical FC Model Using Differential Equations

In two articles F. Standaert, et al.[16,17] brought this analytical approach to fuel cell modelling a few steps further. In the first paper he describes an isothermal fuel cell and in a second paper it was shown that by using an appropriate coordinate transformation a non-isothermal fuel cell can be described as a quasi-isothermal fuel cell.[17] With the previous derivation using sub cells in mind, it is easily understood that the local current density is given by:

$$i(x) = \frac{V_{eq}(u) - V_{cell}}{r(x)} \qquad (76)$$

Recall that the equivalent input current I_{in} is defined as the current that is obtained if all the fuel gas is fully converted. By normalizing I_{in} in the active surface area A of the cell, the equivalent input current density can be defined as:

$$i_{in} \equiv \frac{I_{in}}{A} \qquad (77)$$

The utilization $u(x)$ is now given as the ratio of the current generated from the inlet up to a distance x from the inlet.

$$u(x) = \frac{1}{i_{in}} \int_0^x i(x)\,dx \qquad (78)$$

Or in differential form we have:

$$i(x) = i_{in} \frac{du}{dx} \qquad (79)$$

Combining these equations a differential equation in $u(x)$ is obtained with boundary conditions given as $u(0) = 0$ and $u(1) = u_{tot}$.

$$r(x)\frac{du}{dx} = \frac{V_{eq}(u) - V_{cell}}{i_{in}} \qquad (80)$$

The first step in solving this problem is to multiply the equation by du/dx and integrate the result with respect to x from 0 to unity.

$$\int_0^1 r(x)\left(\frac{du}{dx}\right)^2 dx = \frac{1}{i_{in}} \int_0^1 \left(V_{eq}(u) - V_{cell}\right)\frac{du}{dx} dx \qquad (81)$$

$$\frac{1}{\{i_{in}\}^2} \int_0^1 r(x)\{i(x)\}^2 dx = \frac{1}{i_{in}} \int_0^{u_{tot}} \left(V_{eq}(u) - V_{cell}\right) dx \qquad (82)$$

Rewriting and using $u_{tot} = i_{tot}/i_{in}$ one obtains:

$$V_{cell} = \frac{1}{u_{tot}} \int_0^{u_{tot}} V_{eq}(u) du - \frac{1}{i_{tot}} \int_0^1 r(x)\{i(x)\}^2 dx \qquad (83)$$

Note that we have not solved the differential equation, but merely rewritten it as an integral equation. Secondly, the equation holds for every fuel cell and it is exact within the model assumptions. Yet, it is now in a form where we can clearly recognize two terms and their physical meaning.

Instead of the OCV we see a term appear that can be interpreted as the Nernst potential difference between anode and cathode averaged over the cell; or to be more precise, the term can be averaged from zero utilization at the inlet and to total utilization at the outlet. This leads to the so-called Nernst loss because this first integral is smaller than the OCV. The term has nothing to do with the irreversibilities, which were all included in $r(x)$, and, hence, the term represents reversible (thermodynamic) losses. Secondly, instead of finding a (quasi-) ohmic loss term like $I_{tot}.R_{tot}$ we find that the irreversible losses are proportional to the local irreversible heat dissipation integrated over the cell. In the next two Sections, the two loss terms are discussed in more detail. However, with a few simple assumptions, we can easily solve the integral equation first.

- Assume $V(u)$, $r(x)$ and $i(x)$ to be constant; i.e. independent of u and x than V_{cell} is easily seen to be equal to :

$$V_{cell} \approx V(0) - i_{tot}.r \qquad (84)$$

- If instead we assume that $V(u)$ decreases linearly with u $V(u) = V(0) - \alpha.u$ then V_{cell} is given by:

$$V_{cell} \approx V(0) - \frac{1}{2}\alpha.u_{tot} - i_{tot}.r \qquad (85)$$

So, very simply and in a straightforward manner, the bilinear relation found earlier in Eq. (75), is derived approximately. Furthermore, since within the model assumptions the integral equation is exact, we can calculate the order of magnitude of the deviations introduced by the above approximations. Note that the derivation does not change if $r(x)$ is a function of $i(x)$ as well; i.e., if we want to model fuel cells with a non-linear polarization curve. In particular low temperature fuel cells exhibit a strong increase in polarization at low current density due to kinetic limitations, a linear ohmic region at intermediate current densities, and a strong increase in polarization at high current densities near the limiting current density due to diffusion.

In summary, it is recalled that an integral equation is obtained for the cell voltage which can be solved approximately yielding a bilinear relation also found experimentally for at least the MCFC.[14,18] Correction terms may be introduced to describe the performance of a fuel cell even more accurately. Comparison with experimental results for a MCFC bench cell (100cm^2) shows excellent agreement as will be shown in Section VIII. But perhaps more importantly, the prediction and analysis of good experimental results leads to a better understanding of fuel cells and how they operate. We should now be able to answer questions such as:

- What exactly is Nernst loss and how can it be calculated, avoided or minimized?
- What is the effect of networking of fuel cells and can we minimize the losses in this way?

These and other questions will be answered in the next two sections that refer, respectively, to the two integrals in the integral Eq. (83) with which the cell voltage can be calculated as a function of current density and gas utilization.

V. REVERSIBLE LOSSES AND NERNST LOSS

1. Nernst Loss

We note that in most fuel cells the gas composition changes locally due to utilization of the reactant gas(es) and mixing with product gas(es). Therefore, the entropy term increases locally (in an absolute sense) and more heat is dissipated (as $T\Delta S$). Consequently, the local Nernst potential difference between anode and cathode is lowered. This effect decreases overall fuel cell efficiency and is known as Nernst Loss.

Because the fuel (gas) is consumed, its partial pressure, or, in general, its chemical potential becomes gradually lower towards the outlet. At the same time reaction products are formed diluting the fuel (or oxidant). Both phenomena lead to a decrease in the local Nernst potential difference between anode and cathode as a function of gas utilization.

Since the gas flows of oxidant and fuel can be controlled independently, the local utilization of oxidant and fuel in general is not the same. In fact, often a large oxidant flow is applied for cooling purposes; hence, oxidant utilization is low. For simplicity, only fuel utilization will be described in this section. Oxidant utilization can be described in a completely analogous manner.

In the previous section, the cell voltage was expressed as the sum of two integrals. The first one can be rewritten as the OCV minus a term that is called the Nernst loss. In other words, the Nernst loss is defined as the decrease in cell voltage relative to OCV due to utilization of the fuel.

$$\Delta V_{Nernst} = V_{eq}(0) - \frac{1}{u_{tot}} \int_0^{u_{tot}} V_{eq}(u)du \qquad (86)$$

When the inlet fuel composition is known, the Nernst loss can be simply calculated as the difference between OCV and the averaged Nernst potential. This is illustrated in Fig.10. From the figure we see that the Nernst loss is equal to the surface area between the line $V_{eq}(0)$ and the function $V_{eq}(u)$ up to $u = u_{tot}$.

The figure is reasonably to scale for high temperature MCFC and SOFC fuel cells. It shows that the Nernst loss is a substantial part of the total heat losses as given by the surface areas between χ and area between $V_{eq}(u)$ and V_{cell} represents the irreversible polarization losses. Typical order of magnitude values are $V_{eq}(0) = 1250$ mV; $V_{eq} = 1000$ mV; $V_{cell} = 800$ mV; total reversible losses of 350 mV (including Nernst loss of 100 mV) irreversible (polarization) losses 'only' 100 mV. As shown in this figure, even if the irreversible losses are minimized by a combination of technological improvements, like better catalysts, better porous electrodes, and materials with higher conductivity, lower current density operation or applying the concept of multistage oxidation, the heat losses still amount to a voltage loss of 350 mV. In other words, only the irreversible losses can be reduced by technological improvements, but even an improvement of 50% (100 mV → 50 mV) would only reduce the heat losses by about 10% (50 mV/450 mV).

As shown above in Section III (Nernst law), the Nernst loss is due to entropy changes and, although the loss is dissipated as heat, it is done in a reversible way. In other words, in the reverse reaction this heat is converted into chemical energy again. For example, for an SOFC we can derive:

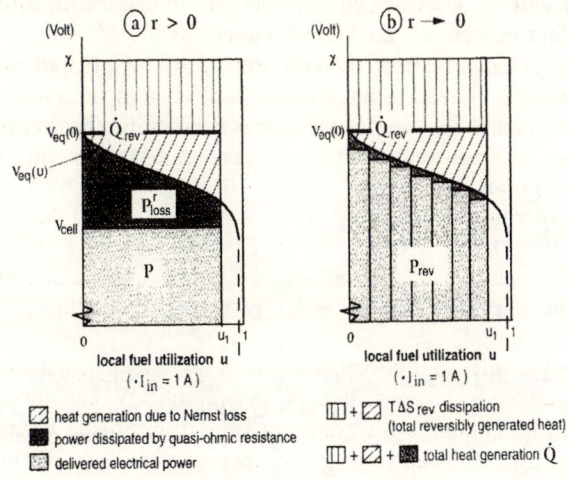

Figure 10. Illustration of the reversible and irreversible heat losses in a fuel cell.

$$\Delta V_{Nernst} = -\frac{RT}{nF}\left(\frac{1}{u_{tot}}\int_0^{u_{tot}}\ln\frac{1-u}{1+u\dfrac{pH_2(0)}{pH_2O(0)}}du\right) \quad (87)$$

In accordance with our conclusion that the Nernst losses are reversible losses associated with entropy terms, we find that the Nernst loss is proportional to the operating temperature (analogous to $T\Delta S_0$). Second, the Nernst loss becomes higher if we apply a hydrogen-rich fuel inlet gas with a low partial pressure of product gas (pH$_2$O(0)). This is because the Nernst loss refers to the difference between OCV and average voltage and the OCV increases strongly for more concentrated fuel gases. However, since the average cell voltage is higher, it is, of course, still advantageous to apply these rich fuel gases with low product concentrations.

Third, the Nernst loss increases more rapidly with utilization for a MCFC than for a SOFC. This is because for a SOFC only H$_2$O is a reaction product emerging at the anode side while for a MCFC also the reaction product CO$_2$ dilutes the hydrogen and builds up its pressure at the anode side. As explained and shown in Figure 5, the higher operating temperature of the SOFC relative to the MCFC counteracts this advantage and in practice Nernst losses of MCFC and SOFC are comparable.

Finally, in spite of the logarithmic nature of the Nernst potential, this potential as a function of utilization shows a large region, which is approximately linear (see Fig. 5 or Fig. 10). Only when we start with a very pure hydrogen inlet gas or reach very high utilization can strong deviations from linear behavior occur. However, for practical purposes the potential can be linearized either by fitting a straight line by a least square fit between $u = 0$ and $u = u_{tot}$, or by approximating $V_{eq}(u)$ by a straight line through say: $u = u_{tot}/2$ and $V = V_{eq}(u_{tot}/2)$ with a slope α equal to $dV(u)/du|_{u\,=\,u1/2}$. When $V_{eq}(u)$ is approximated by a straight line with slope α then the Nernst loss is simply given by $\alpha.u_{tot}/2$. So by determining the slope we can (up to good approximation) determine the Nernst loss.

In Table 2 these slopes are given for a MCFC and a SOFC at 650 °C and 1000 °C, respectively (Standaert)[10]. Inlet fuel gas compositions are as indicated and comparable with gas mixtures produced from reformed methane. Because of the higher operating temperature of a SOFC (around 1000 °C), the Nernst losses in both fuel cells are approximately the same. The lower operating temperature of a MCFC is counteracted by the fact that two product species (H_2O and CO_2) are released at the anode side. Note since α is of the order of 100-200 mV (see Table 2) the Nernst loss for these high temperature fuel cells (MCFC and SOFC) is about 50-100 mV i.e. about 5-10% of the OCV. So reducing the Nernst loss in these high temperature fuel cells is very profitable since this reversible loss is about equal to all the irreversible losses lumped together in the quasi-ohmic resistance at the current densities at which these cells are normally operated. There are some studies reporting that proton conducting ceramics could be used as the electrolyte in a high temperature fuel cell: a PCCFC (proton conducting ceramic fuel cell).[19] Because the reaction products are released at the cathode side, α is relatively small even at higher temperatures. Calculated values are depicted in Table 2.

It is sometimes suggested that by dividing the fuel cell into a number of separate stacks, one could lower the Nernst loss and improve overall performance. The last is true but for other reasons as will be shown in Section VI. However, Nernst loss is determined by thermodynamics and as long as the gases are utilized in the (system of) fuel cell(s) the local V_{eq} varies and Nernst loss appears as defined in Eq. (86). Yet, it is worthwhile to study the possibilities for lowering the Nernst loss. By inspection of the definition of Nernst loss in Eqs. (86)

Table 2
The Slope α (in mV) of the Nernst Potential versus Utilization for an Isothermal MCFC (650 °C), SOFC (1000 °C) PAFC(200 °C) and a PCCFC (Proton Conducting Ceramic Fuel Cell) at 650 and 1000 °C for Various Inlet Gas Compositions (Before the Shift Reaction is Established) (Source F. Standaert)[10]

H_2	CO_2	H_2O	α(MCFC) at 650 °C	α(SOFC) at 1000 °C	α(PAFC) at 200 °C	α(PCCFC) at 650 °C	α(PCCFC) at 1000 °C
0.792	0.008	0.20	216	181	14	27	38
0.64	0.16	0.20	179	162	21	48	80
0.48	0.32	0.20	148	145	28	64	110
0.32	0.48	0.20	122	131	33	75	126

and (87) we see that there are in principle four possibilities to lower the Nernst loss. Either one has to lower the operating temperature of the fuel cell, operate the cell at lower utilization, and lower OCV or in general try to change the Nernst potential as a function of utilization.

2. Minimizing Nernst Loss by Lowering the Operating Temperature

Since the Nernst loss is proportional to temperature, an obvious way to reduce the Nernst loss is by lowering the operating temperature of the fuel cell. Of course, there are limits set by other phenomena. An MCFC, for example, cannot operate below the melting point of the carbonate electrolyte. A more gradual effect is the increase in polarization with lower temperatures. This increase may be exponential in nature due to the Arrhenius behavior of the polarization processes involved, e.g. reaction kinetics and diffusion. Also, conduction in a SOFC electrolyte follows Arrhenius behavior. Hence, ohmic losses increase here exponentially with lowering the temperature. This counteracts the decrease in Nernst loss and other reversible losses ($T \Box S$). So in general there are limits to lowering the operating temperature, and the influence on all the processes inside the fuel cell and on the system as a whole have to be taken into account to determine the optimum temperature.

3. Minimizing Nernst Loss by Decreasing the Utilization

A lower utilization of fuel gas is possible but conflicts with the usual goal of converting all of the fuel that enters the system. Purging of unspent fuel is neither economically nor environmentally advantageous. However, integration of a fuel cell into a chemical process in which an anode effluent gas still containing a relative high amount of H_2 and H_2O (and CO_2 for an MCFC) could be used effectively might be possible. By applying a relative high anode flow the cell is also cooled more compared to standard operation with 80% or 90% utilization. Therefore, the cooling demand in the cathode cycle is lowered for two reasons. Firstly, a lower Nernst loss leads to lower heat production and higher efficiency, and, secondly, the anode effluent gas also carries heat away from the cell. A related operation mode is the use of a high temperature fuel cell as a reformer.[20] The idea is that an internal reforming fuel cell need not only produce hydrogen for its own consumption but the surplus heat generated by the reversible and irreversible processes in the fuel cell can be used to reform a surplus amount of methane. Hence, the anode off-gas is a hydrogen rich gas and utilization is low. Again, the cooling demand of the fuel cell is decreased, saving on auxiliary power in the cathode-cooling loop. (Also see Section XI.1. Tri-Generation Systems.)

4. Minimizing Nernst Loss by Decreasing the OCV by Gas Recycling

Using a leaner fuel gas the OCV will decrease and, in general, the difference of the average Nernst potential over the whole cell will also decrease. So according to the definition, the Nernst loss will become smaller. However, the total reversible losses will increase and reduction of the OCV for instance by mixing a richer fuel with a leaner one, e.g. in a recycling loop, is not beneficial. Although the part of the reversible losses we define as Nernst loss may decrease, the term $T\Delta S$ term increases more as can be seen from Fig.10. In this figure leaner gas composition may correspond to some point half way on the x-axis. The new origin now lies at this point on the x-axis and the new OCV found on the $V_{eq}(u)$ curve is indeed lower. A new u-axis must be defined with u relative to the new (leaner) inlet gas composition. Effectively the u-axis is now stretching out but the decrease from OCV to 80 or 90 % utilization is approximately only half that from the original situation, hence Nernst loss is approximately halved. But the vertically striped area $T\Delta S$ in Fig. 10 is enlarged by a rectangle OCV (old)-OCV (new) approximately twice the new Nernst loss. So although the Nernst loss is halved the total reversible losses ($T\Delta S$ at inlet composition + Nernst loss) are increased by half the old Nernst loss and about equal to the new Nernst loss. So recycling of anode gas will decrease Nernst loss but increase the total reversible heat production. A positive effect of recycling, however, is the more evenly distributed current density. This will be treated in Section VI.

5. Minimizing Nernst Loss by Changing the Relation $V_{eq}(u)$

Now let us look more closely at the reasons for the high Nernst loss in the two high temperature fuel cells that have been developed so far.

Apart from the high temperature, it is the fact the reaction products are produced at the anode side. This is coupled to the fact that the electrolytes conduct anions (negative ions) i.e. CO_3^{2-} and O^{2-} respectively. In fuel cells in which protons or proton carrying cations such as H_3O^+ are conducted in the electrolyte the reaction product H_2O is formed at the cathode side. At the anode only hydrogen is consumed which results in a lower pH_2 unless pure hydrogen is supplied. In the latter case pH_2 must remain constant throughout the cell (except for maybe a small pressure difference due to flow resistance) since there is no other gas available to complement the hydrogen partial pressure to the operating pressure. Hence, Nernst loss (related to the anode) is zero (see Section IX.4. Dead-End Mode). So, although in general the Nernst loss increases (linearly) with increasing utilization, in a proton conducting fuel cell it is possible to obtain zero Nernst loss at 100 % utilization. However, for arbitrary hydrogen/inert gas mixtures, the Nernst loss for a proton conducting electrolyte fuel cell is given by:

$$\Delta V_{Nernst} = \frac{-1}{u_{tot}} \int_0^{u_{tot}} \frac{RT}{2F} \ln \frac{(1-u)}{(1-u\zeta)} du \qquad (88)$$

Note that for $\zeta = 1$; i.e. the fraction of hydrogen in the inlet gas mixture equal to 1, indeed the Nernst loss is zero and that for lean gas mixtures ($x \ll 1$) an upper boundary for the Nernst loss is given by:

$$\Delta V_{Nernst} = \frac{-1}{u_{tot}} \int_0^{u_{tot}} \frac{RT}{2F} \ln(1-u) du \qquad (89)$$

In Fig.11 this upper bound for the Nernst loss is plotted as a function of u for $T = 100$ °C (PEMFC) and 200 °C (PAFC) respectively. So the Nernst loss related to the anode in PEMFC's and PAFC's, is low because of the relatively low temperature and because of reaction products emerging at the cathode, hence not diluting the fuel gas.

Calculation of the Nernst loss related to the cathode proceeds in a completely analogous manner. But since most fuel cells are cooled by a large airflow, oxidant utilization is low hence the contribution of the cathode to the Nernst loss is normally small also for fuel cells with a proton conducting electrolyte. Because of the large oxidant flow also

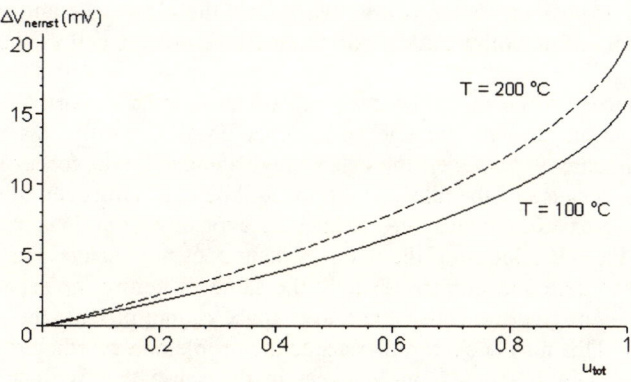

Figure 11. Nernst loss as a function of utilization for proton conducting fuel cells at 100 and 200 °C.

the dilution of the cathode gas by the reaction product (water/steam) in proton conducting fuel cells has a minor influence on the Nernst potential of the cathode and hence on Nernst loss.

6. Can We Minimize the Nernst Loss by Changing the Relation $V_{eq}(x)$?

By changing the geometry of the cell and or gas flow patterns, we may be able to change the local driving force $V(x)$. However, in the definition of the Nernst loss as derived from the model equation, $V(u)$ appears rather than $V(x)$. We may wonder why do we not find the Nernst potential difference integrated with respect to x like in the second integral. After all, V and u are functions of x as well:

$$\Delta V_{Nernst} = V_{eq}(0) - \frac{1}{u_{tot}} \int_0^{u_{tot}} V_{eq}(x) dx \; ??? \qquad (90)$$

The following 'gedanken experiment' will prove the contrary. $V_{eq}(x)$ is a decreasing function because fuel is consumed in the cell. Suppose we cut a fuel cell into two equal parts; one inlet side, one outlet side, and insert a third identical part in the middle, however without drawing current from it. In other words, we isolate the electrodes and current collectors of this third middle part from the rest of the cell as shown in Fig.12.,

Since no current is drawn from this middle part, the gas composition does not change and neither does V_{eq} (x) in this middle part. In general, however, the cell voltage averaged over space 'x' will change because of the additional third middle part. However, since the middle part does contribute, neither in a positive nor in a negative sense, the I-V relation of the whole system does not change. Part 1 and 2 still operate and perform exactly the same as before the separation. Hence averaging the cell voltage over space cannot provide the correct answer. This may seem a rather academic problem. Nevertheless, it has an important practical consequence, in the sense that we can never reduce the Nernst loss by changing the geometry of a fuel cell. Also, the division of a fuel cell into sub cells allowing each sub cell to be operated at a different cell voltage (networking) has no effect on the Nernst loss (all other parameters are kept constant, of course). The Nernst loss is fully determined by the inlet and outlet gas compositions and temperature. In other words it does not matter where exactly the current is produced.

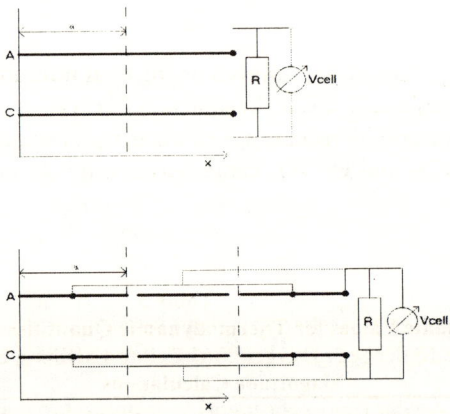

Figure 12. Schematic drawing of the 'gedanken experiment' to show $V(u)$ determines Nernst loss rather than $V(x)$.

7. FC-CE Systems Including Nernst Loss

In Section II we have analyzed fuel cell efficiencies for several values of ΔS for the overall reaction. However, we did not include the entropy changes related to the Nernst loss. If we do include the Nernst loss that occurs in a fuel cell its reversible efficiency is given by:

$$\eta_{fc}(u_{tot}) = \frac{\frac{1}{u_{tot}}\int_0^{u_{tot}} V_{eq}(u)\,du}{\chi} \tag{91}$$

or in terms of thermodynamic quantities:

$$\eta_{fc}(u_{tot}) = \frac{\frac{1}{u_{tot}}\int_0^{u_{tot}} \Delta G(u)\,du}{\Delta H} \tag{92}$$

By incorporating the Nernst loss in the fuel cell efficiency complementary to the result above $T\Delta S$ is replaced by:

$$T\Delta S' = \frac{1}{u_{tot}}\int_0^{u_{tot}} T\Delta S(u)\,du \tag{93}$$

In general $\Delta S' < \Delta S$ and $\Delta S'$ depends on the total utilization just like the Nernst loss. Graphically this is illustrated in Fig. 10.

The area of the rectangle $T\Delta S$ times u_{tot} is enlarged by the area between the OCV and the $V(i)$ curve between $u = 0$ and $u = u_{tot}$. The

Table 3
Approximate Values for Thermodynamic Quantities of a High Temperature Hydrogen / Oxygen Fuel Cell to Facilitate Easy Order of Magnitude Calculations

T (K)	ΔH_o (kJ/mol)	ΔS_o (J/mol.K)	F (C/mol)	ΔV_{Nernst} (mV)
1000	250	50	10^5	100

analysis given in section II for the various FC-CE systems is still valid when replacing ΔS by ΔS'. Therefore, for an H_2/O_2 fuel cell the decrease in OCV with temperature is more rapid when utilization is non-zero and Nernst loss occurs as shown in Fig. 13. For a rough calculation we can use the following approximate values for ΔH, ΔS, T and F as depicted in Table 3.

In Table 4 some important parameter values of a hydrogen-oxygen fuel cell with and without Nernst loss are calculated using the values in Table 3. So assuming 100 mV for the Nernst loss (which can be considered an upper limit for most practical situations as follows from the values in Table 2) an apparent entropy change ΔS' of 70 J/mole.K can be calculate which is an increase of 20 J/mole.K. Consequently, the temperature $T_{\Delta G = 0}$ where the Gibbs free energy is zero, is lowered as well and is now calculated to be approximately 3600 K instead of 5000 K for the standard conditions. So also the fuel cell efficiency as a function of temperature is decreasing more rapidly as shown in Fig. 13. As imposed by the numbers we assumed (which are nevertheless reasonable) we see that at 1000 K the effective reversible cell Voltage is only 900 mV (1250mV – 350 mV) and the reversible fuel cell efficiency taking the Nernst loss into account is only 900/1250 * 100% = 72 %. Assuming another 100 mV for the irreversible (polarization) losses at the anode, cathode and in the electrolyte, the overall fuel cell efficiency is calculated as only 64% (800/1250).

Here is the problem of high temperature fuel cells in a nutshell; they produce a lot of heat! Up to 36% of the heating value of hydrogen in the latter example is converted into heat. Second, the margins to improve fuel cell efficiency by technological means are rather limited because of the total losses of typically 450 mV only 100 mV are irreversible losses that can be minimized by technological breakthroughs such as better electrodes or the use of materials with higher conductivity, etc. or by decreasing current and power density.

Table 4
Some Important Parameter Values of a Hydrogen-Oxygen Fuel Cell with and without Nernst Loss Using the Approximate Values in Table 3.

$T/2F$	$\chi = \mid \Delta H/2F$ (mV)	$T \mid \Delta S \mid/2F$ (mV)	$T \mid \Delta S' \mid/2F$ (mV)	$-\Delta S$' (J/mol.K)	$T = \Delta H/\Delta S$ (K)	$T = \Delta H/\Delta S$' (K)
0.005	1250	250	350	70	5000	3600

Fuel Cells

However, the up side is that as shown in Fig.13 the efficiency of a FC-CE system is hardly effected by the large heat losses because in a second chance the heat losses, reversible as well as irreversible, are converted into power by the CE (Carnot Engine). The FC-CE system efficiency is given by the intersection of the fuel cell efficiency with and without Nernst loss and the vertical line at $T=T_o$ as shown in Fig.13. Again using the values in Table 3 and Table 4 the reversible system efficiency is calculated to be a little less than 92 % i.e. still very close to the 94% without Nernst loss. At 1000 K the Carnot efficiency is 70%. So of the 36% of heat produced in the example above of an irreversible fuel cell with 64% efficiency, theoretically 70 % can be recovered leading to a FC-CE system efficiency of still 89% (64 + 0.7*36). Of course, we should not rely too much on this second chance of heat recovery or bottoming cycle. Firstly, CE's do not exist in practice and also in the heat cycle losses occur. Moreover the heat is produced inside the fuel cell and all this heat must be transferred and exchanged to the Carnot Engine in which processes losses will also occur. Thirdly, the capacity of the Carnot Engine and hence the investment cost will increase with increasing heat losses in the fuel cell. And, last but not least, one must recall that fuel cells are very expensive heaters. All of this indicates that the key to highly efficient fuel cell systems lies in proper heat management.

8. Modeling the Initial Dip in the Nernst Potential

The Nernst potential as a function of the gas utilization can be approximated by a linear relation. However, for low utilization values, especially when using a hydrogen rich inlet fuel gas, the potential decreases steeper than the linear approximation. The curve shows an initial dip (see Fig. 5, Fig. 10). A positive correction term on the cell voltage can be derived to account for this effect, which is found to be relative small namely in the order of a few mV.

It follows directly from the definition of the Nernst loss, Eq. (86), and the linear approximation for $V(u)$ that this correction term is given by the area C between the curve $V(u)$ as given by the Nernst equation and the linear approximation, normalized by u_{tot}:

$$\frac{C}{u_{tot}} \qquad (94)$$

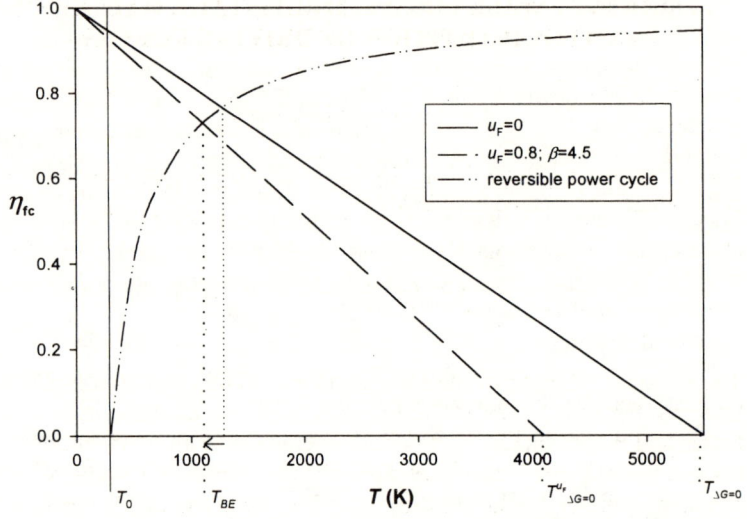

Figure 13. Effect of Nernst loss on fuel cell efficiency and efficiencies of FC-CE systems

This correction term depends on u_{tot} because for small u_{tot} the area between the actual $V(u)$ and the linear approximation is relatively large compared to the whole area under the straight line while for large u including the area of this initial dip becomes less important. If the fuel gas composition at the inlet of the cell is already lean in hydrogen, the OCV lies well within the linear region and there is hardly any initial dip. For fuel gasses with a high hydrogen concentration (and low product gas concentration), the OCV is relatively high and decreases strongly for small u. For these gases the correction term is significant, though in practical situations it will hardly ever exceed 10 mV (i.e. 1% of the cell voltage). For standard fuel gas of an MCFC (H_2/CO_2 80/20% humidified with 60 °C water) Standaert calculates $C = 3$ mV.

VI. IRREVERSIBLE LOSSES, MULTISTAGE OXIDATION AND EQUIPARTITION OF DRIVING FORCES

1. Irreversible Losses

For convenience, integral Eq. (83), the analytical expression for the cell voltage, is repeated here below:

$$V_{cell} = \frac{1}{u_{tot}} \int_0^{u_{tot}} V_{eq}(u)\,du - \frac{1}{i_{tot}} \int_0^1 r(x)\{i(x)\}^2\,dx \qquad (95)$$

The physical meaning of the second integral of this equation is that we integrate the heat losses in the fuel cell to obtain the total irreversible losses and divide that by i_{tot} to obtain the losses in units of volts. If we assume as before, that the local quasi-ohmic resistance is homogeneous throughout the cell, that is to say: $r(x) = r =$ constant, then the second integral represents the average of the square of the local current density times r

$$\frac{1}{i_{tot}} \int_0^1 r(x)\{i(x)\}^2\,dx = \frac{r}{i_{tot}} \int_0^1 \{i(x)\}^2\,dx \geq \frac{r}{i_{tot}} \left\{ \int_0^1 i(x)\,dx \right\}^2 = r.i_{tot} \qquad (96)$$

Since the average of the square is always larger than the square of the average (Cauchy-Schwartz), an upper limit for the cell voltage is given when the 2nd integral is approximated by $r.i_{tot}$:

$$V_{cell} \leq \frac{1}{u_{tot}} \int_0^{u_{tot}} V_{eq}(u)\,du - r.i_{tot} \qquad (97)$$

The equality sign holds if, and only if, the current density is constant throughout the cell. If the current density decreases because the $V(u(x))$ decreases, as is normally the case, the second integral is larger than $r.i_{tot}$. Compared to the ideal situation additional losses occur. Intuitively this is what we expect, because how can a fuel cell operate optimally if most of the current is produced in the first half and the second half near the outlet hardly contributes? In fact it does not and the problem is much more general than might be expected at first sight. This is discussed in the literature and is called equipartition of driving forces [21] or equipartition of entropy production.[22] (see Sections VI.3 and VI.4)

We can approximately calculate the effect of a non-homogeneous current distribution by a 1st order approximation assuming that $i(x)$ is a linear decreasing function (still assuming $r(x) = r$ is constant). Following Standaert's effect of utilization, the principal reason why the current density is not homogeneous, is also taken into account:

$$i(x) \approx i_{tot} + \frac{\alpha . u_{tot}}{r}\left(\frac{1}{2} - x\right) \quad (98)$$

Note that this linear approximation is defined such that the average current density is equal to i_{tot}. By calculating the integral the cell voltage can now be approximated in first order by:

$$\begin{aligned} V_{cell} &\approx \frac{1}{u_{tot}} \int_0^{u_{tot}} V_{eq}(u)du - \left(1 + \frac{1}{3}Z^2\right)ri_{tot} \\ &= \frac{1}{u_{tot}} \int_0^{u_{tot}} V_{eq}(u)du - ri_{tot} - \frac{1}{3}Z^2 ri_{tot} \end{aligned} \quad (99)$$

Where the dimensionless number Z is defined by:

$$Z \equiv \frac{\alpha.}{2r.i_{in}} = \frac{\alpha.u_{tot}}{2r.i_{tot}} \quad (100)$$

The difference between the numerically calculated second integral and the linear approximation is already so small (order of 0.1 mV), that further refinement of the model is not meaningful. The extra losses due to the non-homogenous current distribution are thus summarized in the term:

$$\frac{1}{3}Z^2 r.i_{tot} = \frac{1}{12}\frac{\alpha^2.u_{tot}^2}{r.i_{tot}} \quad (101)$$

For an MCFC at 650 °C with $u_{tot} = 0.8$, $r = 1.08$ Ωcm^2, $\alpha = 180$ mV and $i_{tot} = 150$ mA/cm^2 using Eq. (101), we calculate an extra loss term of 10.6 mV (equivalent to 1,6 mW/cm^2 at 150 mA/cm^2) relative to the optimal situation of equipartition of driving forces. Note that for this case ($r(x) = r$ = constant) equipartition of driving forces means a homogeneous current distribution and is equivalent to equipartition of entropy production, i.e. also $i^2(x).r(x)$ = constant (see Section VI.3.).

2. Multistage Oxidation

Equipartition of driving forces or the isoforce principle refers to the driving force that causes the flux. In our case this driving force is the difference between the cell voltage and the local Nernst potential difference between anode and cathode. In other words it is the (local) overpotential. In a fuel cell one should thus try to keep the local overpotential constant. A cascade of fuel cells each with their own cell voltage can achieve this. This is also called multistage oxidation. This means that the oxidation of the fuel is achieved in several stages (stacks) after one another as shown in Fig.14.

With respect to the gas streams, electrically the stacks are connected in series. If the stacks were to be connected electrically in parallel the situation would be no different from a one-stack operation. However, by connecting the stacks electrically in series, the average current density in each of the identical stacks is the same. The respective cell voltages differ in such a way as to obtain the equal currents through the stacks. And because the stacks are identical, in the first approximation the quasi-ohmic resistances are equal so the average driving forces within the stacks are the same. Moreover within the sub

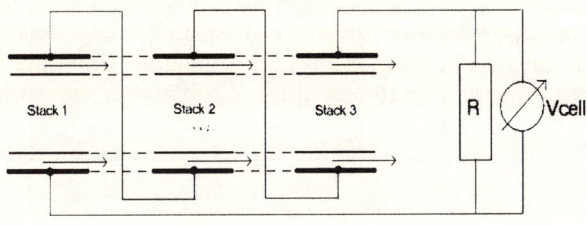

Figure 14. Multistage oxidation in three stacks connected electrically in series.

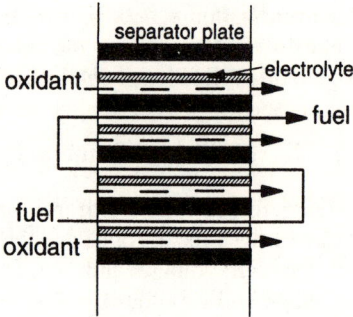

Figure 15. Multistage operation achieved in one stack (Standaert[10])

stacks utilization is low and also the driving forces within the sub stacks are more homogenous and approach equipartition more closely.

Multistage oxidation with 2 or 3 sub stacks can also be achieved in one stack. By a special way of internal manifolding the fuel gas is led back and forth through a number of cells in the stack as shown in Fig. 15.

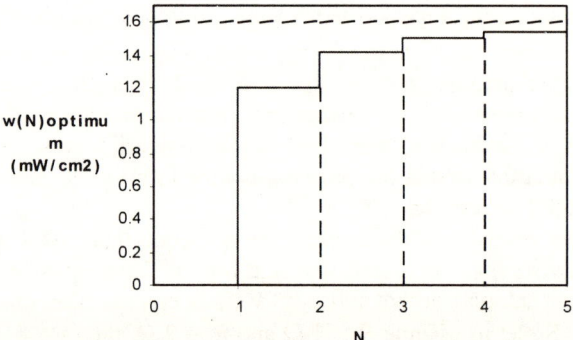

Figure 16. The extra power benefit owing to N-stage operation of an isothermal MCFC, with an average power density of 119 mW/cm^2. (The dotted line at 1.6 mW/cm^2 gives the power for N is infinity, Standaert[10])

Fuel Cells

Since utilization of the oxidant is normally low, the change in driving force due to oxidant utilization is low and there is little advantage to flow the oxidant back and forth through the sub stacks as well. In fact because of the large oxidant flow (to cool the cell), the disadvantage of an increased flow resistance will probably be larger than the small advantage of approaching equipartition of driving forces more closely.

The reduction in losses obtained by multistage oxidation as a function of the number of stages was calculated by Standaert by using Eq. (101) repeatedly for each sub stack. As shown in Figure 16, with 2 or 3 stages most of the 1.6 mW/cm^2 calculated for an infinite number of stacks is already achieved. So in multi-stage oxidation, the Nernst loss is not lowered and the homogenizing of the current distribution causes the positive effect; or, to be more precise and fundamentally correct, it is caused by an equipartitioning of the driving forces.

3. General Formulation of the Equipartition Principle

In the problem above we have just scratched the surface of the much more general problem that has been posed and discussed in literature.[21,23] Given a device that produces a desired product like electricity in the case of a fuel cell or a distillation product for a distillation process, how can we minimize losses given a certain amount of product per unit time? To illustrate the optimization problem with a simple example we consider an electric system with only two paths, i.e., two resistors R_1 and R_2. The problem is to transfer a certain amount of charge; say 1 Coulomb (the yield) through the resistor paths within a finite amount of time; say 1 second, in such a way as to minimize the losses (entropy production). We assume the resistors are constant as a function of time and we will restrict ourselves to a stationary situation so that the problem is how to divide a current of 1 Amp over the two resistors.

The two resistors in the trivial case are equal and the symmetry argument shows that the current through both resistors must be equal. And we have equipartition of forces as well as equipartition of entropy production. So let us assume $R_1 = 1\ \Omega$ and $R_2 = 2\ \Omega$, then we will show that the criterion of equipartition of forces leads to a lower entropy production than the criterion of equipartition of entropy production in accordance with the general proof given in[21] and the references therein.

The constraints and Ohm's law read:

$$I_1 + I_2 = I_{tot} = 1(Amp.)$$
$$I_i = \frac{V_i}{R_i}(i=1,2)$$
(102)

The criterion for equipartition of driving force reads:

$$V1 = V2 \qquad (103)$$

Solving the four equations for the four unknowns yields: $V_1 = V_2 = 2/3$ (V); $I_1 = 2/3$ (A) and $I_2 = 1/3$ (A). Hence, the total entropy production in terms of heat production reads:

$$I_1V_2 + I_2V_2 = \frac{2}{3}(W) \qquad (104)$$

Note that entropy production is proportional to heat production. On the other hand, the equipartition of entropy production principle demands:

$$I_1.V_1 = I_2.V_2 \qquad (105)$$

Solving the equations now yields:

$$I_1 = I_2\sqrt{2} = \frac{\sqrt{2}}{1+\sqrt{2}}(A) \qquad (106)$$

Hence, the total heat production in this case is: $2.I_1.V_1 = 0.6863$ W.

In accordance with the general proof, the equipartitioning of driving forces gives the lowest entropy production, i.e. heat dissipation in this case.[22] The difference in the above example is about 3% and clearly the criterion of constant entropy production does not lead to minimum entropy production. So this one single example disproves the statement that equipartition of entropy production is the optimality criterion.[23] A better solution can be found by division of yields (current) over the two given paths according to the isoforce principle. In fact, this is the optimum distribution, and the total heat production of 2/3 (Watt) is the minimum one can achieve for this example.

4. Equipartition of Entropy Production in a Non-isothermal Fuel Cell

In general, equipartition of driving forces can be proven to yield the optimum solution for producing the minimum amount of entropy given certain constraints. However, in spite of this Standaert could prove in an analogous way for a non-isothermal fuel cell, the very paradoxical conclusion that equipartition of entropy production yielded the optimum solution for his case. He assumed that because of the strong influence of temperature on the local polarization the latter is also a function of utilization: $r = r(u)$. Because now a coupling is made between the distribution of the local resistances and the solution $u(x)$ of the problem an additional (implicit) constraint is imposed. The system is not able to reach the optimum situation of equipartition of driving forces because of this constraint. In other words, suppose $u(x)$ is the solution found by Standaert using the equipartition of entropy production principle. Then $r(u(x))$ is the distribution of local resistances. If we could keep these resistances fixed, then Bedeaux proved that equipartition of driving forces yielded the optimum solution of minimum overall entropy production. And as shown by the simple example above, equipartition of entropy production yields a larger total entropy production than the isoforce principle. Because of the imposed constraint we <u>cannot</u> keep the distribution of resistances fixed. Changing the solution to comply with the isoforce principle would also change $r(u(x))$ again. Hence the isoforce principle does not lead to the optimum in this case. Instead -very surprisingly- the optimum for the non-isothermal fuel cell is found by the equipartition of entropy principle. Although this is a rather academic discussion, the difference in efficiency between the two principles can be a few percent depending on the resistance distribution. The example of multistage oxidation shows that there are practical applications of the principle and that the increase in efficiency compared to a situation that fulfills neither of the equipartition principles can be an additional few percent. The effect on the level of a fuel cell system is even higher because less heat is dissipated in the stack and cooling requirements are decreased. This reduces the energy needed to pump around the cooling fluid (oxidant in most cases) and calculations by Standaert for an MCFC system have shown that the positive effect on a system level may be doubled.[10]

VII. FURTHER ANALYSIS

1. Maximum Power versus Maximum Efficiency

According to Eq. (84) or Eq. (85) assuming u_{tot} is constant, the cell voltage decreases linearly with the output current delivered by the cell. For an MCFC this holds true even if utilization polarization is taken into account. If the input gas flows i_{in} is held constant than u_{tot} is proportional to I_{tot}, the current delivered by the cell. Hence, the Nernst loss is proportional to I_{tot} and the decrease in the cell voltage as a function of I_{tot} is steeper yet still linear. In Fig. 17 this linear I-V relation is shown for typical values of an MCFC (OCV = 1V and r = 1 $\Omega.cm^2$). Also indicated is the specific output power of the cell (in W/cm^2) as a function of current density. It is given by:

$$i_{tot} \cdot V_{cell} = i_{tot} \cdot (OCV - i_{tot} \cdot r) \tag{107}$$

It is easily shown that this parabola has a maximum for $i_{tot} = V_0/2r$. However, it is also easily seen that this current is obtained if the external load resistance is equal to the internal resistance representing the polarization losses. Because the same current flows through both

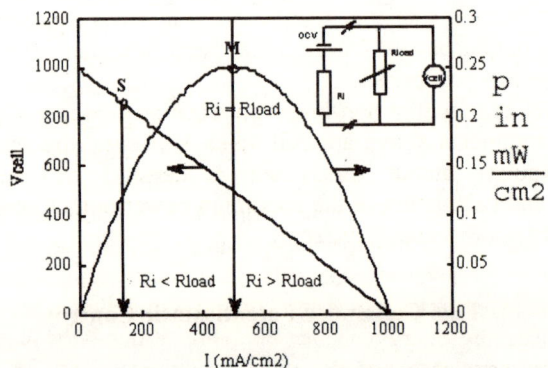

Figure 17. Typical polarization curve and power output for an MCFC according to the simple battery model. Point S indicates the standard operation current density of 150 mA/cm^2, while M indicates maximum output power.

Fuel Cells

resistances, an equal amount of power is dissipated internally as heat losses as is dissipated externally as the desired power output. So, efficiency in this case is limited to 50 % at most. Note that we have to add the irreversible losses, so the efficiency of the fuel cell is even significantly less than 50% at maximum power output. Therefore, fuel cells are normally operated at lower current density to achieve a higher efficiency. The demand for a high efficiency and high power output are contradictory and we have to compromise. In the limit for $i_{tot} \rightarrow 0$, we approach reversibility and the maximum achievable reversible efficiency, while at maximum power output 50% is an upper limit for the efficiency. As a rather arbitrary compromise instead of either optimizing the cell voltage (efficiency) or the power output we can (arbitrarily) choose to optimize their product:

$$p_{cell}.V_{cell} = i_{tot}.(V_0 - i_{tot}.r)^2 \tag{108}$$

The maximum of this third degree polynomial in i is found for $i_{tot} = V_0/3r = 333 mA/cm^2$, i.e., one third of the short-circuiting current of 1000 mA/cm² in our example of Fig. 17.

The standard operating current density of an MCFC is only 150 mA/cm2. So this is a rather low current density and MCFC operators apparently prefer higher efficiency above higher power output. However, the operating current density is mainly a matter of economics. A higher current density and power output will decrease the capital cost because a smaller stack and less material is needed to yield the desired output power. On the other hand, a higher efficiency will decrease the fuel cost per kW.

For fuel cells with a non-linear polarization curve in a qualitative sense the arguments above are still valid. However, the shape of the parabola (power output versus current density) will be skewed somewhat and calculation of the maximum power output point is more difficult, but not essentially different.

2. Fuel Cell under Potentiostatic Control

Under potentiostatic control the cell voltage is kept constant and the current produced by the cell is a function of I_{in} (or u_{tot}). Because $u_{tot} = I_{tot}/I_{in}$, the bilinear relation, Eq. (75), for V_{cell} is directly rewritten to yield a relation between I_{tot} and V_{cell}:

$$V_{cell} = V_{eq}(0) - I_{tot}.R_c - \frac{\alpha}{2}.\frac{I_{tot}}{I_{in}}$$

$$or: \quad I_{tot} \approx \frac{V_{eq}(0) - V_{cell}}{R_c + \frac{\alpha}{2I_{in}}} \qquad (\text{Itot} < \text{Iin}) \qquad (109)$$

We see that the utilization $\alpha/(2.I_{in})$ now appears in the dimensions of a resistance. However, the heat dissipated in this resistance is reversible heat. That is to say, in the reverse process this heat can be converted back into chemical energy. From an engineering point of view this heat is of course no different from the heat ($I_{tot}^2 R_c$) dissipated in the irreversible processes modeled here by the single resistance R_c.

Equation (109) is plotted in Fig. 18 and holds reasonably well for $I_{tot} < I_{in}$. However it loses its meaning of course for $I_{tot} > I_{in}$ (dashed line). If the input flow (equivalent input current) is decreased as to approach the output current, almost 100% utilization is reached. Then the output current I_{tot} will decrease as well, closely following I_{in} along the line $I_{tot} = I_{in}$. This is shown by the thick line in Fig. 18 representing a numerical calculation with slightly different parameters as not to coincide with the line of Eq. (109) potentiostatic control that was already calculated in Section IV.2 and shown in Fig. 8.

3. Fuel Cell Operated at Constant Load Resistance

If the cell or stack is connected to a load with a fixed resistance (R_{load}) such as the oil- or air-cooled resistances sometimes applied in test set-ups then the cell current is calculated as follows. Using Eq. (109) above and noting that the cell voltage equals $V_{cell} \approx I_{tot}. R_{load}$, Eq. (109) is easily rewritten as:

$$I_{tot} \approx \frac{V_{eq}(0)}{R_c + R_{load} + \frac{\alpha}{2I_{in}}} \qquad (\text{Itot} < \text{Iin}) \qquad (110)$$

Again this equation holds reasonably well only if $I_{tot} < I_{in}$. If the input flow or R_{load} are too low then the output current will be cut off to a value very close to the equivalent input current. Decreasing I_{in} (or

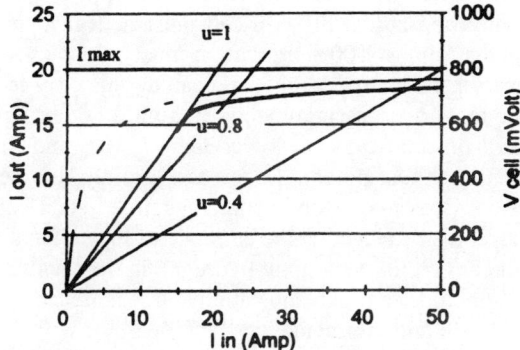

Figure 18. Output current I_{tot} as a function of equivalent input current for a 100 cm^2 MCFC with an OCV of 1 Volt and a specific internal resistance of 1 Ohm.cm^2 or $R_c = 0.01$ Ohm and $R_{load} = 0.04$ Ohm. The solid curve is according to the approximate Eq. (111) and the bold curve represents a numerical calculation. Lines of constant utilization ($u=1$, $u=0.8$ and $u=0.4$) and the maximum achievable current I_{max} are also shown.

R_{load}) further will decrease the cell voltage according to $V_{cell} = I_{tot} \cdot R_{load} \approx I_{in} R_{load}$.

4. Fuel Cell under Galvanostatic Control

Contrary to the two operating modes above, in galvanic control we can get an unstable situation. Operating a fuel cell under galvanostatic control means that I_{out} is kept constant, hence V_{cell} becomes a function of I_{in}.

$$V_{cell} \approx V_{eq}(0) - I_{tot} \cdot R_c - \frac{\alpha I_{tot}}{2 I_{in}} \quad (111)$$

Of course, I_{tot} and R_c should be within reasonable limits such that their product is smaller than OCV (= $V_{eq}(0)$) else V_{cell} would become negative. For high I_{in} the Nernst loss is low and V_{cell} approaches the situation where the losses are only due to polarization. However, V_{cell}

decreases rapidly when I_{in} is lowered such that it approaches I_{tot}. Since the condition imposed is that the fuel cell must deliver a fixed current I_{tot}, this means that almost 100% utilization must be achieved. The fuel cell must try very hard to achieve this and can do this only for very low V_{cell} so that the local overpotential becomes high. In other words, since the cell voltage must always be lower than $V_{eq}(u)$, and since for u approaching 1 the Nernst potential decreases rapidly, the cell voltage will decrease too. Both may even become negative. As a consequence most of the current is produced at the entrance of the cell and the rest of the cell tries to convert the remaining hydrogen in the fuel gas.

One may conclude that galvanostatic control is not the best choice to operate a fuel cell and in practice no stack operator will choose to do so. However, in a stack operated at potentiostatic control or load control one cell of the stack approaches galvanostatic control. This is because the influence of one cell on the whole stack performance is limited (100 cells in a stack is not unusual). The total stack current I_{tot} that also flows through this one particular cell is hardly influenced by the performance of this one cell, but determined by all other cells and the imposed load or stack voltage. Hence, this cell approaches operation under galvanostatic control. If for some reason I_{in} of this cell is lower than for the other cells, e.g., by a statistical variation in flow resistance of the gas channels or part of the flow channel is blocked, I_{in} of this particular cell may approach I_{tot} or even become lower than I_{tot}: a very undesirable situation. The cell will effectively operate as if it were short-circuited ($V_{cell} \approx 0$). It will deliver no power, only heat, and all this heat will predominantly be dissipated at the inlet side of the cell. If $I_{in} < I_{tot}$ then V_{cell} can even become negative decreasing the performance of the stack even further.

5. Heat Effects in Fuel Cells

(i) *Heat Dissipation Due to Irreversible Losses*

Polarization losses are normally separated into ohmic losses, kinetic losses and diffusion losses and can be expressed in a voltage loss. The dissipated heat is than calculated by multiplying the voltage loss by the current (density), but the exact position of the heat dissipation is not trivial. One may argue that from a macroscopic point of view it does not really matter where exactly the heat is dissipated since by good heat conductance the temperature in the cell will at least locally be constant.

- *Ohmic losses.* Local heat dissipation in a conducting medium is calculated as i^2 and most of the heat is dissipated in paths with a low resistance and a high current density. In the electrolyte matrix, calculation of the heat dissipation is straightforward if one assumes that the matrix may be homogenized.
- *Diffusion losses.* Heat is dissipated in those regions where large concentration gradients of reactants and products exist. The local heat dissipation is proportional to flux times the chemical potential gradient. Hence most of the heat is dissipated in the so-called diffusion boundary layer near the electrode surface.
- *Kinetic losses.* Since the electrochemical reactions occur at the surface also heat losses due to limitations in the reaction rate occur at the surface. They are proportional to the kinetic overpotential and the flux (local current density in this case). If chemical reactions are involved heat is dissipated wherever these reactions take place, i.e. in the gas phase (e.g. for the water shift reaction) or in the electrolyte (e.g. the dissociation/recombination reaction of the carbonate ion in MCFC's). If the kinetics is fast the chemical potential differences are small and so is the heat dissipation compared to places in the reaction path where larger potential jumps occur. Note that the term reaction paths is used with both its meaning of a physical path in space (and time) as well as in the conventional meaning of reaction mechanism.

(ii) *Heat Dissipation Due to Reversible Processes*

Bedeaux and Kjelstrup showed that a temperature jump exists across the surface and that the surface temperature is in principle different from the temperature of the surroundings.[24] Bedeaux and Kjelstrup state that: "It is customary to calculate the dissipated energy at the surface from the product of the overpotential and the electric current plus the Joule heat. In this work we show that the overpotential contains reversible parts so-called Peltier effects, in addition to irreversible parts. The Peltier effects do not contribute to the entropy production at the electrode surfaces, but they do effect the temperature profile across the cell." In addition they write: "Local heating or cooling effects in batteries, fuel cells and electrolysis cells during passage of electric current are important in industry." Thermal effects can damage the cell construction in critical cases. Molten electrolytes may be freezing in electrode compartments, and ceramic electrolyte materials may crack. The temperature jumps according to Bedeaux, et al. predict, depending, of course, on the material parameters but can be 20 K over 1.5 nm in some cases. More work is needed to show, for example, if indeed molten carbonate can freeze locally in spite of the fact that the average cell temperature is above the melting point.

In addition to the Peltier effects mentioned above, we recall that in the first sections we discussed the thermodynamics of the fuel cell reactions and the term $T\Delta S$ that equals the heat dissipated in the cell (in a hydrogen oxygen fuel cell). But where exactly is the heat $T\Delta S$ dissipated, at the anode side or at the cathode side? To answer this question the electrochemical reactions of anode and cathode separately must be examined. If the electrolyte is conducting anions then at the cathode side the oxygen reacts into this anion and gas is converted into solid or liquid hence decreasing the number of gas molecules so $\Delta S < 0$ and heat $T\Delta S$ is dissipated in the cathode at the reaction surface.

At the anode side, however, for each mole of hydrogen one mole of water is produced if the operating temperature of the cell is lower than 100 °C (1 atm). So here also $\Delta S < 0$ and heat $T\Delta S$ is dissipated at the anode as well. For higher operating temperatures than 100 °C no reversible heat is dissipated at the anode since the number of gas molecules at the anode remains constant. Note that for each mole of oxygen two moles of hydrogen are converted into two moles of steam.

For an MCFC the situation is more complex since CO_2 is also involved in the reactions. At the cathode for each mole of oxygen two moles of CO_2 are converted into a liquid (molten carbonate) as well. The two moles of CO_2 are released at the anode. So $\Delta S \ll 0$ at the cathode and a lot of heat is dissipated here, whereas at the anode $\Delta S > 0$ part of the heat dissipated at the cathode is consumed in the electrochemical reaction at the anode. So the anode is cooled and more heat than the net dissipated heat for the whole fuel cell is dissipated at the cathode of an MCFC. This asymmetrical local heat production was already described by Jacobsen and Broers in 1977.[25]

VIII. COMPARISON OF TWO ANALYTICAL MODELS WITH EXPERIMENTAL RESULTS ON A 110 CM² MCFC

1. Experimental Results

In a paper by S.F. Au, et. al. measurements are described that were performed on a 110cm² Li/Na MCFC single cell at 650°C at Tohoku university in the laboratory of Prof. I. Uchida.[18]. A comparison is also made with the analytical fuel cell models derived by Standaert.[16,17] Here a slightly different approach is followed to explain some intriguing results further. The MCFC single cell was manufactured, installed, and tested by Ishikawajima-Harima Heavy Industry Co. (IHI) and had been successfully operating for 3330 hours before the measurements described here were performed. The anode was fed with 80% H_2 and 20% CO_2 humidified at 60°C. The cathode was fed with 70% air and 30% CO_2 (not humidified). Measurements were performed under atmospheric conditions. The flow rate of both anode and cathode gases were set according to the current load and desired utilization. The measured cell voltage V_{cell} is tabulated in Table 5 for various current densities and fuel flow rates. The latter is expressed as an equivalent input current density in units of mA/cm². The oxidant input flow rate is held constant at an equivalent of 375 mA/cm² with respect to oxygen. In other words, full conversion of all the oxygen in the MCFC would yield a current density of 375 mA/cm² provided sufficient hydrogen is supplied to the anode as well. Because of the near stoichiometric oxidant gas composition also CO_2 would then be converted almost fully too. Note that oxidant utilization is defined with respect to oxygen utilization and not with respect to CO_2 utilization.

The measured cell voltages in Table 5 are plotted in two 3-D plots in Fig.19a and b as a function of u_f and i_{tot}. In Fig. 19b, the viewing angle is chosen such that it becomes clear that the data points lie in a plane and will thus fit a bilinear relation in u_f and i_{tot} very well except for the data point at the highest current density of 180 mA/cm².

A different representation of the same data of Table 5 is given in Fig. 20. The columns in the table represent data for constant gas input flows of oxidant as well as fuel. So as a function of the current density i_{cell}, u_{ox} and u_f both vary simultaneously and linearly ($u_f = i_{cell}/i_{in}^f$; $u_{ox} = i_{cell}/i_{in}^{ox}$). The data in the columns of Table 5 are plotted as a function of u_f in Fig. 20.

Table 5
Measured Cell Voltage of a 110 cm² Li/Na MCFC Single Cell at 650 °C for Various Current Densities and Fuel Input Flow Rates. The Oxidant Input Flow Rate is Held Constant at an Equivalent of 375 mA/cm².[18]

i_{cell} (mA/cm²)	V_{cell} (mV) i^f_{in} (mA/cm²)			
	750	375	250	188
0	1056	1055	1056	1051
30	1019	1010	1001	989
50	993	979	966	955
80	956	937	918	902
100	927	909	885	868
110	916	894	869	850
120	901	879	852	830
130	888	865	843	811
140	876	848	820	794
150	860	833	803	767
180	801	769	729	---

In addition to the above data, a series of measurements was performed at current densities of 100, 110, 120, 130, 140, 150 mA/cm² while the fuel and oxidant utilization were fixed at 60% and 40% respectively. This is achieved by adjusting the respective input flows for the fuel and oxidant gas to each current density for which the cell voltage is measured. The results are tabulated in Table 6 and shown in Fig. 21.

2. The Simple and Extended Analytical Fuel Cell Model

In the previous sections, a theory has been developed that allows us to construct analytical fuel cell models with a different level of sophistication. In this section two analytical models are constructed: the so-called simple model (SM) and the extended model (EM). Both are compared with the experimental results on a 110 cm2 MCFC presented

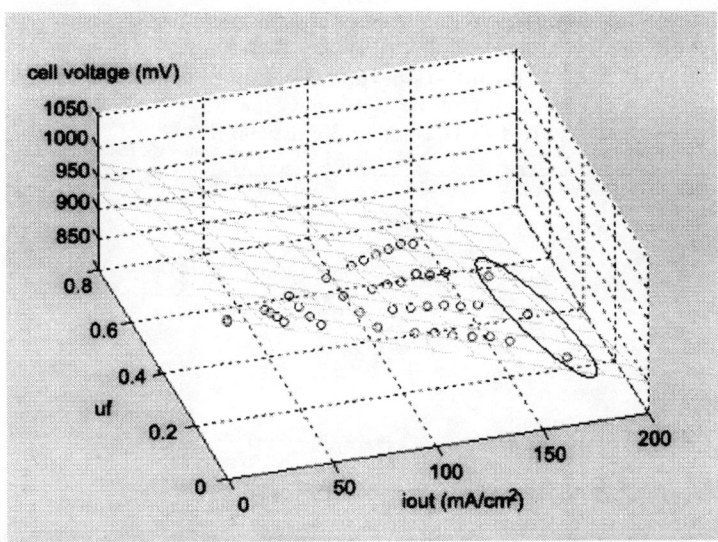
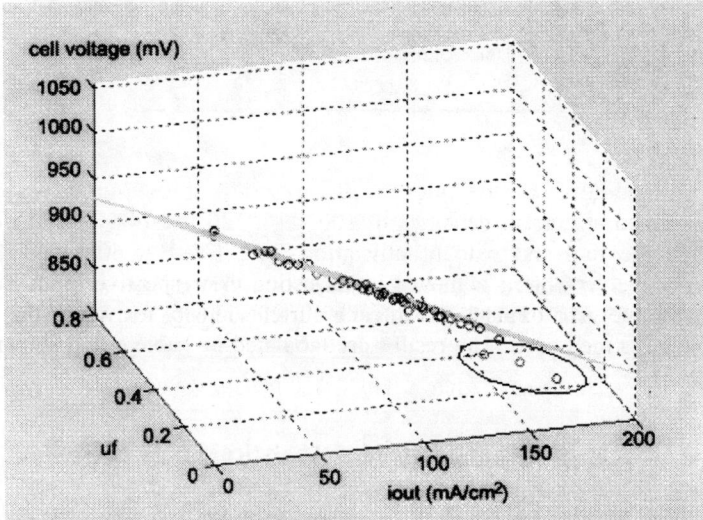

Figure 19. 3D-plots of the results in Table 4 with a plane fitted to the different viewing angles. Three deviating points at 180 mA/cm^2 are encircled and excluded from the fit.

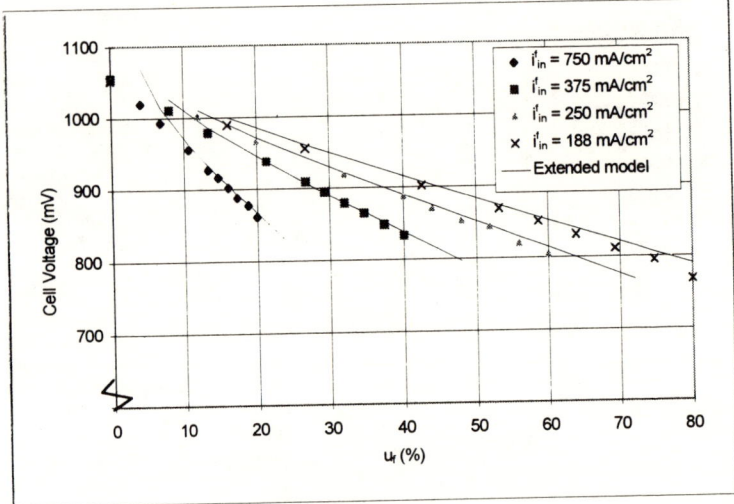

Figure 20. Measured Vcell as a function of fuel utilization for four different equivalent input current densities I_{in}^f. (i_{in}^{ox} = 375 mA/cm^2; $u_{ox} = u_f \cdot i_{in}^f$ / 375 mA/cm^2; data given in the columns of Table 5).

Table 6
Measured V_{cell} as a Function of Current Density at Constant Fuel and Oxidant Utilization of 60% and 40% Respectively

i_{cell} (mA/cm2)	100	110	120	130	140	150
V_{cell} (mV)	850	840	832	821	811	798

Fuel Cells

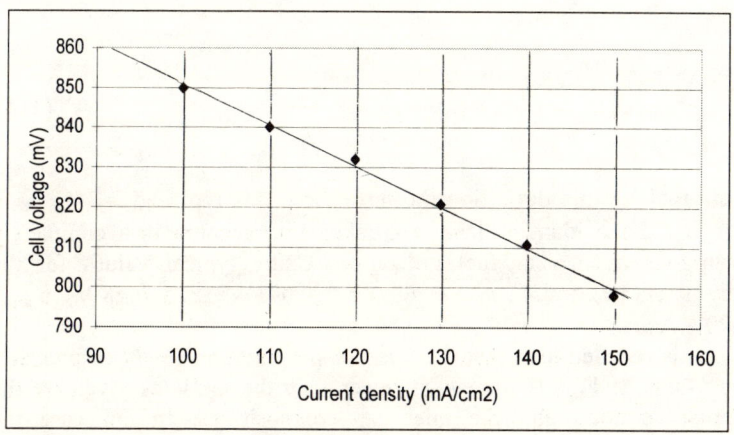

Figure 21. Measured V_{cell} as a function of current density at constant fuel and oxidant utilization of 60% and 40% respectively.

above. The simple model is given already as equation 83 and repeated here below. Note that in the SM (simple model) α and u refer to the fuel utilization only:

$$V_{cell}^{SM} \approx V_{eq}(0) - \frac{1}{2}\alpha_f . u_f - i_{tot} . r \qquad (112)$$

In the extended model three additional terms are included to account for:

1. The initial dip in the V(u) function using Eq. (94) (see Fig. 22).
2. The non-homogenous current density distribution in the cell using Eq. (99).
3. The oxidant utilization.

$$V_{cell}^{EM} \approx V_{eq}^*(0) - \frac{1}{2}\alpha_{tot} u_f + \frac{C}{u_f} - i_{tot} . r - \frac{1}{12}\frac{(\alpha_{tot} u_f)^2}{i_{tot} . r} \qquad (113)$$

The oxidant utilization is taken into account by replacing α by α_{tot} defined as:

$$\alpha_{tot} \equiv \alpha_f + \alpha_{ox}\frac{u_{ox}}{u_f} \tag{114}$$

Note that by this definition the term ½$\alpha_f u_f$ is replaced by ½$\alpha_f u_f$ + ½$\alpha_{ox} u_{ox}$. So oxidant utilization is taken into account in a completely analogous manner as fuel utilization. Using typical values for the parameters (u_{ox} = 0.4 ; u_f = 0.6; α_{ox} = 0.041 V; α_f = 0.18 V) α_{tot} = 0.207 V.[10]

It is recalled here that the linear approximation for $V(u)$ intersects the y-axis at $V_{eq}^*(0)$ somewhat lower than the OCV as given by the Nernst equation and the inlet gas compositions. In our case the theoretical OCV = $V_{eq}(0)$ = 1051 mV. When assuming that the shift reaction is in equilibrium, the anode gas composition contains less hydrogen but also less CO_2 and a little more water and the OCV is calculated as 1060 mV. This is indeed the range in which the experimental values (first row in Table 5) are found (1051-1059mV). As shown in Fig. 22 the intersection of the linear approximation and the y-axis is $V_{eq}^*(0)$ = 1030 mV and the slope of the linear approximation is determined as α_f = 180 mV. For the standard oxidant gas composition used in this experiment we can calculate or obtain graphically that u_{ox} = 41 mV analogous to obtaining α_f from Fig. 22. The constant C in the extended model represents the surface area between the $V(u)$ curve and the linear approximation and is found to be about 3 mV for the fuel gas and temperature used in the experiment (see Fig. 22).

3. The Simple Analytical Fuel Cell Model Compared with Experimental Results

The polarization curve in Fig. 21 resembles a straight line and a least square fit yields an absolute value for the slope of 1.02 +/- 0.03 Ω.cm2. It is very tempting to directly interpret the slope as the quasi-ohmic resistance r. Within the limitations of the simple model this is indeed true. Taking the partial derivative of V_{cell} according to Eq.112 with respect to i_{tot} for constant utilization directly yields minus r.

Fuel Cells

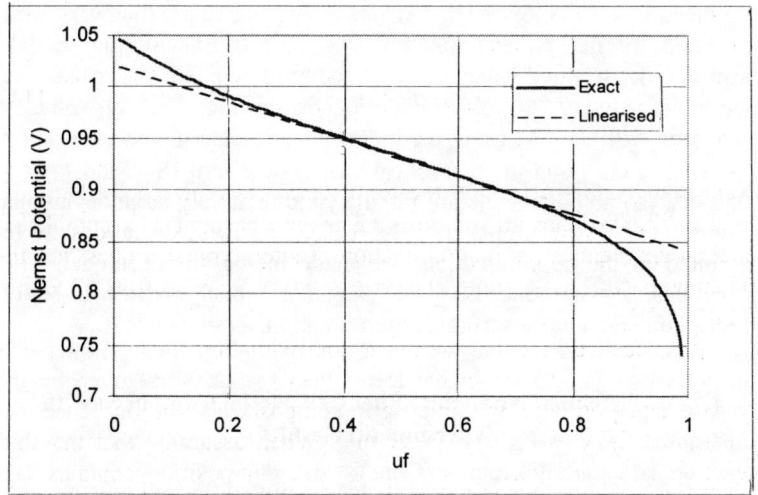

Figure 22. Calculated local Nernst potential of the MCFC anode versus standard cathode as a function of anode utilization u_f.

$$-\frac{\partial V_{cell}^{SM}}{\partial i_{tot}}\bigg|_{u=const.} = r \qquad (115)$$

This value agrees very well with the polarization resistance determined from step measurements on the same cell (r = 1.01 +/- 0.02 Ω.cm2).[18] So using this value for r and the real OCV of 1051 mV calculated with Nernst law and measured (see first row of Table 5) α_f is the only parameter in the simple model for the cell voltage. Fitting the cell voltage to the data in Table 5 (except the data at 180 mA/cm^2) yields the fitted plane as shown in the 3D plots of Fig 19. The plane is described by:

$$V_{cell} = 1.051 - 0.158\, u_f - 1.02\, i_{tot} \qquad (116)$$

Comparing the experimental results with the simple model yields $\alpha_f/2$ = 0.158 V or α_f = 316 mV. So although the data lie almost perfectly in a plane in a 3-D plot of V_{cell} versus i_{cell} and u_f and are thus very well described by the bilinear function for V_{cell} of the simple model. However, the so fitted value for α_f is found to be much larger than the theoretical value of 180 mV as determined from Fig. 22. Cell voltages calculated with the SM using the independently determined value for r ($r = 1,02$ Ω cm²) and the theoretical values of $\alpha_f = 0,18$ V and $V(0) = 1050$ mV (so no parameters are fitted!) yield relatively large deviations of up to 60 mV at 150 mA/cm². The reasons for these deviations are to be found in the approximations we made in the simple model. The terms that we introduced in the extended model provide a better description and a more accurate approximation, as shown next.

4. The Extended Analytical Fuel Cell Model Compared with Experimental Results

According to the extended model the derivative of the cell voltage with respect to the current does not exactly give r but an additional term appears, which is a function of r and i_{tot}. So the extended model predicts a non-linear I-V curve even when assuming r, α, u_{ox} and u_f to be constant

$$-\frac{\partial V_{cell}^{EM}}{\partial i_{tot}}\bigg|_{u=const.} = r(1 - \frac{\alpha_{tot}^2 . u_f^2}{12r^2 i_{tot}^2}) = r - \frac{\alpha_f^2 . u_f^2 + 2\alpha_f u_f \alpha_{ox} u_{ox} + \alpha_{ox}^2 . u_{ox}^2}{12r i_{tot}^2} \quad (117)$$

This additional second term is normally small. It is largely determined by fuel utilization because the oxidant utilization is relatively small (0.4) and because α_{ox} is smaller than α_f (typically 41 mV compared to 180 mV). So without loosing too much accuracy the experimentally determined slope compares to the quasi-ohmic resistance r multiplied by a correction factor:

$$-\frac{\partial V_{cell}^{EM}}{\partial i_{tot}}\bigg|_{u=const.} = r - \frac{\alpha_f^2 . u_f^2}{12r i_{tot}^2} = r\left(1 - \frac{1}{12}\left(\frac{\alpha_f . u_f}{r i_{tot}}\right)^2\right) \quad (118)$$

Table 7
a) Cell Voltage Calculated with the Simple Model Using $r = 1.02$; $\alpha_f = 0.316$ V; OCV = 1.051 V, and b) Deviations Between the Simple model and Experimental Values of Table 5 ($\Delta V_{cell} = V^{SM}_{cell} - V^{exp}_{cell}$).

(a)

i_{cell} (mA/cm²)	V_{cell} (mV) i^f_{in} (mA/cm²)			
	750	375	250	188
0	-5	-4	-5	0
30	-5	-2	0	6
50	-4	0	2	3
80	-3	-1	1	0
100	1	-2	1	-3
110	0	-2	0	-4
120	2	-1	1	-2
130	3	-1	-7	-2
140	3	1	0	-3
150	6	2	0	5
180	28	23	25	-

(b)

i_{cell} (mA/cm2)	V^{SM}_{cell} (mV) i^f_{in} (mA/cm2)			
	750	375	250	188
0	1051	1051	1051	1051
30	1014	1008	1001	995
50	989	979	968	958
80	953	936	919	902
100	928	907	886	865
110	916	892	869	846
120	903	878	853	828
130	891	864	836	809
140	879	849	820	791
150	866	835	803	772
180	829	792	754	716

We see that this correction factor is smaller than 1 and therefore the slope of the I-V curve as predicted by the extended model (EM) is not as steep as the slope predicted by the simple model (SM). This seems

Table 8
a) Cell Voltage Calculated with the Extended Model Using $r = 1.07$ and the Theoretical Values for α_f, α_{ox}, α_{tot}, C and $V_{eq}(0) = 1030$ mV. b) Deviations Between the Extended Model and Experimental Values of Table 5

(a)	V^{EM}_{cell} (mV)			
i_{cell}	i^f_{in} (mA/cm^2)			
(mA/cm^2)	750	375	250	188
0	-	-	-	-
30	1061	1019	1003	992
50	1007	978	963	952
80	954	930	914	899
100	925	900	882	865
110	911	886	867	849
120	897	872	852	832
130	884	858	836	816
140	871	844	821	799
150	858	830	806	783
180	819	789	761	733

(b)	ΔV_{cell} (mV)			
i_{cell}	i^f_{in} (mA/cm^2)			
(mA/cm^2)	750	375	250	188
0	-	-	-	-
30		9	2	3
50	14	-1	-3	-3
80	-2	-7	-4	-3
100	-2	-9	-3	-3
110	-5	-8	-2	-1
120	-4	-7	0	2
130	-4	-7	-7	5
140	-5	-4	1	5
150	-2	-3	3	16
180	18	20	32	-

rather paradoxical, because we know that in the extended model we included the effect of the non-homogenous current density distribution that causes additional losses. However, although the slope of the cell voltage with respect to current density is not so steep, the absolute value of the cell voltage is lower in the extended model. So using the extended model and the experimentally determined slope (1.02 $\Omega.cm^2$) the quasi-ohmic resistance r can be determined by determining the correction factor in the equation above. The slope was determined between $i = 100$ and 150 mA/cm2 so the correction factor lies between: 1- $[(0.18 * 0.8)^2/(0.15)^2]/12 = 0.923$ and 1- $[(0.18 * 0.8)^2/(0.1)^2]/12 = 0.827$ using a first approximation for r of 1 $\Omega.cm^2$. Hence, r is determined between $1.02/0.923 = 1.10$ and $1.02/0.827 = 1.23$ $\Omega.cm^2$. The equation (107) set equal to the experimental value 1.02 $\Omega.cm^2$ can also be solved exactly for r, yielding r between 1.08 $\Omega.cm^2$ at 150 mA/cm^2 and 1.11 $\Omega.cm^2$ at 100 mA/cm^2.

A better way to determine r is by a direct fit of the extended model to the data in Table 5. This can be done because all other parameters are either calculated theoretically ($V^*_{eq}(0) = 1030$ mV; $\alpha_{tot} = 0,18$ V, $\alpha_{ox} = 0,041$ V, $C = 3$ mV) or are experimentally fixed (u_f, u_{ox}, i_{tot}). The result is $r = 1,07$ Ω cm^2 (as for the SM we did not include the data for the highest current density of 180 mA/cm^2). So all variations of calculating r from the experimental data using the extended model yield a value a few percent higher than the slope of the I-V curve. Table 8b depicts the deviations in predicted cell voltage (EM) with the experimental values.

It is emphasized here that r is the only parameter used for fitting the EM to the data set. Furthermore, r is found to be close to the value range obtained from the independent data set of Table 6.

In conclusion the SM yields a perfect fit to experimental data. However, the so determined α_f is much larger than the theoretical value, and the independently determined r. The fitted r agrees well with the value determined by independent step measurements (1,02 versus 1,01 Ω cm^2). Using the theoretical value for α_f and the independently determined r, in other words using the simple model without any fitting, deviations are still restricted to 40 mV maximum (i.e. 5% at most). The EM takes into account three-second order corrections that increase the accuracy to within 1%, i.e.:

1. Initial dip,
2. Oxidant utilization, and
3. Non-homogeneous current density distribution.

The term for the initial dip C/u_f yields an over correction for very low u_f ($\leq 0,2$). Here the constant value for $C = 3mV$ should be replaced by the exact value of C calculated by:

$$C = \frac{1}{u_f} \int_0^{u_f} \{V_{eq}(u) - V_{eq}^*(0) + \alpha_f u_f\} du \qquad (119)$$

5. Conclusions and Application of the Analytical Models to Other Fuel Cell Types

In the previous paragraphs it was shown that the model equations developed in general for any type of fuel cell apply very well for the MCFC.

Like the MCFC, the SOFC is a high temperature fuel cell and as discussed in section V the Nernst loss is a significant loss term for the SOFC as well. Also, like the MCFC, its polarization curve is approximately linear and the polarization losses can be lumped into a quasi-ohmic resistance as for the MCFC. Determination of the function V(u) with the Nernst law is also completely analogous to the MCFC. So there is no reason why the simple and extended analytical fuel cell models should not apply equally well to a planar SOFC. Although planar SOFC's are being developed, the tubular concept by Westinghouse is produced at the largest scale and in units with the largest power output. A 100 kW SOFC unit was demonstrated in Westerfoort, the Netherlands. Furthermore, disk shaped flat plate geometries with radial symmetry are being developed and produced by ECN in the Netherlands in cooperation with Sulzer Hexis Switzerland. However, in the next section it will be shown that with an appropriate coordinate transformation the models are valid for a disk shaped cell as well. It is also argued that the analytical models can describe overall performance of the tubular SOFC provided that losses due to in-plane currents are taken into account.

A large difference between the modeling approach followed here and most other approaches in the literature is the lumping of polarization sources into one parameter; the local (quasi-ohmic) resistance $r(x)$. As explained one can take this resistance constant $r(x) = r$ or include more detailed dependencies $r(u(x))$ or $r(i(x))$. Equation (83) can than still be used, but will be more difficult to evaluate. The determination of the detailed dependencies of r on certain parameters often requires models for the porous electrodes in which the polarization losses predominantly occur. Losses in the electrolyte are ohmic and can be added simply. In addition, temperature differences can be included. Standaert, et. al.[17] have shown that the analytical models that are developed for isothermal fuel cells apply to non-isothermal fuel cells as well using an appropriate conversion of the quasi-ohmic resistance.

IX. FUEL CELL CONFIGURATIONS AND GEOMETRIES

The most common geometry for a fuel cell is the rectangular shape in a flat concept. The gasses are distributed over the cells in the stack either by internal or external manifolding. Ingeniously designed separator plates (also called bipolar plates) separate and distribute the gas streams from the inlet side of the electrode to the outlet side. Also, circular flat plate concepts are proposed and realized (e.g. the Sulzer-Hexis SOFC) where the gasses flow in radial direction. The SOFC of Siemes-Westinghouse consists of tubular cells whereby the oxidant flows through the tube and the fuel gas is supplied in the space between the tubes that are electrically connected. Many more configurations are proposed. For example, U. Bossel compares ten prominent fuel cell (SOFC) configurations.[26] In spite of their differences what they all have in common is that they usually are operated in a 2 in/2 out mode; i.e., both the anode and cathode gasses flow separately through the cell in which they are partly utilized. Hence, the outlet gas composition differs from the inlet composition, which leads to Nernst losses in the fuel cell. An important difference with the conventional burning of fuel gas with the oxidant gas in a reactor is that when using a fuel cell we have two separate output gas flows, whereas there is just one off gas stream from a reactor vessel or combustion chamber. This is why we can use a fuel cell as a gas separator, oxygen pump or active membrane. An MCFC, for example, can be seen as a CO_2 pump or active membrane for separating CO_2 from an oxygen containing gas. However, although the 2 in/2 out concept is the most common and straightforward way to operate a fuel cell, it is not the only one. Next the circular and tubular geometry will be discussed followed by a description of a 1 in/1 out and a 2 in/1 out fuel cell configuration and a recently developed modification of an MCFC that falls in the category 3 in/3 out.

1. Disk Shaped Fuel Cell Geometry

A disk shape flat plate concept with radial symmetry is marketed by Sulzer-Hexis Switzerland based on the technology developed by ECN, the Netherlands. Flat plate circular cells have an axial symmetry as shown in Fig. 23.

It was shown by Standaert[10] that the simple and extended analytical models developed for a rectangular flat-plate concept can be applied to the disc-shaped fuel cells provided an appropriate coordinate transformation is applied. The radii of the inner and outer circles of the cell are denoted as R_1 and R_2 respectively. The radial dimensional coordinate $\hat{R}$ is defined as the distance R to the symmetry axis divided by R_1:

$$\hat{R} = \frac{R}{R_1} \qquad (1 \leq \hat{R} \leq \frac{R_2}{R_1}) \tag{120}$$

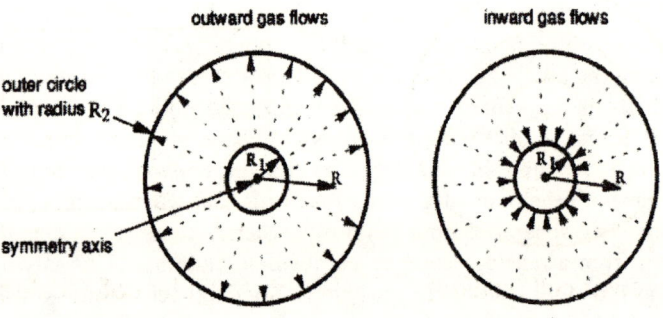

Figure 23. Flat plate circular cells with radial symmetry

Fuel Cells

$$x(R) = \frac{\text{surface between gas inlet and circle with radius R}}{\text{total cell surface between } R_1 \text{ and } R_2}$$

$$x(R) = \frac{R_1^2}{R_2^2 - R_1^2} \cdot ((\hat{R})^2 - 1) \quad \text{(outward gas flows)}$$

$$x(R) = \frac{R_1^2}{R_2^2 - R_1^2} \cdot \left(\frac{R_2^2}{R_1^2} - (\hat{R})^2\right) \quad \text{(inward gas flows)}$$

Standaert has given a mathematical proof of the principle of equivalence between a rectangular and a circular flat plate fuel cell formulated which follows.

Consider a rectangular cell and a circular cell with gas supplies directly proportional to the total active cell areas. If the cells are made of the same materials, they will be operated at the same cell voltage if and only if their average current densities are equal. When operating at the same cell voltage the current distribution in the circular cell can be calculated from the current distribution in the rectangular cell by application of the coordinate transformation given by the equation above.

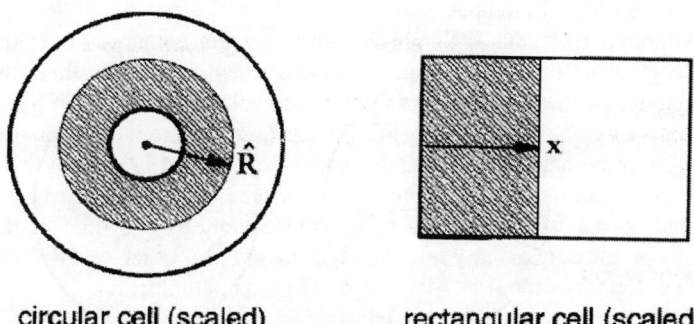

circular cell (scaled)   rectangular cell (scaled)

Figure 24. Coordinate transformation from a circular cell with outward gas flows to a rectangular cell.

The principle of equivalence implies that the analytical fuel cell models also apply to the circular flat plate concept using the given coordinate transformation. This coordinate transformation is visualised in Fig. 24.

2. Tubular Fuel Cell Geometry

While the tubular concept may at first side be very different from the flat plate concept, the analytical models can still be applied. The tubular concept can be obtained from the flat plate by rolling up the cell along the flow direction into a tube. Also co-flow exists between anode and cathode flow. Even if the radial current lines in the electrodes and electrolyte cannot be approximated as being parallel (but radical), still one does find a certain $r(x)$ with x the c-axis in cylindrical coordinates. Nowhere in the derivation of the equations we explicitly used that the fuel cell should be a rectangular flat plate so the equations and the derived model also apply for the tubular geometry, and one can use the analytical model(s) as outlined in this chapter. However, the tubular geometric frequently used for SOFC introduces additional ohmic losses due so-called in-plane currents in the electrodes. Depending on the way the tubes are interconnected in a stack these in-plane currents are either in the tangential direction (Westinghouse concept) or in the longitudinal direction (Boersma & Sammes[27]). The latter describe a new tubular concept with small, extruded tubes that have the advantage of allowing for rapid temperature changes.

Although Boersma & Sammes[27] apply current collectors over the full length of the anode and cathode, they calculate that the conductivity of these collectors is not good enough to prevent in plane losses in the longitudinal direction. Therefore, in their concept, the tubes cannot be made very long. Boersma & Sammes[27] calculate these losses in a stand-alone configuration of a tube where the current is collected by a wire at both the anode and cathode to the inlet of the tube. This introduces in-plane voltage losses. In other words, the current collectors cannot be assumed to be perfect conductors.

They calculate that in this configuration the length of the cells should not be much in excess of 100 mm because of domination of the in-plane losses in longer cells even if current collectors are applied.

In Fig. 26 cross sections are shown of a tubular SOFC showing the tangential in-plane currents in the electrodes. In

Fuel Cells

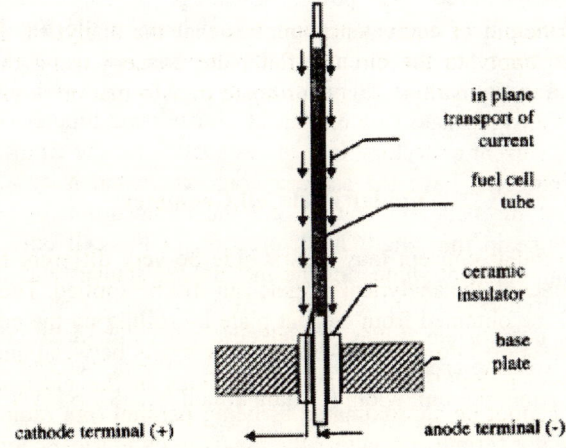

Figure 25. Tubular SOFC with in-plane current losses in the longitudinal direction [Boersma&Sammes[27]]

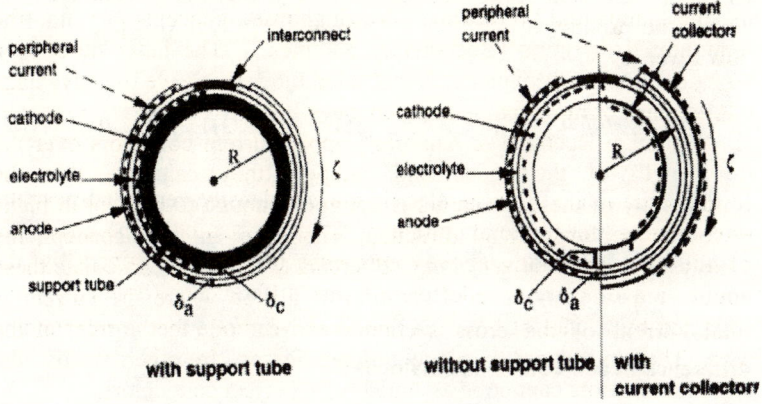

Figure 26. Tubular SOFC with and without support tube and with and without current collectors.[10]

order to calculate the total power advantage of current collectors Standaert[10] considers a cross section somewhere in the middle of a tubular cell. Here his derivation for the in-plane losses and thus the advantage of using (ideal) current collectors will be followed. The cross section on the right hand side of Fig. 26 illustrates a tubular cell with and without current collectors. It is representative for the whole cell, in the sense that both have the same average current density i_{cell}. It is assumed that the cell temperature and the composition of fuel and oxidant change in the longitudinal direction of the cell only. At the cross section, constant values for the internal resistances and the local Nernst potential (V_{eq}) will be used. Hence, if ideal current collectors are applied to both electrodes, the electric current will be uniformly distributed over the cross section. In that case, the power generated by the cross section, per unit width (or tubular cell length), is given by:

$$p_t = 2\pi R_t i_{cell}.V_{cell} = 2\pi R_t i_{cell}(V_{eq} - r.i_{cell}) \quad [W/m] \quad (121)$$

with R_t is the radius of the tubular cell and r (Ω m^2) the quasi-ohmic resistance of the electrode- electrolyte assembly. If the current collectors are omitted, the current density will vary over the cross section as a consequence of the resistances of the current paths in the tangential direction of the electrodes. Analogously the generated power equals the product of terminal current and terminal voltage, which is now given by:

$$p_t = 2\pi R_t i_{cell}.V_{cell} = 2\pi R_t i_{cell}(V_{eq} - r.i(0)) \quad [W/m] \quad (122)$$

In this equation, $i(0)$ is the electric current density generated near the interconnect (not the total current of the cross section). The difference of the two equations above represents the power gained by the application of current collectors. If this difference is divided by the total current of the cross section, i.e., $2\pi R i_{cell}$ the corresponding difference in cell voltage is obtained:

$$(\Delta V_{cell}^{\cdot})_{currentcollectors} = r.(i(0) - i_{cell}) \quad [W/m] \quad (123)$$

The current density $i(0)$ is calculated by Standaert,[10] in a way almost similar to that described by Bossel.[26] The result contains a hyperbolic function:

$$i(0) = \frac{\Upsilon i_{cell}}{\tanh(\Upsilon)} \qquad (124)$$

In this equation, the dimensionless constant Υ represents the square root of the ratio of the resistances in tangential and cross direction:

$$\Upsilon = \pi R_t \sqrt{\frac{\left(\dfrac{\rho_a}{\delta_a} + \dfrac{\rho_c}{\delta_c}\right)}{r}} \qquad (125)$$

Here, the resistivities of the anode and cathode are designated by ρ_a and ρ_c (Ω m), respectively. The thickness of the electrodes is denoted by δ_a and δ_c. For a tubular SOFC without support tube, typical parameter values are: i_{cell} = 300 mA/cm^2 ; R_t = 1 cm , r = 0.5 Ω cm^2 and

$$\frac{\rho_a}{\delta_a} = 0.1\Omega \quad , \quad \frac{\rho_c}{\delta_c} = 0.05\Omega \qquad (126)$$

Based on this data and Eqs. (123)-(125), the potential benefit of current collectors in tubular cells is calculated to be about 125 mV. It is noted that this result depends strongly on the input data, given above. Furthermore, similar calculations are possible if current collectors are applied to either the anode or the cathode alone. The value of 125 mV seems to be relatively large compared to the results by Bossel. Yet the incorporation of good current collectors is a serious option for the improvement of tubular fuel cells and needs further study and experimental verification.

It is emphasized that a current collector will have no influence on the gas sealing at all and has to contact only a relatively small part of the electrode surface. Therefore, current collectors will not necessarily be either expensive or heavy, and will not seriously affect the transport of gases to or from the electrode-electrolyte interface. Therefore, the incorporation of good current collectors is a serious option for the improvement of future tubular fuel cells.

3. 1 in/1 Out or Single-Chamber Fuel Cell Concept

Under certain conditions a mixture of fuel and oxidant does not react spontaneously and can be supplied to a so-called single-chamber fuel cell. Since the same gas is supplied to the anode and the cathode, one might expect the same potential (OCV) for anode and cathode. However, the gas itself is not in equilibrium and neither the anode potential nor the cathode potential is equilibrium potential. At both electrodes a mixed potential is established as is well known from corrosion theory. The mixed potential is closest to the fastest half-cell reaction. So if the fuel oxidation reaction is fastest at the anode catalyst material and the oxygen reduction is fastest at the cathode made of a different material, then a potential difference is established between anode and cathode as in a conventional fuel cell in spite of the same gas conditions at anode and cathode. Van Gool first proposed an inorganic fuel cell of the 1 in/1 out concept also called single-chamber or one-chamber fuel cell.[28] Hibino, et al. reviewed recent developments on this peculiar type of fuel cell. So far only single chamber fuel cells that use SOFC materials and relatively high temperatures are reported.[29,30] Hibino, et al.[31] also studied the use of different hydrocarbons as a fuel below 500 ° C. Asano and Iwahara[32] reported a high power density of 0,17 Wcm² (400 mA/cm² at 420 mV) at 1223 K 273 using a $BaCe_{0,8} Y_{0,2} O_{3-x}$ electrolyte. However, the lifetime of this one-chamber fuel cell was limited because the electrolyte reacted with CO_2 to form $BaCO_3$. YSZ, the electrolyte of the SOFC, is perfectly stable but only relatively low power densities could be obtained: 100m A/cm² x 340mV = 0,035 W/cm² maximum power density.

The advantage of the one-chamber fuel cell would be the simpler stack constructions. It circumvents the need for gas separation once the O_2/CH_4 mixture is available from some source. It seems irrational to deliberately mix CH_4 and air for the purpose of power generation. Application as a gas sensor may be more suitable for the one-chamber fuel cell.

4. 2 in / 1 Out or Dead-End Mode Concept

If either the oxidant or fuel gas is a pure gas and is fully converted by an electrochemical reaction into an ionic form, which is transported to the other electrode through the electrolyte, we can operate one electrode in dead end mode. For example, in fuel cells with a proton-conducting electrolyte (e.g., the PAFC or PEFC) the anode can be supplied with pure hydrogen. The electrons are then separated from the protons in the electrochemical oxidation reaction:

$$H_2 \rightarrow 2H^+ + 2e^- \qquad (127)$$

The electrons as well as the protons are transported away from the anode by conduction. When supplying pure hydrogen the pH_2 must be constant throughout the anode gas channels (except for a small pressure drop between inlet and outlet due to flow resistance). If we would have 100% fuel utilization the gas flow at the inlet would be such that 100% of the gas reacts and the gas flow at the anode outlet would be zero. This means we could just as well close that output, hence, obtain a dead end. Note that in this way we can reach 100% fuel utilization with zero Nernst loss (negating the Nernst loss due to oxidant utilization at the cathode). Because the Nernst loss is proportional with temperature it would be especially advantageous for high temperature fuel cells to operate in dead-end mode. However, in the present two types of high temperature fuel cells (MCFC and SOFC) the reaction product(s) is (are) released at the anode side and, therefore, the anode cannot be operated in dead-end mode. Some studies are reported on proton conducting ceramics that could be used as the electrolyte in a high temperature fuel cell a PCCFC (proton conducting ceramic fuel cell).[19] Also, this cell could be operated in dead and mode when supplied with pure hydrogen. As already mentioned in section VI in this operation mode Nernst loss is not proportional with utilization and 100% utilization can be achieved without any Nernst less at the anode side.

Instead of the anode side, the cathode side of the SOFC and MCFC can be operated in principle in dead-end mode. In an SOFC we have:

$$O_2 + 4e^- \rightarrow 2O^{2-} \qquad (128)$$

and at the cathode of an MCFC:

$$CO_2 + \frac{1}{2}O_2 + 2e^- \rightarrow CO_3^{2-} \tag{129}$$

At the right-hand side only ionic species appear that are transported away from the cathode through the electrolyte. As yet there is little incentive to operate the cathodes of an MCFC or SOFC in dead-end mode since the oxidant streams are used to cool the cells, but in special applications it is an option. In those cases the oxidant cannot be air because it contains a lot of inert gas (nitrogen).

The largest drawback of operating in dead-end mode is, of course, that the supplied gasses have to be very pure. Any inert contaminant will accumulate in concentration and finally fill the electrode and gas chambers. In principle, there are three ways to circumvent this problem:

1. Selective diffusion of the inert gas through a membrane or the electrolyte itself to the other electrode.
2. Back diffusion of the inert gas.
3. Discontinuous flushing.

However, this effect can also be used advantageously if the inert contaminant has economic value. For instance, noble gases in air are an inert contaminant of which the partial pressure can be increased significantly in a dead-end fuel cell cathode. Only separation of the noble gas from nitrogen needs to be accomplished then.

5. 3 in/3 Out Concept; MCFC with a Separate CO_2 Supply

Recently, a feasibility study was performed on a modified MCFC with a separate CO_2 supply.[33] It was noted that in principle the cathode reaction of an MCFC need not involve CO2 but could be the same as for the SOFC (see previous section). The O^{2-} ions formed in this reaction diffuse to the anode where they can react with hydrogen forming water. However, O^{2-} is a minority species in molten carbonate while the carbonate ions are present in excess ($[CO_3^{2-}] = 1$). It is found in present MCFC's that unless the cathode gas contains CO_2 with a partial pressure of about 0.05 atm or higher large cathode polarization losses occur. The minority charge carrier O^{2-} is converted into a majority species (CO_3^{2-}) by the well-known carbonate recombination reaction:

$$CO_2 + O^{2-} \Leftrightarrow CO_3^{2-} \qquad (130)$$

The transport process now becomes conduction of carbonate ions analogous to conduction of electrons in the electron-sea of a metal and molten carbonate is found to be a good ionic conductor.

Although the exact oxygen reduction mechanism in MCFC's is not yet resolved, it is found that in every possible mechanism the overall electrochemical reaction is the reduction of oxygen into oxide ions followed by the chemical recombination reaction (130). Although recent publications provide convincing evidence that the recombination reaction must be fast, for a long time it was reported and believed to be slow.[34] So based on this probably erroneous belief it was conceived that the MCFC would operate in a sense analogous to a SOFC. Only somewhere on its way to the anode the O^{2-} ion is recombined to carbonate ions. In an i-MCFC the notion of somewhere on its way is taken literally and the CO_2 is supplied only through a separate gas channel in the matrix.

In Fig. 27 a laboratory scale MCFC of 3 cm² (1.92 cm Ø) is shown in which three matrix tiles are placed such that a gas channel is formed through which the CO_2 rich gas can be flown. It is noted that in the preliminary experiments performed with this cell, neither for the i-MCFC nor for the same cell under standard gas composition, the cell performance is optimal. This is because the start-up procedure of the cell was not optimized for minimum polarization but focussed primarily on preventing the matrix from cracking. Nevertheless, this 3 in/3 out concept was found to work surprisingly well. The cell was operated under three different gas conditions shown in Table 9. The polarization curve of the oxygen cell is shown in Fig. 28. For comparison also a polarization curve is shown of the same cell when it is operated under standard conditions i.e. the cathode is supplied with a 70/30% air/CO_2 mixture. The air cell and the standard cell showed comparable performance.

In the i-MCFC several phenomena oppose each other in determining the overall polarization losses because of the extra matrix tile for the gas channels the matrix is thicker resulting in a higher ohmic polarization. Also, a larger diffusion distance exists for the O^{2-} ions but

Table 9
Gas Compositions of the Three Cell Configurations Used in the Work of Hemmes & Peelen, et al. Reported Here. Total Pressure 1 atm.; All Balances N_2 [33]

	Oxygen Cell		Air Cell		Standard Cell	
	pCO_2 atm	pO_2 atm	pCO_2 atm.	pO_2 atm	pCO_2 atm	pO_2 atm
cathode gas	0	1.0	0	0.21	0.30	0.15
matrix gas	1.0	0	1.0	0	0.20	0
anode gas	$pH_2 = 0.8$ and $pCO_2 = 0.2$ humidified at 60 °C					

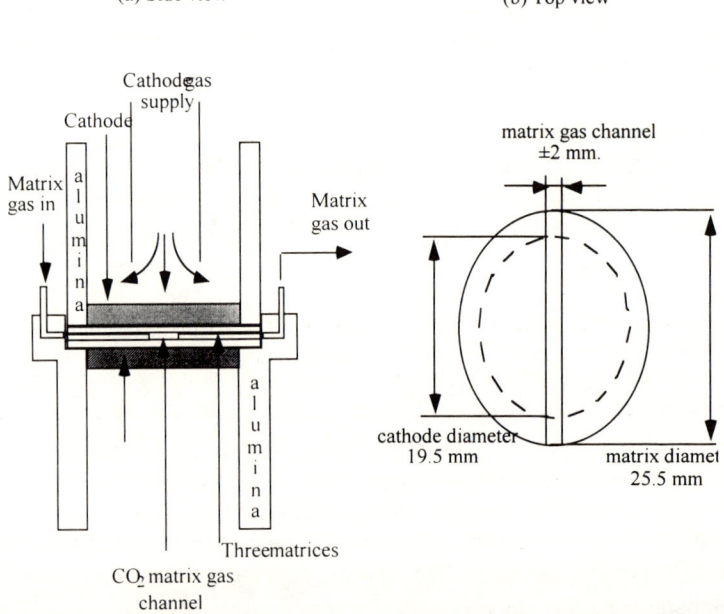

Figure 27. Schematic representation of the i-MCFC laboratory cell with separate CO_2 matrix gas channel, (a) side view (matrix gas channel is drawn 90° rotated with respect to the other components) and (b) top view.[33]

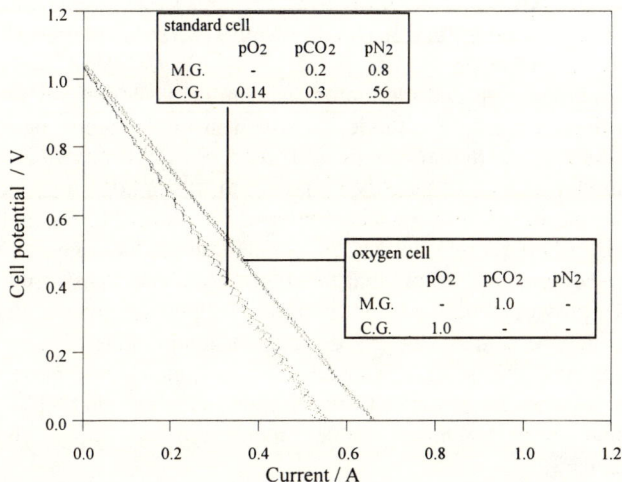

Figure 28. *I-V* curves of the oxygen cell and the standard cell defined in Table 9.[33]

In the i-MCFC several phenomena oppose each other in determining the overall polarization losses because of the extra matrix tile for the gas channels the matrix is thicker resulting in a higher ohmic polarization. Also, a larger diffusion distance exists for the O^{2-} ions but this is counteracted by a larger O_2 solubility in the carbonate since this is a chemical dissolution process in which CO_2 is released:

$$CO_3^{2-} + O_2 \rightarrow O_2^{2-} + CO_2 \tag{131}$$

Since no, or very little, CO_2 is supplied to the cathode, this reaction shifts to the right-hand side leading to a large peroxide concentration. Moreover, since we are no longer diluting the air with CO_2 we have a higher pO_2 at the cathode, leading to a higher OCV and lower polarization losses. On the other hand, we have a high pCO_2 (up to 100% of the total pressure) in the matrix where we want to recombine O^{2-} to carbonate ions; hence, this reaction also is promoted into the desired direction. In fact, the last two arguments lead to the idea of an i-MCFC, that is to say: "The pCO_2 should be low where oxygen is dissolving, and it should be high where it is needed for the recombination reaction."

It is noted here that the major problem of the MCFC, which is the dissolution of the NiO cathode, is solved in this concept, because the dissolution is minimal at low pCO_2 (order 10^{-2} to 10^{-3} atm.).[35] Such a low amount of CO_2 can easily be added to the cathode gas. Also, the transformation of the matrix material Li-aluminate is solved or minimized in the 3 in/3 out MCFC concept due to the high pCO_2 in the matrix. Terada, et al. found that the allotropic phase transformation and particle growth occurs via a dissolution-deposition mechanism. The results they obtained suggest that lower temperature, higher partial pressure of CO_2 or lower basicity of carbonates are preferable to control the particle growth of lithium aluminate, and that alpha-lithium aluminate appears more stable under typical MCFC operating conditions.[36]

Another advantage lies in the extra degree of freedom we now have in designing FC systems. Since the CO_2 and oxidant flows are decoupled one can optimize and utilize them separately in a fuel cell system. The functionality of the matrix channel is the selective extraction of CO_2 from the matrix gas stream. This can be used advantageously if the fuel for the i-MCFC is CO_2-rich, such as in the cases of reformed methane, biogas or landfill gas. By feeding the fuel gas, first through the matrix channel, CO_2 is removed, and an enriched fuel is obtained at the outlet that can be fed directly to the anode of the i-MCFC because of the lower pCO_2 and higher pH_2 a higher OCV is obtained in the i-MCFC. On the other hand, the outlet of the anode with its high pCO_2 can be used as matrix gas or combinations of both operation modes. These examples are to illustrate a number of new and interesting possibilities that might form the incentive to study and develop the i-MCFC concept further.

In the cathode of the i-MCFC there exists a large driving force for removing (small factions of) CO_2 that might be contained in the oxidant. This yields possible applications whenever one wants to remove CO_2 from the air; for example, in closed life support systems as in manned space flights. By applying the dead-end mode concept to the cathode and/or CO_2 matrix channel we can create also 3 in/2 out or even 3 in/1 out systems.

X. NUMERICAL FUEL CELL MODELING

1. Introduction

The analytical models derived in the previous sections apply to every type of fuel cell. However, the accuracy depends on the accuracy of the assumptions for the particular fuel cell. As explained in section IV.1 numerical methods are sometimes indispensable for studying the processes in side a fuel cell in sufficient detail to be able to engineer and optimize the particular fuel cell under study. In this section the more detailed models developed for the various types of fuel cells are briefly reviewed while highlighting the crucial factors and problems for each type. With a few exceptions all the models require numerical algorithms to solve the set of equations of the model.

2. AFC: Alkaline Fuel Cell

In low temperature fuel cells such as the AFC proper water management is essential for good functioning of the cell. Detailed models are needed to describe the processes inside the cell. For example, Yang and Bjornborn[37] developed a mathematical model for an AFC based on concentrated electrolyte solution transport theory to study the effect of humidity content in the reactant gas, and electrode operating temperature on water flux transport in gas and liquid phase as well as its effect on electrolyte concentration distribution. This is specific for an AFC, which uses a concentrated aqueous electrolyte. The goal was a better understanding of humidity content management during electrode operation. However, essentially the developed model is a porous electrode model namely the flooded agglomerate model. In a second paper Yang and Bjornborn[38] extended their model and refined the water and hydroxide ions transport mechanisms. It was shown that for the transport of water it is important to take into consideration the effect of the combination of gas phase and liquid phase transport and the transfer between the two phases.

A rigorous mathematical model of an AFC is provided by Kimble and White.[39,40] To model the performance of the cell as a whole, porous electrode models for the anode and cathode are incorporated. Specifically for the AFC, water is produced in the overall reaction dilutes the KOH electrolyte. Kimble & White assume that the electrochemically-produced water evaporates into the gas streams. In a steady state this, of course, must be true otherwise the electrolyte would continuously be diluted with water. Like Yang & Bjornborn[37,38] they also take into account the influence of electrolyte concentration on the ion-conductivity of the electrolyte.

These models are essentially one-dimensional in the direction perpendicular to the cell incorporating gas diffusion layer - anode - electrolyte - cathode - gas diffusion layer. Because of the complex nature of the porous electrode models that form the basis of the cell model of the AFC, a solution can only be found by numerical methods. The difference in gas composition at the inlet and outlet of the cell is not taken into account. Even so Kimble, et al., have to solve 25 governing equations with 36 outer and internal boundary conditions. By combining anode, cathode, electrolyte and gas diffusion layers Kimble and White[39,40] found that an interaction between anode and cathode also exists. They conclude that the diffusion of dissolved oxygen contributes the most to the polarization losses at low potentials while the electronic resistance contributes the most resistance at high cell potentials. Gas diffusion in the gas diffusion layers was not found to be a limiting factor.

Jo, et al.[41] conducted a computational simulation using a one-dimensional isothermal model for an AFC single cell developed earlier[42] to investigate influences of the thicknesses of the separator, catalyst layer, and gas-diffusion layer in an AFC. The parameters and operating conditions of the model are based on the Obiter Fuel Cell, which is employed as a power source for NASA space shuttles. The cell polarizations were predicted at various thicknesses and their influences were also analyzed. Thickening the separator layer decreased the limiting current density and increased the slope of the ohmic polarization region. Investigation of the thickness of the anode catalyst layer showed that the optimum thickness varied between 0.04-0.15 mm depending on cell voltage. The thickness of the cathode catalyst layer significantly influenced the cell performance presumably due to mass transfer limitations in the electrolyte. Their investigation of the influence of initial electrolyte concentration shows that the performance of the AFC is maximized at a concentration of 3.5 M. Finally, it is found that increasing the operating pressure steadily enhances cell performance. This is generally true for all fuel cells.[2]

Another example to show the need for numerical methods is the thermal modeling of an AFC by Baumann, et al.[43] They state that an essential problem connected with the operation of the fuel cell systems in space is the rejection of waste heat. The intention of their investigation was to gain a better understanding of the heat generation and heat rejection mechanism in alkaline fuel cells by performing detailed thermal modeling of a single cell stack. In particular, spatial temperature profiles within the fuel cell stack and the start-up behavior of the cells were predicted. Furthermore, a model simulation of an emergency situation due to a partial failure of the coolant circuit was performed and theoretical temperature versus time curves were given for restarting the cooling.

3. PEMFC: Polymer Electrolyte Membrane Fuel Cell

Research and development on the polymer fuel cell has increased rapidly in recent years predominantly due to the possible application of polymer fuel cells in automotive applications.[1] The overall goal of the modeling efforts is the same as that for high temperature fuel cells namely optimizing fuel cell performance through a better understanding of limiting processes in the fuel cell and a better engineering of the cell construction and its components. Although some researchers follow an empirical approach, most of the modeling of the PEMFC reflects the complex interaction of phenomena mainly caused by the presence of water in liquid form. Water is formed in the overall reaction and water is dragged along with the protons in an electro-osmotic process. So particularly at high current densities one predicts a two-phase flow of water and gas in the porous cathode catalyst layer. The water content strongly influences the proton conductivity of the membrane. Therefore, proper water and heat management is found to play an important role in maintaining a high performance in the polymer fuel cell.

Modeling approaches vary from single porous electrode modeling (mainly the cathode because of its higher polarization losses) and one dimensional modeling for complete MEA's to pseudo 2-dimensional models accounting for the utilization of the gases along the flow path.

Also, full 2-D models and, recently, a 3-D stack model using CFD (computational fluid dynamics) were reported in literature.[44,45]

Because of the complex nature of the processes and interactions, usually one falls back on solving the problem using numerical methods. The drawback of the numerical methods is evident from the large number of independent attempts reported in literature to set up a model for the PEMFC. It appears to be very difficult to transfer knowledge and know-how on numerical models as also noticed by Okada[46] who states that the unavailability of most of the fuel cell models prevent its full use by other researchers. In this modeling a tendency is observed that one should take into account all possible processes and interaction that take place. Also for high temperature fuel cells one started modeling the fuel cell (MCFC for example) by taking into consideration porous electrode models to predict the local performance as a function of gas composition and temperature. Although this is fundamentally correct, this is not always necessary, and much simpler approaches can be followed sometimes allowing for analytical solutions and providing better understanding of the fuel cell principles as shown in this chapter.

Occasionally, one finds analytical solutions for a fuel cell (sub) model for a PEMFC for example Springer and Gothesfeld[47] and Eikerling and Kornyshev.[48] Also, Gurau, et al.[49] provide analytical solutions for a half-cell domain including cathode channel, gas diffuser (i.e. the porous carbon backing), catalyst layer and the membrane of a PEMFC. On the other hand Gurau, et al.[44] also present a full 2D-numerical model for the entire sandwich of a PEMFC including MEA, the backings and the gas-channels.

Excellent overviews of the modeling efforts for PEMFC's are given by Okada[46] and Janssen.[50] Their modeling efforts fall into two categories. The first models are essentially porous electrode models which model the polarization behavior in the catalyst layers and the second are transport models that deal with the issue of water management in the PEMFC. Okada[46] provides a comprehensive overview in the form of two tables; one of the transport models and one of the models dealing with the polarization behavior in the catalyst layer (i.e., the porous electrode models). In the tables he briefly indicates the various modeling methods applied in the papers and the results obtained.

Janssen further divides the transport-oriented models into two categories, which are the so-called diffusion models and the hydraulic models that find their origin in two articles that stand at the beginning of developments in PEMFC modeling. Bernardi and Verbrugge[51] propose a hydraulic model for the transport of water counteracting the electro-osmotic drag of water from anode to cathode. While Springer, et al. [47] balance the electro-osmotic drag by a back diffusion of water. Many authors such as Okada et al.,[52,53] Wohr, et al.,[54] Yi and Nguyen[55,56] and Futerko and Hsing[57-59] have adopted the diffusion approach by Springer, et al.[47,60] Authors that adopted the hydraulic model are Singh, et al.[61]

Singh, et al.[61] could identify 5 different water flow regimes in a co flow stack revealed by their 2D-model. Anode and cathode water fluxes are found to vary considerably along the oxidant and fuel flow channels and can reverse sign depending on the average current density of the cell. At low current density back diffusion of water from the cathode to the anode overrules the electro-osmotic effect of water transport from anode to cathode (in 200mA/cm2). At intermediate current density liquid water flows out of both sides of the fuel cell throughout the whole length of the channels (from inlet to outlet). At very high current densities one can predict it to be completely reversed with respect to the first regime and the electro-osmotic effect overrules the back diffusion of water, so that humidification is required along the entire anode side to prevent membrane dehydration. In intermediate regimes a transition is predicted at either the cathode (50-150 A/cm2) or anode (200-2000 mA/cm2) where part of the electrode (near inlet respectively outlet) needs humidification whereas in the other part a surplus of water is found. Although no direct evidence was presented for the existence of these regimes, their work illustrates the importance of detailed modeling to understand the water household in a PEMFC and to find appropriate solutions for this problem.

Because a great number of phenomena occur at the same time in a PEMFC and influence each other the modeling becomes quite complex, whereas comparison with experiment is usually done only through the well known S-shaped polarization (I-V) curves. As G. Janssen[50] states, and supported by the work of Sena,[62] it is for example often very difficult to distinguish between the effects of drying–out the membrane and the effects of mass transport limitations as a result of flooded pores. Therefore, she uses the effective drag coefficient to validate her model that is based on the transport theory of irreversible thermodynamics as applied to PEMFC by Kjelstrup and Bedeaux[63]. Data on the effective drag coefficient proved relatively easy to measure and are reported in a separate paper by Janssen and Overvelde[64]. In order to avoid modeling of complex phenomena that cannot be validated by direct measurements some authors (for example Kim, et al. [65] Squadrito[66] and Amphlette, et al.[67-69] and Lee and Lalk[70,71]) use empirical expressions for all losses. In the latter references complete stacks are modeled and it is evident that simpler approaches for local cell performance are needed.

Two dimensional (2-D) modeling is necessary to take into account the changing conditions along the flow direction of the channels as concluded by Janssen[50] as well as by Singh, et al.[61] Because the conductivity of the membrane depends strongly on the water content in the membrane, the transport of the water in the membrane has received much attention. It is often modeled as part of the fuel cell model but sometimes it is the sole subject of a paper. For example Eikerling[72] compares the Darcy flow model for porous media with the diffusion model for water transport in the polymer membrane. In spite of the fact that the distribution of water in the PEMFC is so important for the proper functioning of the cells only two attempts were reported to measure the water content. A direct measurement using neutron scattering is reported by Bellows[73] and an indirect method by Buechi[74] that applies the measurement of the local membrane resistance.

The conductivity of the Nafion 117 membrane in equilibrium with water vapor of various relative humidity (RH) was modeled by Thampan, et al.[75] as a function of RH and compared with experimental results of Sone, et. al.[76] The conductivity depends on RH increasing from 0.001 S/cm at 20% RH to about 0.007 at 40% RH and 0,02 at 80% RH. Most models are isothermal models. However, if temperature effects are implicitly modeled for example by Berning, et. al.[45] one finds small deviations from the isothermal situation of only a few degrees, so PEMFC are essentially isothermal.

4. DMFC: Direct Methanol Fuel Cell

Although the performance of DMFC's is poor compared to a PEMFC that is fed with hydrogen they attracted attention because they circumvent the need for a reformer when using MeOH as a fuel e.g. in automotive applications. The reason for the low performance is twofold firstly the electrochemical oxidation of MeOH on presently known catalysts is slow and secondly a cross over of methanol from anode to cathode through the polymer membrane occurs. Cross over reduces fuel efficiency and also lowers the cell voltage. Meyer and Newman[77-79] extensively modeled DMFC behavior to optimize its performance with presently used materials. They describe their work in a series of three papers. As in other low temperature fuel cells the role of water is important. And again the model is essentially one-dimensional describing the transport phenomena in the direction perpendicular to the cell. Hereby the authors have implicitly implemented the notion that there are no large temperature differences across the cell (from inlet to outlet) and that influence of gas composition on performance is small and that the Nernst loss is small.

Baxter, Battaglia and R.E. White[80] include species movement in what they call pseudo y-direction into account for water, methanol and carbon dioxide by use of an effective mass transfer coefficient. But otherwise also this model is essentially one-dimensional.

Kulikovsky[81,82] and Scott, et al.[83-103] developed 2-D models while especially Scott, et al. contributed with a great number of papers to the modeling of a DMFC. They included the dynamics of a DMFC in their model, which is very important in automotive applications.

5. PAFC: Phosphoric Acid Fuel Cell

It was only very recently that a two-dimensional model for a PAFC was presented in literature by Choudhury, et. al.[104] As the authors already mention, most of the published literature deals with one-dimensional models for PAFC, which are essentially porous electrode models. For example Yang[105] and Maggio[106] focus on the cathode while Cutlip, et al.[107] develop a detailed model for the anode of a PAFC. Choudhury, et al.[104] state that a two-dimensional model is necessary in cases where especially the oxygen concentration changes substantially in the flow direction due to depletion of oxygen and back diffusion of product water. Their model was validated by: (i) testing in one-dimensional mode for verification of the basic parameters through a micro setup known as "unit cell", and (ii) evaluation of the two-dimensional model through an experimental setup of a PAFC stack with four cells. Furthermore, they illustrate the utility of their model for the PAFC design and in its humidity management. They present the results of a parametric sensitivity studies using the model as well.

6. MCFC: Molten Carbonate Fuel Cell

As early as 1983 a two-dimensional non-isothermal model for an MCFC was developed by Wolf and Willemski,[108] the so-called PSI-model (PSI stands for Physical Science Incorporated). Apart from the water management, which does not play a role in these high temperature fuel cells, the modeling is analogous to the approach followed in 2-D models for the low temperature fuel cells. The local current density is calculated as outlined in Section IV.2. However the local quasi-ohmic is determined using a porous electrode model for each electrode that takes into account the influence of gas composition and local temperature. The model takes account of gas stream utilization due to the electrochemical reactions leading to Nernst loss, conductive heat transfer between cell hardware and gas streams, energy transfer accompanying mass addition to the bulk streams, convective heat transfer by the bulk streams and in-plane heat conduction through the cell hardware. In an iterative procedure the solution is found in terms of the local current density as a function of applied cell voltage. From this the external I-V curve can be predicted as a function of the gas utilization. Moreover because of the 2D nature of the model, the local current density and temperature at every position in the cell can be calculated which allows for detailed study of these distributions as a function of, for instance, co-flow or cross-flow of anode and cathode gas streams. Because of the great number of parameters, the model could be tuned to fit particular experiments. This is always necessary because as every fuel cell developer knows, the performance of a particular fuel cell depends on many sometimes poorly controllable parameters and cannot be predicted by *ab-initio* calculations. The model proved very useful in the development of the MCFC for example at ECN in the Netherlands. When comparing this approach with our simpler analytical approach, it becomes very clear that the level of detail in the modeling should relate to the modeling goals. The PSI model provides enough detail to study the internal processes in the MCFC and allows for a more detailed engineering whereas the analytical approach gives a more general picture and a better understanding of the essential processes and phenomena. Sometimes the differences are found to be very small and results may contradict. For example Wolf and Willemski[108] conclude that the best performance was achieved with counter-flow conditions. Although the analytical approach is not suitable for a cross-flow cell still a comparison can be made between co-flow and counter-flow cells based on the analytical expressions derived by Standaert[10]. He concludes that for example in co-flow a more homogenous current density exists over the cell resulting in a somewhat higher cell voltage at sufficiently high current densities. The difference between the analytical approach and the numerical approach is that Standaert[10] is able to note that it makes a difference whether one operates at high or low average current densities, whereas this only follows from numerical calculations if one somehow got the idea that this might make a difference and starts investigating the influence of this particular parameter. This lesson learned in MCFC modeling might apply as well to other types of fuel cells.

7. SOFC: Solid Oxide Fuel Cell

Bistolfi, et al.[13] explain the different modeling approaches needed to solve different R&D problems illustrated by the example of a SOFC. Steady-state modeling of a single planar cell was used by these authors to study cross-flow and co-flow configurations and design operating conditions for internal reforming. Zooming in to a single unit cell a commercial finite element package was used to study current flow fields and losses in the anode and inter connector of an SOFC. To study gas flow patterns in the gas distribution system a CFD (computational fluid dynamics) software package is most appropriate to optimize the gas manifolding construction and assure a proper gas supply to all cells in a stack.

In order to avoid stack damages during dynamic load conditions dynamic models are developed that account for the different processes in the cell such as heat transfer. In studying the effects of indirect internal reforming one does want to calculate local temperature profiles and therefore one needs to know local current densities. Due to a rapid endothermic reforming reaction the local temperature at the inlet of a cell strongly decreases. This effect is enhances itself because due to lower temperature the local cell resistance increases and current density decreases. Therefore less heat is produced locally. This is only partly counteracted by a higher Nernst voltage difference at lower temperature. Aguiar, et. al.[109] present results of such more detailed modeling including the modeling of heat exchange between SOFC and the internal reformer chamber. They suggest the use of catalysts with reduced activity (e.g. oxide-based instead of Ni based catalyst) and show in their modeling that smooth temperature profiles can be obtained in SOFC's with internal reforming.

The SOFC with its all-solid components allows for the construction of monolithic fuel cell stacks avoiding the use of separator plates between the cells in a stack. Vayenas et al. constructed and tested a cross-flow monolith electrochemical reactor that can be operated as a SOFC but can also be used in tri-generation systems (see Section XI.1.). Also a mathematical model was derived by defining a unit cell with oxygen flowing in the X direction and fuel in the Y-direction. The reactor is considered to be composed of these unit cells.[110,111]

XI. NEW DEVELOPMENTS AND APPLICATIONS

1. Fuel Cells in Trigeneration Systems

The examples at the end of the previous section, in relation to a new yet experimental type of fuel cell, show that fuel cells need not only be seen as devices that generate electricity as their main purpose, but can also be seen as devices that separate certain gas-components from a mixture or enhance their concentration. These properties of fuel cells can be used to improve the chemical process into which they can be incorporated making appropriate use of the specific fuel cell characteristics.

An example is the use of a fuel cell as an oxygen depletion device. Oxygen is removed from the air in the cathode of for instance a polymer fuel cell. Therefore the fuel cell can be used to de-aerate for example a storage hall for agricultural products, thereby prohibiting or delaying rotting processes. The cathode off gas is cooled and fed back to the storage hall. Because often the storage space must be cooled as well the use of high temperature fuel cells is less suitable. Also combustion in e.g. a gas motor is less suitable for this reason. Moreover the latter also suffers from the disadvantage that the oxygen depleted exhaust gas contains combustion products from the fuel. So here the feature of a fuel cell having two separate outlets is a valuable advantage and only clean air with a lower oxygen pressure re-enters the storage hall.

The use of exhaust gas form the anode that contains a high amount of CO_2 in greenhouses is another well-known application for fuel cells increasing the yield of agricultural products. Here the fuel cell is an alternative for a gas motor that is used nowadays for this purpose.

Instead of looking at ad-hoc solutions or possibilities for fuel cells in the chemical industry, Dijkema, et al. propose a systematic method to analyse the possibilities for innovations in the chemical process industry.[112] The systematic method starts with looking at a fuel cell as one of the unit operations in a chemical plant. An inventory of the characteristics of the mass and energy flows is made and composed with what is needed in certain chemical processes such as the production of ammonia or methanol. It was shown that in some cases a leverage effect can be brought about with a relative small fuel cell (~10 MWatt) increasing the efficiency of production of a methanol production plant on a hundreds of MWatt-equivalent scale [112]. These innovative systems are often referred to as tri–generation systems because not only electricity and heat are produced but also chemicals. We can distinguish between three sorts of tri-generation with fuel cells:

1. Integration of present type of fuel cells using common fuels and oxidants.
2. Integration of present type of fuel cells using uncommon fuels or oxidants.
3. Integration of new modified or specially designed types of fuel cells using uncommon fuels and/or oxidants.

Vayenas et al have pioneered this research area. They investigated the co-generation of electric power and NO using NH_3 as a fuel in a SOFC.[113] The SOFC was modified in the sense that Pt electrodes were used. Vayenas et al. also studied the (electrochemical-) oxidation of H_2S to SO_2 and the oxidation of methanol to formaldehyde in SOFC's. [114-116]

2. Coal/Biomass Fuel Cell Systems

In the literature a number of system studies have appeared on integrated biomass or coal gasifier-fuel cell systems.[117-120] The off gas from a coal or biomass gasification unit is particularly suitable for the high T fuel cells SOFC and MCFC because the off gas is also of high temperature, and no pre-heating is necessary. Secondly, CO is not a poison but a fuel for MCFC and SOFC. Thirdly, the caloric value is not so high and pH_2 is relatively low, due to the large N_2 content. This makes the gas less suitable for combustion while the open cell voltage hardly decreases (see Section III.1.). In particular for the MCFC the high content of N_2 even increases the Nernst potential as derived in Section III.2.

An important problem in using coal-biomass gas or landfill gas is the large contamination of the gas due to the contaminations in the feedstock. In particular in combination with the MCFC a renewed effort in developing MSO (molten salt oxidation) technology would be worthwhile.[121-123] In molten salt oxidation frequently molten carbonate is used. The gasification of coal and biomass etc. in molten carbonate is reported to yield a relatively clean gas because a lot of the contaminants react with or will be physically contained in the melt (e.g. dust particles). Presumably also the problem of carbonate evaporation from the MCFC stack is solved when the fuel gas is derived from a carbonate melt.

Because air is normally used as oxidant, a large amount of N_2 is contained in the biogas, unless an oxygen plant is added to the gasification unit that removes the nitrogen from the air. The same holds for coal gas. In Table 10 typical composition of biomass gas from a gasifier is given. Note that the nitrogen content is higher than 50%. A third source of low caloric gas is gas from landfills that also contains a relatively high concentration of N_2.

3. Indirect Carbon Fuel Cell (IDCFC)

Although the presence of a large amount of the inert gas N_2 in the fuel gas from a coal/biomass gasifier has a positive effect on the open cell voltage of an MCFC it does not mean that it is not favorable to prevent that it is mixed into the fuel gas. Also the cathode gas consists of a major fraction of inert N_2 gas and all this inert gas has to be pumped through the fuel cell to carry with it the essential reactants to the electrodes. This decreases the total system efficiency. Using the features of a fuel cell one can prevent the mixing of N_2 into the fuel

Table 10
Typical Composition of Gas from a Biomass Gasifier
(Vol-% Dry Basis) (Source ECN the Netherlands)

CO	H_2	CH_4	CO_2	N_2	C_2H_4	C_2H_6
14.7	14.5	2.5	15.5	52.3	0.3	0.07

gas. M. Ishida and N. Nakagawa depicted the principle in Fig. 29. Instead of feeding the oxidant (air) directly to the coal or biomass it is done through a so-called concentration cell for which an SOFC can be used. Oxygen is separated from the solid fuel by the SOFC electrolyte YSZ, which is capable of conducting O^{2-} ions, having electrodes on both sides as in an ordinary SOFC.

The cathode reaction is the same as in an SOFC and the oxide ions are conducted to the anode where they react with CO:

$$CO + O^{2-} \rightarrow CO_2 + 2e^- \qquad (132)$$

At the solid fuel (carbon) CO_2 is regenerated into CO by the Boudouard equilibrium:

$$C + CO_2 \rightarrow 2\,CO \qquad (133)$$

Assuming that, ideally speaking, all of the CO_2 has reacted with the carbon, then half of the so-formed CO goes to the exhaust and the other half is recycled to the anode and converted into CO_2 again. This concept can be called an indirect carbon fuel cell because one of the

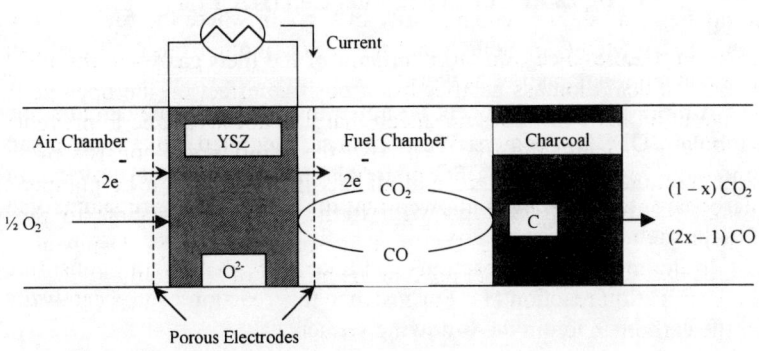

Figure 29. Nakagawa and Ishida tested indirect conversion of elemental carbon through generation of electrochemically reactive CO using the Boudouard reaction and electrochemical conversion of CO to CO_2.[124]

reaction steps is electrochemical and producing electricity and the overall reaction is:

$$C + \tfrac{1}{2} O_2 \rightarrow CO \qquad (134)$$

However, the carbon is not in direct contact with the anode and does not directly deliver electrons to an electrode. We can also call it electrochemical gasification. As explained in section II.3, it has large exergetic advantages to perform the gasification reaction electrochemically. A detailed exergy analyses of this indirect carbon fuel cell has been given by Nakagawa [124]. The regeneration principle via CO_2 and the Boudouard reaction can be assisted via H_2O and the reactions:

$$C + H_2O \rightarrow H_2 + CO \qquad (135)$$

and:

$$H_2 + O^{2-} \rightarrow H_2O + 2e^- \qquad (136)$$

Therefore, if steam is present (as is normally always the case with biomass) a parallel reaction path is provided decreasing polarization losses. This regeneration concept can also be realized with existing components as shown schematically in Fig. 30, where the fuel cell can either be an MCFC or SOFC, and the MSO (molten salt oxidizer) can be a conventional gasifier as well.[125]

A more integrated concept is shown in Fig. 31.[125] One can imagine a tubular SOFC by Siemens Westinghouse integrated into a molten salt oxidation vessel. If the SOFC materials are stable in the molten salt (carbonate) then it is even allowed that the SOFC tubes are submerged into the melt.

In the melt various reactions can take place that overall constitute the gasification reaction (1). For instance this reaction can be catalyzed by the carbonate ion in the following sequence:

- At the anode oxygen is produced in the reverse cathode reaction

$$O^{2-} \rightarrow \tfrac{1}{2} O_2 + 2e^- \qquad (137)$$

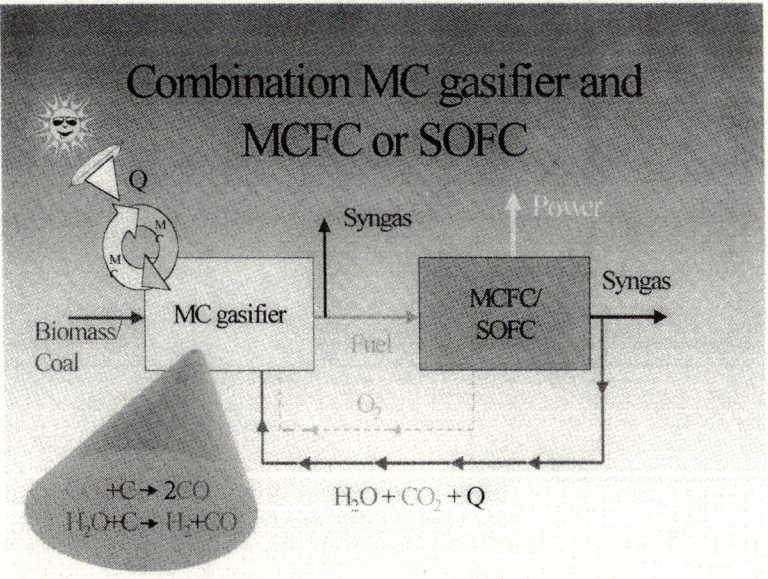

Figure 30. Gasifier - fuel cell concept with regeneration, simulating an indirect carbon fuel cell.

- The oxygen dissolves chemically:

$$\tfrac{1}{2} O_2 + CO_3^{2-} \rightarrow CO_2 + O_2^{2-} \tag{138}$$

- The so-formed peroxide reacts with the carbon:

$$C + O_2^{2-} \rightarrow CO + O^{2-} \tag{139}$$

- Also CO in its turn can react with peroxide:

$$CO + O_2^{2-} \rightarrow CO_2 + O^{2-} \tag{140}$$

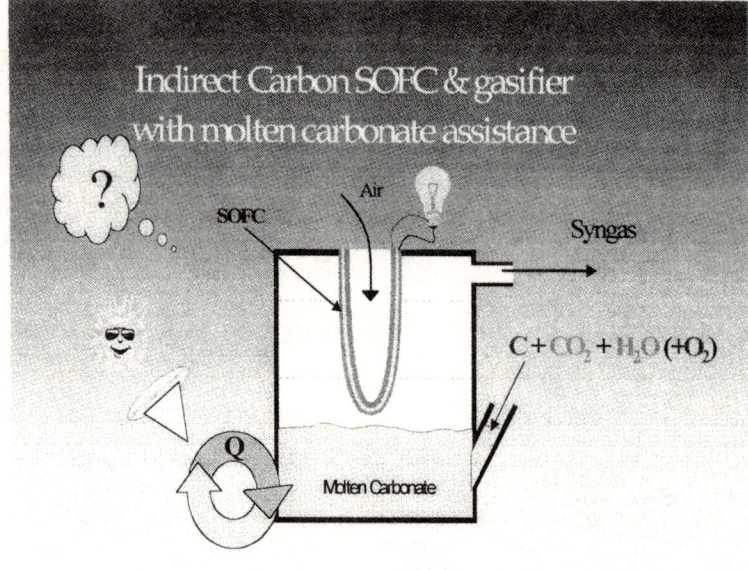

Figure 31. Indirect Carbon Fuel Cell concept with SOFC technology.

- Finally oxide ions and CO_2 produced in these reactions recombine into carbonate again in the well-known recombination reaction:

$$CO_2 + O^{2-} \rightarrow CO_3^{2-} \qquad (141)$$

This is just one of many possible reaction paths allowed by bringing the SOFC anode in contact with the melt containing the solid fuel. However, experiments are needed to prove the concept and to determine the dominant reaction path and speed.

4. Direct Carbon Fuel Cell (DCFC)

In an indirect Carbon Fuel Cell, which we have seen a few concepts above, the carbon is oxidized chemically, but the oxidant is introduced into the system via a fuel cell/concentration cell. In a direct carbon fuel cell the electron transfer from the carbon is directly at the electrode, although indirect reaction-path always remain possible as well.

By direct contact between the electrode, the carbon and the electrolyte, the following anode reaction can occur if molten carbonate is used as electrolyte:

$$C + CO_3^{2-} \leftrightarrow CO + CO_2 + 2e^- \tag{142}$$

$$C + 2\,CO_3^{2-} \leftrightarrow 3\,CO_2 + 4e^- \tag{143}$$

If the electrolyte is YSZ as in a SOFC:

$$C + O^{2-} \leftrightarrow CO + 2e^- \tag{144}$$

$$C + 2\,O^{2-} \leftrightarrow CO_2 + 4e^- \tag{145}$$

In general a mixture of CO and CO_2 will be the product gas. The ratio between the two will depend on thermodynamics (mainly temperature) and kinetics of the respective reactions. At higher temperature the formation of CO is preferred since the Boudouard equilibrium shifts towards CO at higher temperature.

$$C + CO_2 \leftrightarrow CO \tag{146}$$

Since the gas is produced at the anode it is in direct contact with the carbon anode and the Boudouard reaction can take place. Therefore, the off gas will probably be largely determined by the Boudouard equilibrium, hence temperature. This is reflected in the ΔG (=ΔH - $T\Delta S$) of the two overall fuel cell reactions:

$$C + \tfrac{1}{2}\,O_2 \leftrightarrow CO \tag{147}$$

for which $-\Delta H_{CO} = 110\text{-}113 \text{kJ/mol}$ and $\Delta S_{CO} = +90$ to 97 J/mol.K in the range $T = 300\text{-}1200$ K and the full conversion to carbon dioxide:

$$C + O2 \leftrightarrow CO_2 \qquad (148)$$

with $\Delta H_{CO2} = -395$ kJ/mol and $\Delta S_{CO2} \sim 0$ in the range $T = 300\text{-}1200$ K

The standard potentials for both overall reactions are equal for:

$$T_B = \frac{\Delta H_{CO} - \dfrac{\Delta H_{CO_2}}{2}}{\Delta S_{CO}} \qquad (149)$$

The factor 2 appears because in reaction (148) four electrons are transferred while in reaction (147) only two. Using respectively $\Delta H_{CO2} = -395$ kJ/mol; $\Delta H_{CO} = -95$ kJ/mol and $\Delta S_{CO} = +95$ J/mol.K, we find that $T_B = 1097$ K as shown in Fig.32.

So DCFC's operating above this temperature are predicted to mainly produce CO and to deliver a higher open cell voltage than the

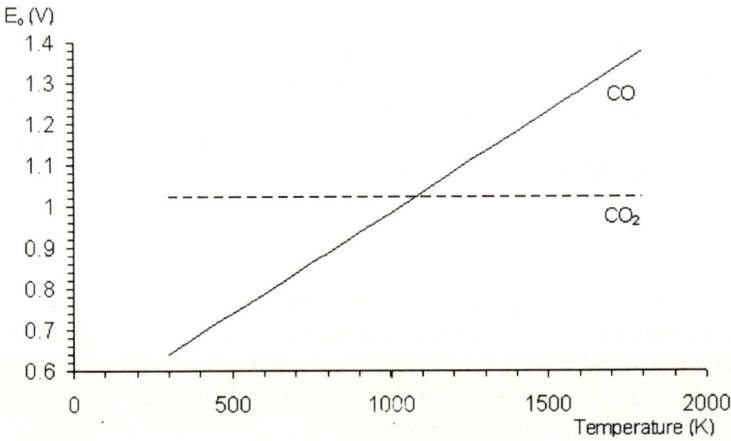

Figure 32. E_o for the conversion of carbon to respectively CO_2 and CO as a function of temperature.

Fuel Cells

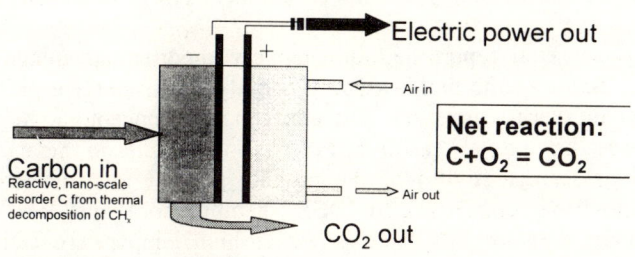

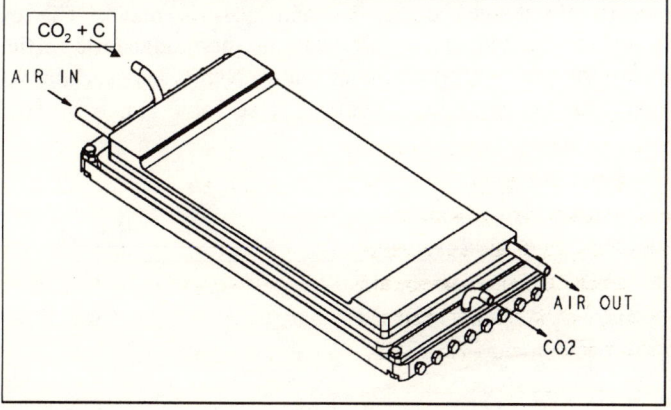

Figure 33. Schematics and construction of a 100W Direct Carbon Fuel Cell test set-up used at LLNL.[126]

1.02 Volt for the conversion to CO_2. Lower operating temperatures will favor CO_2 production. Cooper, et al. from Lawrence Livermore National Laboratory located in California, USA, built and operated a DCFC at 750-850° C (1023-1123 K) shown in Fig.33.[126,127]

Although the temperature is higher than T_B they report full conversion to CO_2 at $T = 800°C$ (1073 K). They used molten carbonate slurry containing carbon nano particles as fuel for the anode in an MCFC like cell. The cathode reaction is the same as for the MCFC. An optimum morphology of the carbon nano particles a current density was obtained that approaches the standard for an MCFC within a factor of two, as shown in Fig. 34 (r = 2 Ohm.cm^2 vs. 1 Ohm cm^2 for an MCFC).

Cooper, et al. report the following very important advantage of the DCFC. Since a solid fuel is used, its activity is constant irrespective of the amount that is converted. Therefore, in their concept in which pure CO_2 is used as a driver gas to feed the carbon particles to the anode the partial pressure (activity) of the product gas, CO_2 is also constant throughout the cell. Hence, the DCFC exhibits no Nernst loss, which constitutes a major loss factor in other high temperature cells (see Section V). A very interesting comparison between the electrochemical conversion efficiencies of C, H_2 and CH_4 is made. The overall efficiency is treated as a product of thermodynamic efficiency ($\Delta F/\Delta H$), the fuel utilization factor due to Nernst loss and the voltage efficiency due to irreversible losses η_{irr}. (see Table 11):

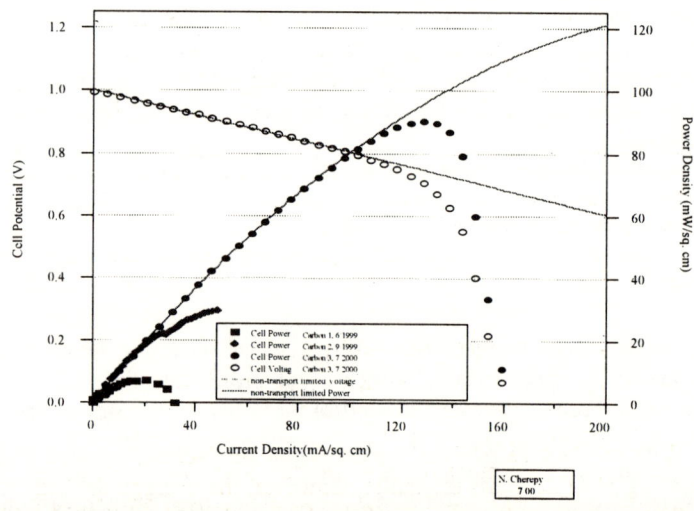

Figure 34. Polarization curves and power density obtained with the DCFC test set-up at LLNL [126]

Table 11
Comparison between the electrochemical conversion efficiencies of C, H_2 and CH_4 [127]

Fuel	$\eta_{fc} = \Delta G / \Delta H$	$\eta_{\text{Nernst loss}}$	η_{irr}	η_{tot}
C	1.0	1.0	0.8	0.8
H_2	0.7	0.8	0.8	0.45
CH_4	0.89 (1.0)*	0.8	0.8	0.57 (0.64)*

*Cooper, et al. report $\eta_{fc} = 0.89$ but using data from the CRC Handbook one finds that the entropy change in the overall reaction is nearly zero so that $\Delta G = \Delta H$ and $\eta_{fc} = 1$.

$$\eta_{tot} = \eta_{fc} \cdot \eta_{\text{Nernst loss}} \cdot \eta_{irr} \tag{150}$$

Although in principle it is not correct to multiply efficiencies, because the loss terms are additive in nature, the calculation does correctly indicate the advantage of direct conversion of carbon, and the limitations in the efficiencies of electrochemical conversion of gases due to Nernst loss and $T\Delta S$ losses. However, the numbers are a bit exaggerated.

Another important advantage of the DCFC is that the anode off gas is pure CO_2 that can either be used or stored (CO_2 sequestration). On the other hand when using an inert carrier gas, for example nitrogen, the OCV will increase as can be seen from the Nernst equation (assuming standard conditions at the cathode).

$$V_{eq} = E_0 + \frac{RT}{4F} \ln\left(\frac{[C]}{pCO_2}\right) \tag{151}$$

With an inert carrier gas we do encounter Nernst loss due to the increasing pCO_2 towards the outlet. However, the average driving force and thus also the cell voltage will be higher than starting with 100% CO_2 because pCO_2 is always smaller than 1 atm. in the whole cell. The drawback is that now the off gas is not pure CO_2 but a N_2/CO_2 mixture.

Separation of carbon from methane (CH_4) to produce the required carbon nano particles is an interesting possibility since in the process hydrogen is formed that can be used in other types of fuel cells, or it can be mixed into the natural gas pipeline system to decarbonize the gas and thus reduce diffuse CO_2 emissions at the end users.

ACKNOWLEDGEMENT

The author would esspecially like to acknowledge the contributions of Frans Standaert[10] and Siu Fai Au[128] who as PhD students significantly contributed to the work reported here. Furthermore Gaby Janssen of the Dutch Energy Research Center (ECN) is greatly acknowledge for sharing her knowledge on polymer fuel cell modeling

LIST OF SYMBOLS

A	Cell area	(m^2)
α	Slope of the linear part of $-V_{eq}(u)$ or $-dV(u)/du$ assuming $V(u)$ linear	(V)
α_f	Slope of the linear part of $-V_{eq}(u_f)$	(V)
α_{ox}	Slope of the linear part of $-V_{eq}(u_{ox})$	(V)
α_{tot}	α calculated with both fuel and oxidant utilization (def Eq. 114)	(V)
β	Dilution factor of fuel by an inert gas $0 < b < 1$	-
C	Area between $V(u)$ and its linear approximation in the initial dip.	(V)
δ a	Electrode thickness of the anode (tubular SOFC)	(m)
δ c	Electrode thickness of the cathode (tubular SOFC)	(m)
ΔG	Gibbs Free energy change of a reaction	(kJ/mol)
ΔH	Enthalpy change of a reaction	(kJ/mol)
ΔH_{CO}	Enthalpy change of the reaction C + 1/2 O2 = CO	(kJ/mol)
ΔH_{CO2}	Enthalpy change of the reaction C + O2 = CO2	(kJ/mol)
ΔS	Entropy change of a reaction	(J/mol.K)
ΔS_{CO}	Entropy change of the reaction C + 1/2 O2 = CO	(J/mol.K)
ΔS_{CO2}	Entropy change of the reaction C + O2 = CO2	(J/mol.K)
ΔV_a	Deviation of local cell voltage with respect to the cell voltage with all anode gases in their standard state.	(V)
E_o	Standard potential	(V)
F	Faraday's constant (= 96485.34 C/mol)	(C/mol)
γ	Parameter (approximately equal to fuel utilization)	-
η	Efficiency	-
η_c	Carnot efficiency	-

Symbol	Description	Units
η_{fc}	Fuel Cell efficiency	-
η_{irr}	Efficiency factor accounting for the Nernst loss	-
$\eta_{Nernstloss}$	Efficiency factor accounting for the irreversible fuel cell losses	-
η_s	System efficiency	-
η_s^{max}	Maximum system efficiency	-
i	Index of hypothetical sub-cell	-
i	(local) current density	(A/m2)
$I(i)$	Current through the ith sub-cell	(A)
$I_{1.2}$	Current through resistance $R_{1.2}$	(A)
I_{in}	Equivalent input current density; i.e. the current that the FC would deliver if all the input gas is fully converted	(A/m2)
i_{in}	Equivalent input current density: $i_{in} = I_{in}/A$	(A/m2)
i_{in}^f	Equivalent input current density explicitly referring to fuel gas	(A/m2)
i_{in}^{ox}	Equivalent input current density referring to oxidant gas	(A/m2)
i_{tot}	Total current delivered by the fuel cell	(A)
i_{tot}	Average current density of the cell: $i_{tot} = I_{tot}/A$	(A/m2)
i_{cell}	Average current density of a tubular fuel cell	(A/m2)
k	Index in summation	-
n	Number of electrons transferred	-
N	Number of hypothetical sub-cells	-
OCV	Open Cell Voltage	(V)
p	Pressure	Pa
p_0	Standard pressure	Pa
p_1	Loss term accounting for irreversible fuel cell losses	-
p_2	Parameter in temperature dependent irreversible FC losses	-
pX	Partial pressure of component X (e.g. pH_2, pO_2, etc.)	Pa
q_1	Loss factor accounting for irreversible losses in a CE	-
q_{comp}	Heat of compression	J
r	Specific quasi-ohmic cell resistance (assumed constant in the cell)	(Ω m2)
$r(x)$	Local specific quasi ohmic cell resistance	(Ω m2)
R_c	Total quasi ohmic cell resistance of the whole fuel cell	(Ω)
R	Gas constant (8.31451 J/mol.K)	(J/mol.K)
R	Radial coordinate of circular cell	(m)
R_1	Inner radius of circular cell	(m)
R_2	Outer radius of circular cell	(m)
$\hat{R}$	Scaled radial coordinate of circular cell. (Def.: $\hat{R} = R/R_1$)	-
ρ_a	Electronic resistivity of the anode (tubular SOFC)	(Ω m)
ρ_c	Electronic resistivity of the cathode (tubular SOFC)	(Ω m)
R_i	Internal cell resistance	(Ω)
R_{load}	External load resistance	(Ω)
R_t	Radius of tubular cell	(m)
Rt	Radius of tubular cell	(m)

T	Temperature	K
T_1	Fuel cell temperature where $\eta_{fc}=1$ for a fuel cell with $\Delta S > 0$	K
T_{BE}	Break even temperature where $\eta_{fc}=\eta_c$	K
T_B	Boudouard temperature. Def. Eq. (149)	K
$T_{\Delta G=0}$	Spontaneous combustion temperature. Def.: $T_{\Delta G=0} = H/\Delta S$	K
T_{max}	Temperature where η_s is maximum	K
T_o	Temperature of the environment	K
u	Utilization	-
$u(i)$	Utilization after the ith sub-cell	-
u^*	Utilization with respect to pure H_2 where $V_{eq}(u^*) = \chi$	-
u^{**}	Utilization with respect to pure H_2 where $V_{eq}(u^{**}) = E_o$	-
u_f	Fuel utilization	-
u_{ox}	Oxidant utilization	-
u_{tot}	Utilization of the whole fuel cell (if referring to the fuel gas: $u_{tot} = u_f$)	-
V	Gas volume	(m^3)
V^*_{eq}	Slope of the of the linear empirical function	(V)
$V_{1,2}$	Voltage across resistance $R_{1,2}$	(V)
V_{cell}	Cell voltage	(V)
V_{eq}	Equilibrium cell voltage or OCV	(V)
V_{eq}	Theoretical Nernst potential	(V)
$V^*_{eq}(0)$	OCV in the linear approximation for $V_{eq}(u)$	(V)
$V_{eq}(i)$	Local equilibrium cell voltage the ith sub-cell	(V)
W	Work	(J)
W_{comp}	Compression work	(J)
χ	Voltage equivalent of enthalpy change (Def.: $\chi = \Delta H/nF$)	(V)
x	Dimensionless space coordinate; scaled distance to fuel cell inlet	-
Υ	Dimensionless parameter (Def. Eq. 125)	-
Z	Dimensionless parameter (Def. Eq. 100)	-
ζ	Fraction of hydrogen in the inlet gas mixture	-

AFC	Alkaline Fuel Cell
CE	Carnot Engine
DCFC	Direct Carbon Fuel Cell
DMFC	Direct Methanol Fuel Cell
FC	Fuel Cell
FC-CE	Fuel Cell – Carnot Engine system
IDCFC	Indirect Carbon Fuel Cell
MCFC	Molten Carbonate Fuel Cell
PAFC	Phosphoric Acid Fuel Cell
PEMFC	Polymer Electrolyte Membrane Fuel Cell

RH	Relative Humidity
SOFC	Solid Oxide Fuel Cell
YSZ	Yttrium Stabilized Zirconium Oxide

REFERENCES

[1] Perry,M.L. & Fuller,T.F. A historical perspective of fuel cell technology in the 20th century. *Journal of the Electrochemical Society* **149** (2002) S59-S67.
[2] Hirschenhofer,J. H., Stauffer,D. B. and Engleman,R.R. *Fuel Cells, A Handbook* (5th edition). US Department of Energy, (2000).
[3] Appleby,A. J. & Foulkes, F. R. *Fuel Cell Handbook*. Van Nostrand Reinhold, New York :_ (1989).
[4] Reif, F. Berkeley Physics Course, *Part V: Statistical physics*. Mc Graw Hill, (1964).
[5] Massardo,A.F. & Lubelli, F., *Journal of Engineering for Gas Turbines and Power* **122** (2000) 27-35.
[6] Blomen, L.J.M.J. & Mugerwa,M.N. *Fuel Cell Systems*. Plenum, New York (1993).
[7] Au,S.F., Hemmes,K. & Woudstra, N. *J. of Power Scources*, accepted (2003).
[8] Park,S., Craciun, R., Vohs, J.M. & Gorte, R.J., *Journal of the Electrochemical Society* **146** (1999) 3603-3605.
[9] Lide, D.R. *CRC Handbook of chemistry and physics : a ready-reference book of chemical and physical data*, CRC Press (2001).
[10] Standaert, F.R.A.M., *Analytical Fuel Cell Modeling and Exergy Analysis of Fuel Cells*, Delft University of Technology ISBN 90.9012330-X, 1998.
[11] Suski.L., *Thermochimica Acta* **245** (1994 57-67.
[12] Bessett, N.F. & Wepfer, W.J, *Journal of Energy Resources Technology-Transactions of the Asme* **117** (1995) 307-317.
[13] Bistolfi,M., Malandrino,A. & Mancini,N., *Computers & Chemical Engineering* **20** (1996) S1487-S1491.
[14] Machielse, L.A.H. in: *Modeling of Batteries and Fuel Cells*. The Electrochemical Society Proceedings PV 91-10, 166-174. 1991.
[15] Hemmes,K., in: *Computer algebra applied in electrochemistry and fuel cell modeling*, The Electrochemical Society Proceedings PV 91-10 Modeling of Batteries and Fuel Cells. 1991. Phoenix, Az, USA.
[16] Standaert,F.R.A.M., Hemmes,K. & Woudstra,N. *Journal of Power Sources* **63** (1996) 221-234.
[17] Standaert, F.R.A.M., Hemmes,K. & Woudstra, N., *Journal of Power Sources* **70** (1998) 181-199.
[18] Au, S.F., Peelen,W.H.A., Standaert,F.R.A.M., Hemmes,K. & Uchida,I., *Journal of the Electrochemical Society* (2001).
[19] Browning, D.; Weston, M.; Lakeman, J. B.; Jones, P.; Cherry, M.; Irvine, J. T. S.; Corcoran, D. J., *Journal of New Materials for Electrochemical Systems* **5** (2002)25-30.
[20] Vollmar,H.E., Maier,C.U., Nolscher,C., Merklein,T. & Poppinger, M., *Journal of Power Sources* **86** (2000) 90-97.
[21] Bejan,A. & Tondeur,D., *Rev. Gen. Therm.* **37** (1986) 165-180.
[22] Bedeaux,D., Standaert,F., Hemmes,K. & Kjelstrup,S., *Journal of Non-Equilibrium Thermodynamics* **24**(1999) 242-259.

[23] Tondeur, D. & Kvaalen,E. *Ind.Eng.Chem.Res.* **26** (1986)26-50.
[24] Bedeaux,D. & Kjelstrup Ratke,S,. *J. Electrochem. Soc.* **143**(1996) 767-779.
[25] Jacobsen & Broers G.H.J. *J. Electrochem. Soc.* **124** (1977) 297.
[26] Bossel U.G. in: *Comparative Evaluation of the performance Potentials of Ten prominent SOFC Configurations.* The Electrochemical Society. PV 93-4 : Proceedings of the 3rd international symposium on SOFC's., 1993.
[27] Boersma,R.J., Sammes,N.M. & Fee,C.J,. *Solid State Ionics* **135**(2000) 493-502.
[28] Van Gool,W. Philips Res.Repts. **20**(1965)81.
[29] Hibino,T., Wang,S., Kakimoto,S. & Sano,M., *Electrochem. Solid State Lett.* **2** (1999) 317-319.
[30] Hibino,T., Hashimoto,A. & Kakimoto,S., *Journal of the Electrochemical Society.* **148** (2001) H1-H5.
[31] Hibino, T.; Hashimoto, A.; Inoue, T.; Tokuno, J.; Yoshida, S.; Sano, M., *Journal of the Electrochemical Society* **148** (2001) A544-A549.
[32] Asano,K. & Iwahara,H., *Journal of the Electrochemical Society.* **144** (1997) 3125-3129.
[33] Hemmes,K., Peelen,W.H.A. & deWit,J.H.W., *Electrochemical and Solid State Letters* **2** (1999) 103-106.
[34] Peelen,W.H.A., Hemmes,K. & Lindbergh,G., *Journal of the Electrochemical Society* **147** (2000) 2122-2125.
[35] Doyon,J.D., Gilbert,T., Davies,G. & Paetsch,L., *Journal of the Electrochemical Society* **134** (1987) 3035-3038.
[36] Terada,S., Nagashima,I., Higaki,K. & Ito,Y,. *Journal of Power Sources* **75** (1998) 223-229.
[37] Yang,S.C. & Bjornbom,P, *Electrochimica Acta* **37** (1992) 1831-1843.
[38] Bjornbom,P. & Yang,S.C., *Electrochimica Acta* **38** (1993) 2599-2609.
[39] Kimble,M.C. & White,R.E., *Journal of the Electrochemical Society* **139** (1992) 478-484.
[40] Kimble,M.C. & White,R.E. *Journal of the Electrochemical Society* **138** (1991) 3370-3382.
[41] Jo,J.H., Moon,S.K. & Yi,S.C., *Journal of Applied Electrochemistry* **30** (2000) 1023-1031.
[42] Jo,J.H. & Yi,S.C. A computational simulation of an alkaline fuel cell. *Journal of Power Sources* **84** (1999) 87-106.
[43] Baumann,A., Hauff,S. & Bolwin,K,. *Journal of Power Sources* **36** (1991) 185-199.
[44] Gurau,V., Liu,H.T. & Kakac,S, *Aiche Journal* **44** (1998) 2410-2422.
[45] Berning,T., Lu,D.M. & Djilali,N. *Journal of Power Sources* **106** (2002) 284-294.
[46] Okada,T. *Journal of New Materials for Electrochemical Systems* **4** (2001) 209-220.
[47] Springer,T.E., Zawodzinski,T.A. & Gottesfeld,S, *Journal of the Electrochemical Society* **138** (1991) 2334-2342.
[48] Eikerling,M. & Kornyshev,A.A. *Journal of Electroanalytical Chemistry* **453** (1998) 89-106.
[49] Gurau,V., Barbir,F. & Liu,H.T. *Journal of the Electrochemical Society* **147** (2000) 2468-2477.
[50] Janssen,G.J.M. *Journal of the Electrochemical Society* **148** (2001) A1313-A1323.
[51] Bernardi, D.M. & Verbrugge, M.W *Journal of the Electrochemical Society* **139** (1992) 2477-2491.
[52] Okada, T.; Xie, G.; Gorseth, O.; Kjelstrup, S.; Nakamura, N.; Arimura, T.. *Electrochimica Acta* **43** (1998) 3741-3747.
[53] Okada,T., Xie,G. & Meeg,M.. *Electrochimica Acta* **43** (1998) 2141-2155.

[54] Wohr, M.; Bolwin, K.; Schnurnberger, W.; Fischer, M.; Neubrand, W.; *International Journal of Hydrogen Energy* **23** (1998) 213-218.
[55] Yi,J.S. & Nguyen,T.V. *Journal of the Electrochemical Society* **146** (1999) 2000.
[56] Yi,J.S. & Nguyen,T.V. *Journal of the Electrochemical Society* **145** (1998) 1149-1159.
[57] Futerko,P. & Hsing,I.M. *Electrochimica Acta* **45** (2000) 1741-1751.
[58] Futerko,P. & Hsing,I.M. *Journal of the Electrochemical Society* **146** (1999) 2049-2053.
[59] Hsing,I.M. & Futerko,P. *Chemical Engineering Science* **55** (2000) 4209-4218.
[60] Springer,T.E., Wilson,M.S. & Gottesfeld,S. *Journal of the Electrochemical Society* **140** (1993) 3513-3526.
[61] Singh,D., Lu,D.M. & Djilali,N. *International Journal of Engineering Science* **37**(1999) 431-452.
[62] Sena,D.R., Ticianelli,E.A., Paganin,V.A. & Gonzalez,E.R. *Journal of Electroanalytical Chemistry* **477**(1999) 164-170.
[63] Kjelstrup Ratke,S., Vie P.J.S. & Bedeaux,D. in: *Surface chemistry and electrochemistry of membranes*, Ed. by Sorensen,T.S., Marcel Dekker,1999, pp. 483.
[64] Janssen,G.J.M. & Overvelde,M.L.J. *Journal of Power Sources* **101** (2001) 117-125.
[65] Kim,J., Lee,S.M., Srinivasan,S. & Chamberlin,C.E. *Journal of the Electrochemical Society* **142**(1995) 2670-2674.
[66] Squadrito,G., Maggio,G., Passalacqua,E., Lufrano,F. & Patti,A.. *Journal of Applied Electrochemistry* **29** (1999) 1449-1455.
[67] Amphlett, J. C.; Baumert, R. M.; Mann, R. F.; Peppley, B. A.; Roberge, P. R.; Harris, T. J. *Journal of the Electrochemical Society* **142** (1995) 1-8.
[68] Amphlett, J. C.; Baumert, R. M.; Mann, R. F.; Peppley, B. A.; Roberge, P. R.; Harris, T. J, *Journal of the Electrochemical Society* **142** (1995) 9-15.
[69] Amphlett, J. C.; Baumert, R. M.; Mann, R. F.; Peppley, B. A.; Roberge, P. R.; Rodrigues, A.. *Journal of Power Sources* **49** (1994) 349-356.
[70] Lee,J.H. & Lalk,T.R. *Journal of Power Sources* **73** (1998) 229-241.
[71] Lee,J.H., Lalk,T.R. & Appleby,A.J. *Journal of Power Sources* **70** (1998) 258-268.
[72] Eikerling,M., Kharkats,Y.I., Kornyshev,A.A. & Volfkovich,Y.M. *Journal of the Electrochemical Society* **145** (1998) 2684-2699.
[73] Bellows,R.J., Lin,M.Y., Arif,M., Thompson,A.K. & Jacobson,D.. *Journal of the Electrochemical Society* **146** (1999) 1099-1103.
[74] Buechi F.N., Huslage J. & Scherer G.G. Paul Scherrer *Scientific Report 1997* Vol V General Energy. Daum C. and Leuenberger J. Vol V, 48. 1998. Villigen, Paul Scherrer Institute.
[75] Thampan,T., Malhotra,S., Tang,H. & Datta,R. *Journal of the Electrochemical Society* **147**(2000) 3242-3250.
[76] Sone,Y., Ekdunge,P. & Simonsson,D. *Journal of the Electrochemical Society* **143** (1996) 1254-1259.
[77] Meyers,J.P. & Newman,J. *Journal of the Electrochemical Society* **149** (2002) A710-A717.
[78] Meyers,J.P. & Newman,J. *Journal of the Electrochemical Society* **149** (2002) A718-A728.
[79] Meyers,J.P. & Newman,J. *Journal of the Electrochemical Society* **149** (2002) A729-A735.
[80] Baxter,S.F., Battaglia,V.S. & White,R.E. *Journal of the Electrochemical Society* **146** (1999) 437-447.
[81] Kulikovsky,A.A. Gas dynamics in channels of a gas-feed direct methanol fuel cell: exact solutions. *Electrochemistry Communications* **3** (2001) 572-579.

[82] Kulikovsky,A.A. *Journal of Applied Electrochemistry* **30** (2000) 1005-1014.
[83] Argyropoulos,P., Scott,K. & Taama,W.M. *Journal of Applied Electrochemistry* **31** (2001) 13-24.
[84] Argyropoulos,P., Scott,K. & Taama,W.M. *Journal of Power Sources* **87** (2000) 153-161.
[85] Argyropoulos,P., Scott,K. & Taama, *Electrochimica Acta* **45** (2000)1983-1998.
[86] Argyropoulos,P., Scott,K. & Taama,W.M. *Chemical Engineering Journal* **73** (1999) 217-227.
[87] Argyropoulos,P., Scott,K. & Taama,W.M. *Chemical Engineering Journal* **73** (1999) 229-245.
[88] Argyropoulos,P., Scott,K. & Taama,W.M. *Journal of Applied Electrochemistry* **30** (2000) 899-913.
[89] Argyropoulos,P., Scott,K. & Taama,W.M. *Chemical Engineering Journal* **78** (2000) 29-41.
[90] Argyropoulos,P., Scott,K. & Taama,W.M. *Journal of Applied Electrochemistry* **29** (1999) 661-669.
[91] Argyropoulos,P., Scott,K. & Taama,W.M.. *Journal of Power Sources* **79** (1999) 184-198.
[92] Argyropoulos,P., Scott,K. & Taama,W.M. *Chemical Engineering & Technology* **23** (2000) 985-995.
[93] Argyropoulos,P., Scott,K. & Taama,W.M. *Journal of Applied Electrochemistry* **30** (2000) 899-913.
[94] Scott,K., Argyropoulos,P. & Sundmacher,K. *Journal of Electroanalytical Chemistry* **477** (1999) 97-110.
[95] Scott,K., Argyropoulos,P. & Taama,W.M. *Chemical Engineering Research & Design* **78** (2000) 881-888.
[96] Scott,K., Taama,W. & Cruickshank,J. *Journal of Power Sources* **65** (1997) 159-171.
[97] Scott,K., Argyropoulos,P. & Taama,W.M. *Chemical Engineering Research & Design* **78** (2000) 881-888.
[98] Scott,K., Taama,W. & Cruickshank,J. *Journal of Power Sources* **65** (1997) 159-171.
[99] Simoglou, A.; Argyropoulos, P.; Martin, E. B.; Scott, K.; Morris, A. J.; Taama, W. M. *Chemical Engineering Science* **56** (2001) 6761-6772.
[100] Sundmacher, K.; Schultz, T.; Zhou, S.; Scott, K.; Ginkel, M.; Gilles, E. D.*Chemical Engineering Science* **56** (2001) 333-341.
[101] Sundmacher,K. & Scott,K. *Chemical Engineering Science* **54** (1999) 2927-2936.
[102] Sundmacher, K.; Schultz, T.; Zhou, S.; Scott, K.; Ginkel, M.; Gilles, E. D.. *Chemical Engineering Science* **56** (2001) 333-341.
[103] Sundmacher,K. & Scott,K.. *Chemical Engineering Science* **54** (1999)2927-2936.
[104] Choudhury,S.R., Deshmukh,M.B. & Rengaswamy,R. *Journal of Power Sources* **112** (2002) 137-152.
[105] Yang,S.C. *Journal of the Electrochemical Society* **147** (2000) 71-77.
[106] Maggio,G. *Journal of Applied Electrochemistry* **29** (1999) 171-176.
[107] Cutlip,M.B., Yang,S.C. & Stonehart,P. *Electrochimica Acta* **36** (1991)547-553.
[108] Wolf,T.L. & Wilemski,G. Molten *Journal of the Electrochemical Society.* **130** (1983) 48-58.
[109] Aguiar,P., Chadwick,D. & Kershenbaum,L. *Chemical Engineering Science* **57**(2002) 1665-1677.
[110] Michaels J.N., Vayenas,C.G. & Hegedus,L.L *J.Electrochem.Soc.* **133**(3)(1986)522-525.

[111] Vayenas,C.G., Debenedetti,P.G., Yentekakis,I. & Hegedus,L.L. Cross-Flow, *I&EC Fundamentals* **24(3)** 1985 316-324.
[112] Dijkema,G.P.J. & Luteijn,C.P.W.M.P.C. *Chemical Engineering Communications* Gordon Breach Sci. Publ. Ltd **168**. 1998. , 111-125. [113] Vayenas,C.G. & Farr R.D.. *Science* **208** (1980) 593-594.
[114] Yentekakis,I.V. & Vayenas,C.G. *Journal of the Electrochemical Society* **136** (1989) 996-1002.
[115] Vayenas,C.G., Bebelis,S.I. & Kyriazis,C.C. *Chemtech* **21** (1991) 422-428.
[116] Neophytides,S. & Vayenas,C.G. *Journal of the Electrochemical Society* **137** (1990) 839-845.
[117] Alderucci,V., Antonucci,P.L., Maggio,G., Giordano,N. & Antonucci,V. *International Journal of Hydrogen Energy* **19** (1994) 369-376.
[118] McIlveen-Wright,D.R., Williams,B.C. & McMullan,J.T. *Renewable Energy* **19** (2000)223-228.
[119] Mathur,A., Bali,S., Balakrishnan,M., Perumal,R. & Batra,V.S. *International Journal of Energy Research* **23** (1999) 1177-1185.
[120] Kivisaari,T., Bjornbom,P. & Sylwan,C. *Journal of Power Sources* **104** (2002) 115-124.
[121] Hsu, P. C.; Foster, K. G.; Ford, T. D.; Wallman, P. H.; Watkins, B. E.; Pruneda, C. O.; Adamson, M. G. *Waste Management* **20** (2000) 363-368.
[122] Canning,K. *Pollution Engineering* **32** (2000) 11-12.
[123] Adamson, M. G.; Hsu, P. C.; Hipple, D. L.; Foster, K. G.; Hopper, R. W.; Ford, T. D. *High Temperature Material Processes* **2** (1998) 559-580.
[124] Nakagawa,N. & Ishida,M. *Industrial & Engineering Chemistry Research* **27** (1988) 1181-1185.
[125] Peelen,W.H.A., Hemmes,K. & De Wit,J.H.W. *High Temperature Material Processes* **2** (1998) 471-482.
[126] Cooper,J.F., Cherepy,N., Berry,G., Pasternak,A. & Surles,T. in: *Direct Carbon Conversion: Application to the Efficient Conversion of Fossil Fuels to Electricity*, Paper 50. 2000. Phoenix, the Electrochemical Society. Fall Meeting of the Electrochemical Society, Electrochemistry and Global Warming. 2000.
[127] Cooper,J.F., Cherepy,N., Upadhye,R., Pasternak,A. & Steinberg,M. in: *Direct Carbon Conversion: Review of Production and Electrochemical Conversion of Reactive Carbons. Economics and Potential Impact on the Carbon Cycle*. UCRL-ID-141818. 12-12-2000. CA, USA, Lawrence Livermore National Laboratory.
[128] Au, S.F. *Innovative High Temperature Fuel Cell Systems*. 2003. Delft University of Technology; ISBN 90-407-2375-3. 20.

5

Trace-Anion Catalysis of Outer-Sphere Heterogeneous Charge-Transfer Reactions

Zoltán Nagy

Argonne National Laboratory, Materials Science Division, Argonne, Illinois, 60439, USA

I. INTRODUCTION

That anions can affect the rate of metal redox reactions has been known for a long time. More often than not, this was caused by formation of complex ions in the solution, thereby completely changing the nature of the reaction, although other causes (electrostatic effects and surface ligand bridging) have also been considered. This general field has been summarized in a review by Weaver.[1] Much of the early work was carried out on mercury electrodes, usually with polarographic techniques, reviewed by de Levie.[2] The anion concentration and the ratio of anion to reactant concentration were so high in practically all these reports that bulk solution complex formation could not be ignored.

In contrast, the catalytic effect of trace anions is a relatively little-known and not-well-documented area of electrode kinetics in spite of the fact that it can cause several orders-of-magnitude change in the rate constant of an outer-sphere heterogeneous charge-transfer reaction. More important, this effect can be observed at such low anion concentrations (ppb level) that are below the analytical detection limit for most anions, making it rather difficult to carry out quantitative measurements. Standard purification techniques for the removal of these impurities from the source chemicals or their solutions has not been established, and the effectiveness of a purification procedure often can only be established by measuring the catalytic effect because of the above mentioned analytical difficulties.

II. SUMMARY OF OBSERVATIONS

The effect was probably first reported by Gerischer[3] in 1950. He observed an increase of the exchange current density of the ferrous/ferric redox reaction in sulfuric acid solution by chloride ions at concentration levels as low as 10^{-4} molar (~ 4 ppm).

Very few papers were published during the following fifty years on this subject, and Table 1 gives a summary of the reported observations. With the exception of the ferrous/ferric redox reaction system (see below) many of the reported observations were, at best, only

Table 1
Trace-Anion Catalytic Effects Reported in the Literature

Reaction	Electrode	Anion	Maximum concentration $\times 10^6$ (mol^{-1})	Rate increase	Ref.
Fe^{2+}/Fe^{3+}	Pt	Cl^-	100	1.2	3
Sb^{3+}/Sb^{5+}	Pt	I^-	2	200	4,5
Fe^{2+}/Fe^{3+}	Pt	Cl^-,Br^-,I^-	17	qual.	6
Fe^{2+}/Fe^{3+}	Au,Pt	Cl^-,Br^-,SO_4^{2-}	100	1,000	7
Fe^{2+}/Fe^{3+}	Au,Pd,Pt	Cl^-	500	250	8,9
V^{2+}/V^{3+}	Ag,Au,Pt	anions	traces	~10,000	10
Ru^{2+}/Ru^{3+}	Ag,Au	anions	traces	~1,000	10
Eu^{2+}/Eu^{3+}	Ag	anions	traces	~5,000	10
Sb/Sb^{3+}	Hg	halides,SCN^-	100	qual.	11
Fe^{2+}/Fe^{3+}	Au	Cl^-	400	5,000	12
Cr^{2+}/Cr^{3+}	Hg	N_3^-	1,620	~1,000	13
Fe^{2+}/Fe^{3+}	Au,Pt	anions	traces	5,000	14
Cu^+/Cu^{2+}	Cu	Cl^-	2,000	10	15
Ti^{3+}/Ti^{4+}	Au	Cl^-	100	qual.	16
$V^{3+}/V^{4+}/V^{5+}$	Au	Cl^-	100	qual.	16

semi-quantitative in nature. The type and concentration of the trace anions present were often not known. Even if known concentrations of anions were added to the test solution, the starting concentration (the naturally occurring impurity concentration) was usually not known. The electrochemical measurements were often such as to make the quantitative effect of the impurity on the rate constant difficult to determine.

Even with these uncertainties, it is quite clear, at least qualitatively, that the catalytic effect is considerable. Catalytic effects starts to manifest at very low concentrations, and the effect can produce many orders-of-magnitude changes in the apparent standard rate constant. The catalytic effect was observed for a variety of reactions, on a variety of electrode materials, caused by a variety of anions.

It has been generally recognized.[4,5] that the catalytic effect of these anions must be the result of their adsorption on the electrode surfaces, because the catalytic effects were observed at such low anion concentrations in the presence of often overwhelmingly large reactant concentrations that any complex formation between the anions and the reactant in the bulk solution must be negligible. Two possible explanations have been suggested early on[4,5] for the catalytic effect of the trace anions adsorbed on electrode surfaces: (i) electrostatic (double-layer) effects and (ii) the "ligand-bridging" mechanism, which is a well-known mechanism in homogeneous kinetics. While there is no definitive resolution of the problem, the electrostatic effect has been considered to be too small to account for the occasionally observed large differences in the rate constant.[4, 5] Consequently it cannot, by itself, be the full explanation, requiring the necessity to assume the ligand-bridging mechanism as at least a partial explanation.

Strong, albeit indirect, evidence for the bridging mechanism is the fact the presence of some excess chromous ions in the solution can eliminate the anion catalysis[10] (this chromous "gettering" will be discussed in more detail below) and this can only be explained by assuming that the chromous ions themselves are oxidized through a bridging mechanism and thereby removing the catalytic ions permanently from the electrode surface because of the strongly "substitutionally inert" nature of the resulting chromic complexes. By analogy, one could expect other, rather very similar, redox reactions also to proceed through the "ligand-bridging" mechanism. This mechanism is also supported by some preliminary theoretical calculations of the electronic coupling for different possible reaction paths.[15] It is worth mentioning some related "surface-species" catalysis of similar reactions that are not being caused by trace-anion impurities. Very strong catalysis of the Fe^{2+}/Fe^{3+}, Eu^{2+}/Eu^{3+}, and V^{2+}/V^{3+} reactions was observed by surface oxidation of glassy carbon and highly orientated pyrolytic graphite electrodes.[17,18] The increase of rate constants was found to be as high as two-to-three orders-of-magnitude, and the bridging mechanism was suggested as the explanation of the catalysis. On the other hand, the ligand-bridged pathway was questioned for some systems, based on kinetic data about the stability of the involved surface complexes arriving at maximum possible rate constants smaller than the measured ones.[10]

The best-documented case is the catalytic effect of chloride ions on the ferrous/ferric redox reaction studied in perchloric acid solutions on platinum and gold electrodes. Quantitative measurements were reported as a function of chloride concentration by more than one group, several completely different purification approaches were tried, and the "chloride-free" rate constant has been reliably determined. This system will be discussed in detail in the following section and it will be used to demonstrate the experimental difficulties associated with these studies.

III. THE FERROUS/FERRIC REACTION

The kinetics of the ferrous/ferric redox electrode reaction has been investigated by many workers as a simple, uncomplicated charge-transfer reaction, which seems ideal for testing experimental techniques and charge-transfer theories. The kinetics of the reaction has been measured using numerous types of electrodes and solutions, and the results have been summarized in several reviews.[19-21] The apparent standard rate constant was typically between 10^{-2} and 10^{-4} cm s^{-1}, which could be considered a reasonable range of values considering the variety of electrodes and solutions used in the experiments. In spite of its seeming simplicity, many complications were encountered. In solutions containing sulfate or chloride ions, complex formation with both ferrous and ferric ions occurs; consequently, the reaction is comprised of a series of elementary steps with the charge transfer not necessarily occurring to/from a simple hydrated cation. To avoid complex formation, many workers have used perchlorate solutions because in this medium complex formation has never been shown to occur conclusively.[22,23] Furthermore, the specific adsorption of perchlorate ions is also very weak or nonexistent on noble metal electrodes.[24,25] Therefore, the reaction indeed can be expected to be an uncomplicated single-step, outer-sphere charge transfer:

$$Fe(H_2O)_6^{3+} + e = Fe(H_2O)_6^{2+} \tag{1}$$

Most measurements were carried out with platinum electrodes and the apparent standard rate constant was found to be between 10^{-2} and 10^{-3} cm s^{-1} in this solution, which could be considered a reasonable reproducibility for a solid electrode surface.

Some complications are encountered, however, even in perchloric acid due largely to changes of the electrocatalytic properties of the electrode surface caused by adsorbed species. For example, on platinum surfaces oxygen adsorption (or oxide formation) can start to occur at potentials as low as 0.75 V,[26-29] that is uncomfortably close to the standard potential of the ferrous/ferric reaction of 0.771 V,[21] especially for anodic oxidation. Oxide formation on gold occurs at much more positive potentials,[30-32] removing this difficulty at the price of possible slight adsorption of the perchlorate anions.[25]

However, in none of these cases was the trace-anion content of the test solution measured, and, although the normal purification procedures customarily used in electrode kinetics were generally followed, special efforts were not made to eliminate the trace anions completely from the solution. As will be shown later, these early results can be explained by the presence of ppm order-of-magnitude or less chloride content in the test solutions. This should not be surprising, since it is difficult even today to obtain chemicals of such purity as to produce solutions with higher purity without extra purification.

1. Gerischer

The accelerating effect of chloride was first demonstrated on platinum electrode in sulfuric acid electrolyte by Gerischer[3] in 1950. He found a 20% increase of the exchange current density of the reaction when chloride was added to produce a 10^{-4} molar (~ 4 ppm) solution. However, the chloride content of the original solution was not known, and it is not unreasonable to assume that it was at least the same order-of-magnitude as the addition.

2. Johnson and Resnick

It was not until the work of Johnson and Resnick[6] in 1977 that it was realized that very small traces of chloride could have a considerable effect on the reaction rate. They applied the technique of hydrodynamic voltammetry using a rotating platinum disc electrode in perchloric acid solutions. They concluded that "the electrocatalysis of Fe(III) reduction in $HClO_4$ solutions results from adsorption, at a mass transport limit rate, of Cl^- present in the solution as impurity." The catalytic effect was observed even without addition of any chloride to the test solution, which was estimated to be less than 10^{-6} molar (< 0.04 ppm) in chloride. This work also demonstrated the difficulties with platinum oxide formation overlapping the investigated reaction. At some potentials the measured current included both reactions, and the redox reaction (and anion adsorption) could occur either on platinum or on oxide, complicating the interpretation of the results. The addition of bromide and iodide were also investigated and the catalytic effect was found to decrease as $Cl^- > Br^- > I^-$.

3. Weber et al.

The effect was first measured quantitatively by Weber et al.[7] in 1978 on platinum and gold electrodes in perchloric acid. They have applied a rigorous purification procedure to produce "impurity-free" solutions of ferric perchlorate in perchloric acid. The purification consisted of adsorbing the impurities at a large surface area platinum sponge electrode at positive potentials, followed by cleaning the electrode in another solution by negative polarization. This procedure was repeated until a somewhat arbitrary purity criterion was achieved. Hydrodynamic voltammetric curves were measured with rotating disc electrodes, and the solution was considered "impurity-free" when the voltammetric curve measured at a stationary electrode coincided with one measured at high rotation rate. The justification of this criterion was the observation of Johnson and Resnick,[6] which they confirmed, that the catalytic effect of small impurities is mass-transport controlled, and it would take a much longer equilibration time to obtain the same voltammogram at stationary electrodes than at a high rotation rate if impurities were present. Subsequently, they measured current-potential curves by observing the stationary currents at a series of potentials on rotating disc electrodes. The electrode was negatively polarized before each measurement "so that adsorption of remaining impurities in the solution was eliminated as much as possible at the beginning of the measurement." The apparent standard rate constant for the ferrous/ferric redox reaction in the "impurity-free" solutions was found to be 9×10^{-6} and 5×10^{-5} cm s^{-1} for platinum and gold electrodes, respectively. These values are much smaller (in the extreme case, three orders-of-magnitude smaller) than the values typically reported for this reaction in the past.

They have also investigated quantitatively the catalytic effect of chloride by additions to their "impurity-free" solutions. A linear relationship was found between the logarithm of the apparent standard rate constant and the logarithm of the chloride concentration over several orders-of-magnitude (between 10^{-1} and 10^{-4} cm s^{-1}, and 10^{-4} and 10^{-7} molar), and it was found that even at a low chloride concentration of 10^{-6} molar (~ 0.04 ppm) the rate constant on both platinum and gold electrodes was still about two orders-of-magnitude larger than that in "impurity free" solutions (a comparison of some of these results to other observations will be given later in Fig. 3). The logarithm of the rate constant was also found to be linearly related to the estimated surface coverage by chloride (based on radiotracer measurements).

They also measured the effect of bromide and sulfate ions and found that "bromides increased the rate of the ferrous/ferric reaction almost to the same degree as chlorides; sulfates, however, were less effective."

4. Smith and Wadden

The catalytic effect of chloride ions was also confirmed by Smith and Wadden.[8,9] They used a simple model to estimate the "chloride-free" rate constant from electrode kinetic measurements combined with surface coverages determined by radiotracer technique. Assuming that the measured rate is a linear combination of the catalyzed rate occurring on the covered area of the surface and the uncatalyzed rate occurring on the uncovered area of the surface, they estimated the "chloride-free" apparent standard rate constant to be 2.6×10^{-5} cm s^{-1} on platinum electrode in perchloric acid.

5. Nagy et al.

The purification technique of Weber et al.[7] was also used by Nagy et al.[14] to obtain the "impurity-free" rate constant in perchloric acid solutions, with the only difference that the rate was measured after every purification (using the potential-step relaxation technique) and the solution was considered "impurity-free" when the rate remained constant after repeated purification cycles. The apparent standard rate constant was found to be 1.3×10^{-5} and 3.3×10^{-5} cm s^{-1}, on platinum and gold electrodes, respectively.

6. Hung and Nagy

Hung and Nagy[12] attempted to use techniques other than rigorous purification of the solution to obtain "impurity-free" rate constants. The experiments were carried out in perchloric acid solutions using gold electrodes to avoid the complications resulting from platinum oxidation (see above). The solution purification of Weber et al.[7] is very time consuming, consisting of several cycles of adsorption of impurities on the platinum sponge and desorption and cleaning of the electrode in a separate solution. Furthermore, the question always remains whether or not new chloride will be introduced in time from the perchlorate electrolyte. Perchlorate has long been considered a very stable ion, but recent results indicate that it can be cathodically reduced under certain conditions;[24, 33-36] consequently, trace amounts of chloride ions, sufficient to increase the measured rate constant, can conceivably be generated in the cell. Therefore, it seemed desirable to develop methods for the measurement of kinetic data free of anion effects in solution containing traces of anions and using an *in situ* cleaning of the measuring electrode surface immediately before the kinetic measurements. Two such methods were examined. In the first method the anions were desorbed from the surface prior to measurements, and in the second method the anions were "gettered" by chromous ions. The test solutions were prepared from the highest purity chemicals available commercially, and the chloride content, which was the main anionic impurity, was determined using an ultra-sensitive specific-ion electrode. The cleanest solution prepared this way was found to contain 3×10^{-6} mol l^{-1} (~ 0.1 ppm) chloride.

In the desorption method the measuring electrode was potentiostatted at negative potentials to desorb the anions from the surface followed by return to the equilibrium potential and a potential-step relaxation measurement to determine the kinetics of the ferrous/ferric reaction. A typical experiment is shown in Fig. 1. Increasingly negative desorption potentials were applied and the rate constant decreased with negative potentials, leveling off before the potential of hydrogen evolution was reached. Assuming that at the most negative potentials practically all of the anions were desorbed from the surface, the "impurity-free" apparent standard rate constant was found to be 4×10^{-5} cm s^{-1}.

In the "chromous-gettering" method, anions were removed from the surface before and during the kinetic measurement by continuous oxidation of chromous ions added in small concentration to the test solution. This method was originated by the work Weaver et al.,[10,37] and it was based on the observation that addition of chromous ions to the test solution caused a several orders-of-magnitude decrease of the

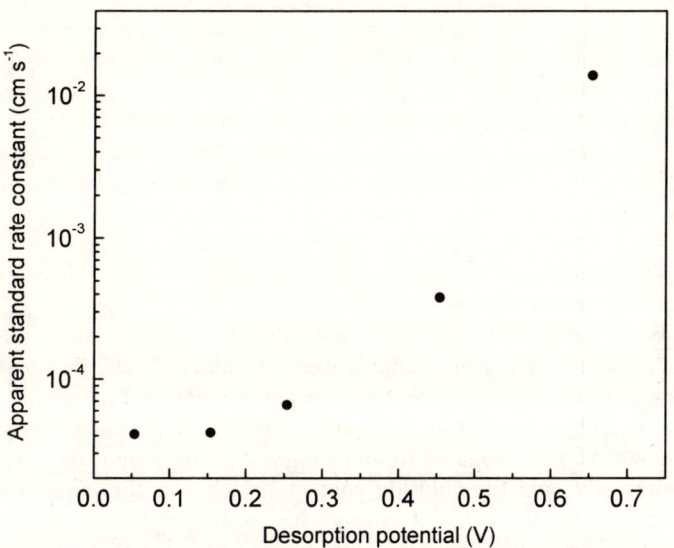

Figure 1. Dependence of the rate constant on the desorption potential. Chloride ion concentration: 1.5×10^{-5} mol l^{-1}.

rate constant of several redox reactions. Weaver, et al. have suggested that the ligand-bridged path provides a much faster rate for some redox reactions than the outer-sphere reaction path of a hydrated metal ion, and that normally measured rate constants of redox systems are influenced by trace amounts of anions adsorbed on the electrode surface. The chromous ions are also preferentially oxidized through a ligand-bridged mechanism, and, since the resulting chromic complex is substitutionally inert,[38] this process can thereby remove and permanently "getter" the adsorbed catalytic anions from the electrode surface. This forces the other redox reactions to occur predominantly on a "clean" surface.

Hung and Nagy[12] used various chromous and chloride concentrations and the kinetic measurements were carried out using the potential-step relaxation technique. The results are summarized in Figure 2, plotting the logarithm of the apparent standard rate constant of the ferrous/ferric reaction as a function of the logarithm of the

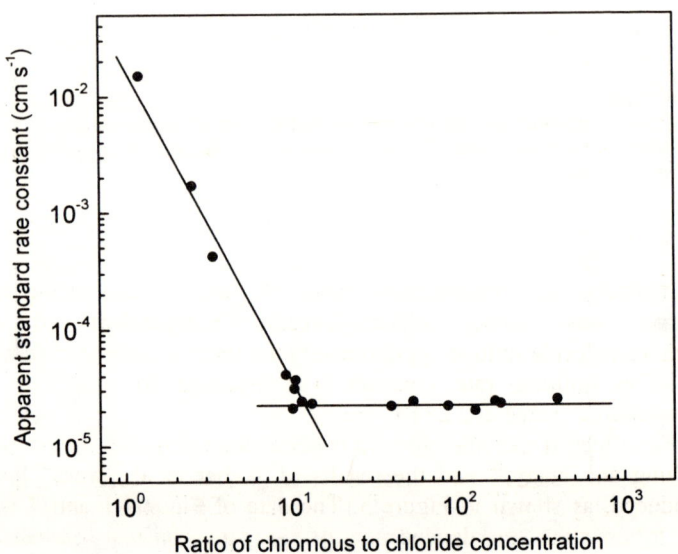

Figure 2. Dependence of the rate constant on the ratio of the concentration of chromous ions to that of the chloride ions.

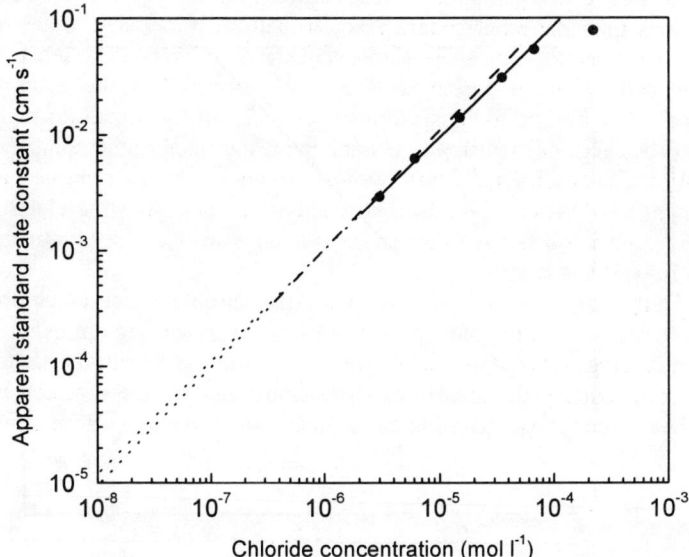

Figure 3. Dependence of the rate constant on the chloride ion concentration. Points and solid line: Hung and Nagy.[12] Dashed line: Weber et al.[7] Dotted lines: extrapolations.

chromous/chloride concentration ratio. A linearly decreasing rate constant was found which became independent of the chromous/chloride ratio at approximately ten times excess of chromous ions. The limiting rate constant was assumed to represent the "impurity-free" value at 2.2×10^{-5} cm s^{-1}.

The effect of variable chloride concentration was also investigated by Hung and Nagy,[12] and the results of Weber et al.[7] were closely reproduced, as shown in Figure 3. The data of Figures 1 and 3 were used to estimate the dependence of the rate constant on surface coverage by adsorbed chloride, utilizing the radiotracer adsorption results of Horanyi et al.[39] (Figure 4). Linear dependence of the logarithm of the rate constant on the coverage was found, confirming the results of Weber et al.,[7] and an additional estimate of the "impurity-

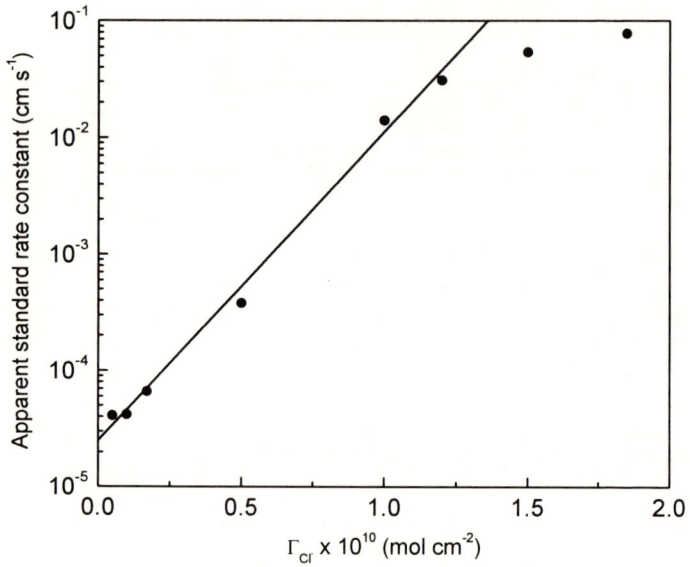

Figure 4. Dependence of the rate constant on the surface excess of chloride ions, for bulk chloride concentration of 1.5×10^{-5} mol l^{-1} and larger at various potentials.

free" rate constant could be obtained by extrapolating to zero coverage as 2.5×10^{-5} cm s^{-1}.

7. Discussion

The chloride catalysis of the ferrous/ferric reaction has been thoroughly investigated, by a number of researchers, using different purification and kinetic measuring techniques, as summarized above. The agreement amongst the different researches is indeed good as illustrated in Tables 2 and 3.

Table 2
"Impurity-Free" Apparent Standard Rate Constant of the Ferrous/Ferric Reaction on Platinum Electrodes in Perchloric Acid

Technique	$k^o \times 10^5$ (cm s^{-1})	Reference
Solution Purification	0.9	7
Radiotracer Coverage	2.6	8, 9
Solution Purification	1.3	14

Considering that the experiments were carried out with polycrystalline electrodes, with their source, purity, and surface-treatment history largely unknown, the agreement for the "impurity-free" rate constants can probably be even considered excellent. What is even more reassuring is that the procedures used to achieve "impurity-free" results are drastically different from each other, every one of the four techniques used is based on totally different physical principles, and while each of them may be questioned to a certain degree, it is very unlikely that they would result in similar errors. Consequently, it is reasonable to state that the apparent standard rate constant has been reliably established as $2\pm1 \times 10^{-5}$ cm s^{-1} for platinum electrodes and as $4\pm2 \times 10^{-5}$ cm s^{-1} for gold electrodes in perchloric acid solutions. These values are orders-of-magnitude smaller than the historically reported values between 10^{-2} and 10^{-4} cm s^{-1}.[19-21]

The change of the rate constant with chloride concentration has also been well reproduced, albeit by only two groups (Figure 3). This plot can also be used to estimate the degree of cleanliness required to

Table 3
"Impurity-Free" Apparent Standard Rate Constant of the Ferrous/Ferric Reaction on Gold Electrodes in Perchloric Acid

Technique	$k^o \times 10^5$ (cm s^{-1})	Reference
Solution purification	5.0	7
Solution purification	3.3	14
Desorption	4.0	12
Chromous gettering	2.2	12
Radiotracer coverage	2.5	12

produce "impurity-free" results. Although a long (and maybe questionable) extrapolation is needed, one can get at least an order-of-magnitude estimate of 5×10^{-8} mol l^{-1} (~2 ppb) or lower chloride content that has to be achieved to avoid catalysis by the chloride ions. Similarly, one can estimate the chloride content of the solutions used in earlier experiments that produced much larger rate conctants.[19-21] This turns out to be in the range of ppm concentrations of chloride ions, which is not surprising, since it is difficult even today to obtain chemicals of such purity as to produce solutions with higher purity without extra purification.

This leads us to the question of the experimental difficulties to ensure that the measured rate constant is indeed an "impurity-free" value, and at what impurity concentration is that value achieved. I assume that it is obvious that "impurity-free" value does not mean that the concentration of the impurity is absolute zero, rather it means that the concentration of the impurity is sufficiently low as not to cause any measurable catalysis. One of the major problems is that the analytical techniques available are not sensitive enough to determine the impurity concentrations even at levels where catalysis occurs. (The analytical problem is made especially difficult by the presence of usually large supporting electrolyte concentrations in the test solutions.) So one can purify and purify, and measure lower and lower rate constants, but the chemical analysis will just report "less than" some value even though the rate constant continues to decrease with repeated purifications. The question then remains as follows: if further purifications do not change the rate constant any more, did we reach the "impurity-free" value, or did we just reach the limitation of the given purification process to remove any more impurity? There is no easy answer. In the case of the ferrous/ferric system, a reassuring answer is that four different kind of "purifications" (either removing the impurity from the bulk solution, or at least from the surface of the electrode) gave very closely agreeing rate constant values. This, however, required a considerable effort. And even this provided only a rate constant, but the limiting concentration of the impurity where it does not cause catalysis any more still remained unanswered, at least quantitatively.

IV. CONCLUDING REMARKS

The results summarized above, and especially the detailed discussion of the ferrous/ferric redox reaction case, gives a compelling picture that trace-anion catalysis of outer-sphere charge-transfer reactions may be a very general phenomenon. It is reasonable to conclude that it may be present in many other systems even though it may not have been noticed in prior measurements. The effect may not be easily detectable since removal of the catalytic anions to concentrations below parts-per-billion level may be required in some cases, in solutions where this level of concentration cannot even be detected by available analytical techniques. These difficulties explain the scarcity of reported results (Table 1) and the fact that most results are only qualitative. The catalytic effect is, more often than not, considered to be caused by a ligand-bridging charge-transfer pathway through the anions adsorbed on the electrode surface, but this mechanism has not been proven conclusively.

While we have a reasonably reliable "impurity-free" rate constant for the ferrous/ferric reaction, this value is still not known for the other reactions listed in Table 1, and they indeed may be orders-of-magnitude smaller than typically reported in the literature. This may be the case for a number of other redox reactions, for which the effect has not yet been observed. Consequently, the reported rate constant values for outer-sphere charge-transfer reactions may, in general, be suspect and probably too high. This is especially important to realize when the measured rates are compared to those predicted by charge-transfer theories.

ACKNOWLEDGMENT

This work at Argonne National Laboratory was supported by the U.S. Department of Energy, Basic Energy Sciences-Materials Sciences, under Contract #W-31-109-ENG-38.

REFERENCES

[1] M. J. Weaver in *Comprehensive Chemical Kinetics*, Ed. byR.G. Compton, Vol. 27, Elsevier, NY, 1987, 1.
[2] R. de Levie, *J. Electrochem. Soc.* **118** (1971) 185C.
[3] H. Gerischer, *Z. Elektrochem.* **54** (1950) 362.
[4] R. J. Davenport, D. C. Johnson, *Anal. Chem.* **45** (1973) 1755.
[5] L. R. Taylor, B.A. Parkinson, D.C. Johnson, *Electrocim. Acta* **20** (1975) 1005.
[6] D. C. Johnson, E.W. Resnick, *Anal. Chem.* **49** (1977) 1913.
[7] J. Weber, Z. Samec, V. Marecek, *J. Electroanal. Chem.* **89** (1978) 271.
[8] J. T. Wadden, MS Thesis, Memorial University of Newfoundland, Canada (1978).
[9] F. R. Smith, J. T. Wadden, *Prepr. Can. Symp. Catal.* **6** (1979) 263.
[10] S. W. Barr, K. L. Guyer, M. J. Weaver, *J. Electroanal. Chem.* **111** (1980) 41.
[11] H. Verplaetse, P. Kiekens, E. Temmerman, F. Verbeek, *Talanta* **28** (1981) 431.
[12] N. C. Hung, Z. Nagy, *J. Electrochem. Soc.* **134** (1987) 2215.
[13] P. K. Wrona, *Polish J. Chem.* **65** (1991) 687.
[14] Z. Nagy, L. A. Curtiss, N. C. Hung, D.J. Zurawski, R. M. Yonco, *J. Electroanal. Chem.* **325** (1992) 313.
[15] Z. Nagy, J. P. Blaudeau, N. C. Hung, L.A. Curtiss, D.J. Zurawski, *J. Electrochem. Soc.* **142** (1995) L87.
[16] N. C. Hung, Z. Nagy, unpublished results.
[17] C. A. McDermott, K. R. Kneten, R. L. McCreary, *J. Electrochem. Soc.* **140** (1993) 2593.
[18] P. Chen, M. A. Fryling, R. L. McCreary, *Anal. Chem,* **67** (1995) 3115.
[19] N. Tanaka, R. Tamamushi, *Electrochim. Acta* **9** (1964) 963.
[20] R. Tamamushi, *Kinetic Parameters of Electrode Reactions of Metallic Compounds*, Butterworths, London, 1975.
[21] K. E. Heusler, in *Encyclopedia of Electrochemistry of the Elements*, Ed. by A.J. Bard, Vol. IXA, Marcel Dekker, New York,1982, 229.
[22] M. R. Rosenthal, *J. Chem. Ed.* **50** (1973) 331.
[23] L. Johansson, *Coord. Chem. Rev.* **12** (1974) 241.
[24] S. Ya. Vasina, O. A. Petrii, *Soviet Electrochem.* **6** (1970) 231.
[25] J. Clavilier, C. N. Van Huong, *J. Electroanal. Chem.* **80** (1977) 101.
[26] W. Boeld, M. Breiter, *Electrochim. Acta* **5** (1961) 145.
[27] M. W. Breiter, *Electrochim. Acta* **8** (1963) 447.
[28] M.W. Breiter, *Electrochim. Acta,* **8** (1963) 925.
[29] S. Gilman, in *Electroanalytical Chemistry,* Ed. by A.J. Bard, Vol. 2, Marcel Dekker, New York,1967, 111.
[30] D. A. J. Rand, R. Woods, *J. Electroanal. Chem.* **35** (1972) 209.
[31] T. Takamura, Y. Sato, K. Takamura, *J. Electroanal. Chem.* **41** (1973) 31.
[32] C. M. Ferro, A. J. Calandra, A. J. Arvia, *J. Electroanal. Chem.* **55** (1974) 291.
[33] G. Horanyi, G. Vertes, *J. Electroanal. Chem.* **64** (1975) 252.
[34] F. Colom, M. J. Gonzalez-Tejera, *J. Electroanal. Chem.* **190** (1985) 243.
[35] M. Sanchez Cruz, M. J. Gonzalez-Tejera, M. C. Villamanan, *Electrochim. Acta* **30** (1985) 1563.
[36] G. M. Brown, *J. Electroanal. Chem.,* **198** (1986) 319.
[37] K.L. Guyer, S.W. Barr, R.J. Cave, M.J. Weaver, in *Electrode Processes 1979,* Ed. by S. Bruckenstein, J.D.E. McIntyre, B. Miller, E. Yeager, The Electrochemical Society Softbound Proceedings Series, Vol. 80-3, Pennington, NJ, 1980, p. 390.

[38] H. Taube, *Electron Transfer Reactions of Complex Ions in Solution,* Academic Press, New York, 1970, p. 4.
[39] G. Horanyi, E. M. Rizmayer, P. Joo, *J. Electroanal. Chem.* **152** (1983) 211.

6

Thermodynamic and Transport Properties of Bridging Electrolyte-Water Systems

Maurice Abraham and Marie-Christine Abraham

Departément de Chimie, Université de Montréal, Montréal, Canada

I. INTRODUCTION

For a long time, most efforts in the physical chemistry of solutions were made to obtain information on the structure of aqueous electrolytic solutions. But, at the same time as our knowledge of these solutions was expanding, especially regarding dilute solutions, another field of research attracted attention, the physical chemistry of molten salts, with increasing interest in new technologies. Due to various reasons of scientific, technological and even historical nature, most investigations in these two fields of research were made almost independently.

According to Braunstein classification,[1,2] between dilute aqueous solutions and molten salts lie the hydrates, with complete or incomplete water shells around the ions, covering a very large water mole fraction range, from about 0 to 0.9. Obviously, investigations on hydrates are very important for the knowledge of transition properties between dilute aqueous solutions and molten electrolytes. As a matter of fact, these electrolyte-water systems became gradually the object of intense attention and their study is now seriously expanding as much for technical applications as for theoretical reasons. Hydrates are used, or planned to be, in technologies such as ore leaching and extraction processes, waste-water treatment, chemical and electrochemical manufacturing, absorption-type

Modern Aspects of Electrochemistry, Number 37, Edited by Ralph E. White *et al.* Kluwer Academic/Plenum Publishers, New York, 2004.

refrigeration machines... Due to concern about pollution and security problems, areas of technological interest include energy storage and generation. For example, solar energy storage, exploitation of geothermal energy sources, molten salts based fluids in nuclear reactors which could contain more or less water, deliberately introduced or not. The knowledge of thermodynamic and transport properties, for example water vapor pressure, heat of vaporization, viscosity, electrical conductance..., as well as the influence of water concentration, temperature... on these properties, over the ranges of anticipated operating conditions, are essential in the design and operation of technical systems in which they are utilized.

With regard to scientific interest, the suggestion was made, now and then,[1-8] that more progress in the understanding of very concentrated aqueous solutions could come from the consideration of solutions obtained by adding water to fused electrolytes rather than concentrating dilute aqueous solutions. Reciprocally, one would expect the structure of solutions where water plays the role of the solute, its mole fraction being less than about 0.5, be akin to that of the anhydrous molten electrolytes so that in the limit of vanishing water mole fraction the properties of the solutions would tend to those of the anhydrous electrolytes. From this point of view, any theoretical bringing-in regarding those solutions could contribute to more progress concerning the anhydrous electrolytes themselves.

Recently, electrolyte-water systems were investigated in the liquid phase over practically the whole water concentration range. Since these systems bridge the gap between anhydrous electrolytes and dilute aqueous solutions, they are designated by the expression "bridging electrolyte-water systems", the electrolyte being a single one or a mixture of several components. Bridging systems lie at the heart of the present chapter. Electrolyte-water systems which do not cover the entire concentration range will also be examined or taken into account inasmuch as they cover a sufficiently large water mole fraction region, starting especially near the anhydrous electrolyte, so that they can give information on the transition of properties between those at the two extremes of the concentration scale.

It now appears that experimental and theoretical investigations on bridging electrolyte-water systems have produced results with regard to

II. THERMODYNAMIC PROPERTIES

One important aim of the thermodynamics of electrolyte-water systems is to obtain information on the water activity a_w. The values of a_w in the liquid system are generally obtained from water vapor pressure measurements by the equation

$$f_w = f_w^o \, a_w \tag{1}$$

where f_w^o is the fugacity of pure water and f_w, the fugacity of the water vapor in equilibrium with the electrolyte-water system.

The water activity a_w is related to the water mole fraction on an unionized basis x_w by

$$a_w = \gamma_w \, x_w \tag{2}$$

where γ_w is the water activity coefficient.

In the case of a mixed electrolyte, x_w is defined by

$$x_w = \frac{n_w}{n_w + \sum_i n_{ei}} \tag{3}$$

with n_w, the number of moles of water and n_{ei}, the number of moles of the electrolyte i.

The corresponding mole fraction of the electrolyte x_e is given by

$$x_e = \frac{\sum_i n_{ei}}{n_w + \sum_i n_{ei}} \tag{4}$$

The composition of an anhydrous mixed electrolyte is expressed by

$$X_{ei} = \frac{n_{ei}}{\sum_i n_{ei}} \tag{5}$$

where X_{ei} is the mole fraction of the electrolyte i in the anhydrous melt.

The first ever determinations of a_w values for a salt-water system over the whole water mole fraction range were made on the system [0.533 $AgNO_3$–0.467 $TlNO_3$ + H_2O], at 372 K, by Trudelle, Abraham and Sangster.[7] In this abbreviated notation, the number before a chemical formula represents X_{ei}. Figure 1 gives the Raoult's law plot of the data; the system shows positive deviations over almost all water concentration range. This was unusual behavior since the vast majority of electrolyte-water systems previously studied, though over limited water concentration ranges, showed strong negative deviations,[2,9] except for such salts as $AgNO_3$,[10] $RbNO_3$,[11] and $CsNO_3$[11] showing slight positive deviations.

From further investigations on the bridging systems [$AgNO_3$–$TlNO_3$–$M(NO_3)_2$ + H_2O],[12,13] where X_{AgNO_3} / X_{TlNO_3} is fixed at 1.06, $X_{M(NO_3)_2}$ varying from 0 to 0.125, with M = Cd and Ca, at 372 K, it has been established that three cases of deviations from Raoult's law may appear, as il-

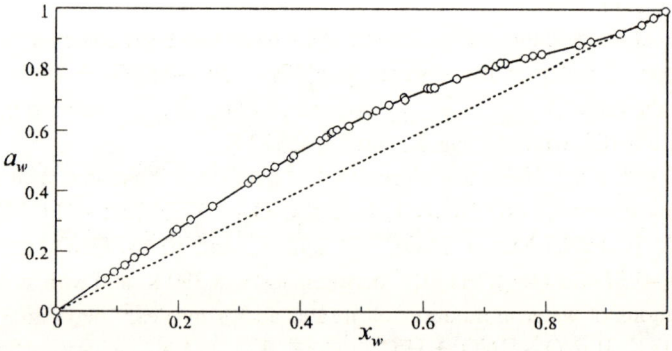

Figure 1. Water activity a_w as function of water mole fraction x_w for the [0.533 $AgNO_3$–0.467$TlNO_3$ + H_2O] system, at 372 K. Dashed line: Raoult's law. (From Ref. 7, with kind permission from Canadian Journal of Chemistry).

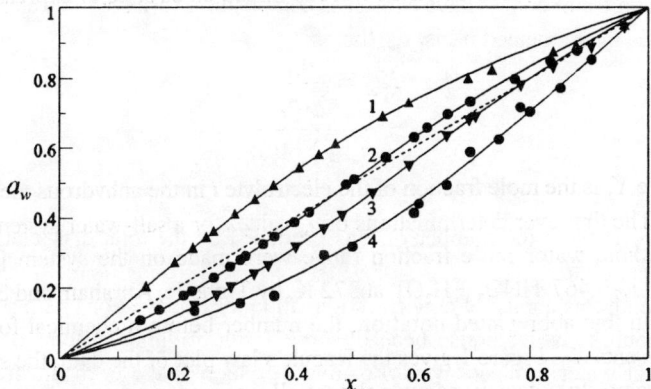

Figure 2. Water activity a_w as function of water mole fraction x_w. 1: [0.515AgNO$_3$–0.485 TlNO$_3$ + H$_2$O],[12] at 372 K; $r = 0.49$, $\epsilon = 1.33$. 2: [0.476AgNO$_3$–0.449TlNO$_3$–0.075 Cd(NO$_3$)$_2$ + H$_2$O],[12] at 372 K; $r = 0.72$, $\epsilon = 2.29$. 3: [0.451AgNO$_3$–0.424TlNO$_3$–0.125 Cd(NO$_3$)$_2$ + H$_2$O],[12] at 372 K; $r = 0.88$, $\epsilon = 2.69$. 4: [0.467TlNO$_3$–0.214CsNO$_3$–0.319 Cd(NO$_3$)$_2$ + H$_2$O],[19] at 383 K; $r = 1.42$, $\epsilon = 2.35$. ϵ in kJ mol^{-1}. Dashed line: Raoult's law. Solid line: calculated curve from Eq. (7).

lustrated in Fig. 2: positive deviation from salt to dilute aqueous solutions, negative deviation from salt to dilute aqueous solutions and S-shaped curve with a cross-over point for which $a_w = x_w$ and where the deviation changes sign.

Positive deviations from Raoult's law have also been observed with solutions of water in the systems [AgNO$_3$–TlNO$_3$–MNO$_3$ + H$_2$O][14–16] where the ratio X_{AgNO_3} / X_{TlNO_3} is fixed at 1.06, X_{MNO_3} varying from 0.025 to 0.100, with M = Na, K, and Cs, at 372 K.

Negative deviations from Raoult's law have been illustrated with the bridging systems [0.667 NH$_4$NO$_3$–0.258LiNO$_3$–0.075NH$_4$Cl + H$_2$O],[17] at 366 K, [C$_2$H$_5$NH$_3$NO$_3$ + H$_2$O],[18] at 298 K, and [0.467TlNO$_3$–0.214 CsNO$_3$–0.319Cd(NO$_3$)$_2$ +H$_2$O],[19] between 360 and 390 K, and with several solutions of water in mixtures of nitrates and/or nitrites,[20] between 383 and 403 K: [LiNO$_3$– KNO$_3$ + H$_2$O], where X_{LiNO_3} / X_{KNO_3} varies from 1.00 to 2.33, [0.52 LiNO$_3$–0.31KNO$_2$–0.17NaNO$_2$ + H$_2$O], [0.41 LiNO$_3$–0.41KNO$_3$– 0.18 MNO$_3$ + H$_2$O], with M = Na and Cs, and [0.472 LiNO$_3$–0.472KNO$_3$– 0.056M(NO$_3$)$_2$ + H$_2$O], with M = Ca and Mg.

Some authors[20-31] have chosen as composition variable the water mole fraction on an ionized basis, y_w, that is

$$y_w = \frac{n_w}{n_w + \sum_i \nu_i n_{ei}} \tag{6}$$

in which ν_i is the number of ions produced by the total dissociation of the electrolyte i.

With this choice of concentration, all kinds of deviations with respect to Raoult's law have also been observed with bridging systems or solutions of water in electrolytes. Positive deviations have been illustrated[21, 22, 26, 27, 29, 30] with the systems: [0.533AgNO$_3$–0.467TlNO$_3$ + H$_2$O],[7] at 372 K, [CH$_3$CO$_2$Na + H$_2$O],[26] at 590 K, [C$_6$H$_5$CO$_2$Na + H$_2$O],[26] at 590 K, [KNO$_3$ + H$_2$O],[27] between 425 and 492 K, [NaNO$_3$ + H$_2$O],[29] at 492 K, [0.600AgNO$_3$–0.400RbNO$_3$ + H$_2$O],[29] at 492 K, [0.550NaNO$_3$–0.450CsNO$_3$ + H$_2$O],[29] at 492 K, [AgNO$_3$–CsNO$_3$ + H$_2$O],[30] with X_{AgNO_3} = 0, 0.400, 0.600 and 1, at 492 K; negative deviations[21-23, 25, 26] with the systems: [0.500LiNO$_3$–0.500KNO$_3$ + H$_2$O],[9, 23, 32] between 373 and 436 K, [0.245LiNO$_3$–0.755NH$_4$NO$_3$ + H$_2$O],[23] at 373 K, [KOH + H$_2$O],[26] at 590 K, [0.500KOH–0.500KCl + H$_2$O],[26] at 590 K; alternance of positive and negative deviations[26] with the systems: [NaOH + H$_2$O],[26] at 590 K, [0.500 NaOH–0.500CH$_3$CO$_2$Na + H$_2$O],[26] at 590 K.

1. Adsorption Theory of Electrolytes

(i) Component Activities

The adsorption theory of electrolytes was initiated, in 1948, by Stokes and Robinson[33] as a conceptual adaptation to electrolyte solutions of the theory of gas adsorption on solid surfaces developed ten years earlier by Brunauer, Emmet, and Teller[34] (BET model). This approach is related to the concept of liquid quasi-lattice. From this point of view, at high electrolyte concentrations, a solution may be pictured as containing ions in various stages of hydration. Some ions have a complete hydration shell forming a monolayer around them, other ions have incomplete shell, and others have several layers, with the second and higher layers much less strongly bound than the first. All the layers are in equilibrium, the relative amounts of each varying with concentration. Since this model bears strong resemblance to that from which the gas adsorption isotherm is derived, Stokes and Robinson[33] obtained the following equation giving the water activity a_w as function of the water mole fraction x_w by modification of the notation of the BET isotherm:

$$\frac{a_w(1-x_w)}{x_w(1-a_w)} = \frac{1}{cr} + \frac{(c-1)}{cr}a_w \tag{7}$$

where c is an energetic parameter and r, a structural parameter, both independent of x_w.

The parameter c is given by

$$c = \exp\left(\frac{\epsilon}{RT}\right) \tag{8}$$

in which

$$\epsilon = E_L - E \tag{9}$$

where E is the molar binding energy of water on sites close to the ions, E_L, the molar binding energy of water in pure water, R, the gas constant, and T, the Kelvin temperature.

The parameter r is expressed by

$$r = \frac{N_S}{N_A} \tag{10}$$

where N_S is the number of available sites with the molar binding energy E for the water molecules, per mole of electrolyte, and N_A, the Avogadro constant.

Stokes and Robinson[33] have tested Eq. (7) with 1:1 and 2:1 electrolytes: LiCl, LiBr, HCl, HClO$_4$, NaOH, Ca(NO$_3$)$_2$, ZnCl$_2$, ZnBr$_2$, CaCl$_2$, and CaBr$_2$, at 298 K, over concentration ranges between 6 and 30 M, by plotting its left-hand side against a_w and obtained good straight lines. For all these electrolyte-water systems, values of ϵ lie between 4 and 12 kJ mol^{-1} which are reasonable values of energy of adsorption.[33] Values of r for most of these systems lie between 3 and 4, but for CaCl$_2$ and CaBr$_2$, they are 6.7 and 7.1 respectively. The authors were not satisfied by the non-integral values of r which could not, from their point of view, correspond to any physical reality and had arisen as a result of approximations in the BET theory and its application to electrolyte solutions, the most drastic approximation of the theory being that of treating all water molecules beyond the first layer around the ions like pure water. Therefore, taking into account the modification introduced into the BET model by Anderson,[35] they transformed Eq. (7) into

$$\frac{a_w(1-x_w)}{x_w(1-ba_w)} = \frac{1}{bcr} + \frac{(c-1)}{cr}a_w \qquad (11)$$

in which

$$b = \exp\left[\frac{(E_L - E_b)}{RT}\right] \qquad (12)$$

and

$$c = \exp\left[\frac{(E_b - E)}{RT}\right] \qquad (13)$$

where E_b is the molar binding energy of water in the multilayer hydration sphere of the ions where the nearest neighbors are only water molecules.

Fixing r at integral values (4 or 8), Stokes and Robinson[33] have tested Eq. (11) and obtained good straight lines as with Eq. (7), values of $(E_L - E_b)$ lying between 0 and -0.6 and $(E_b - E)$, between 5 and 10 kJ mol^{-1}.

Recently, Abraham et al.,[19] making a non-linear least-squares fitting of Eq. (11) to the data of the system [0.467TlNO$_3$–0.214CsNO$_3$–0.319 Cd(NO$_3$)$_2$ + H$_2$O] over the large water mole fraction range $x_w \approx 0.2 - 0.9$, at 365, 372, 380, and 388 K, found non-integral values of r (≈ 1.4) and observed that E_b is very close to E_L ($E_L - E_b \approx -0.02$ compared to $E_b - E$

≈ + 2.5 kJ mol^{-1}), confirming that the parameters c and r of Eq. (7) are sufficient to compute the water activity, as is usually done with other hydrate melts.

It must be pointed out that the concept of liquid quasi-lattice would rather be in favor of non-integral values of r. Indeed, Abraham and Abraham,[36, 37] in discussions of electrolyte-water systems properties, have considered a simplified picture in which quasi-lattices consist of submicroscopic parts representing, in a statistical sense, the most probable groupings of all entities involved: ions, hydrated ions, ion pairs, complex ions, water molecules, structural defects..., by analogy with the cybotactic theory.[38] In some submicroscopic parts, salt-like parts, containing ions and adsorbed water molecules, these water molecules would be at the average energy level E. Other submicroscopic parts, water-like parts, would have water molecules at the average energy level E_L (or E_b). Structure and size of the salt-like parts vary with time while the boundaries between the salt-like and water-like parts of the liquid fluctuate. Even if the picture is simplified so that ions and water molecules are assumed to be the only existing entities in the salt-like parts, the maximum number of available sites close to an ion will vary with time and location so that it is the average of integral values which is measured. Since it is an average value, there is no reason for r to be necessarily an integer.

Application of Eq. (7) to various bridging electrolyte-water systems has shown that its validity range extends from anhydrous electrolyte to an x_w upper limit situated between about 0.7 and 0.9, depending upon the nature of the electrolyte,[7, 12, 13, 18, 19, 39, 40] as illustrated in Figs. 2 and 3. Equation (7) has also been successfully tested with solutions of water in fused electrolytes.[2, 9, 14–16, 41, 42] It is remarkable that whatever the features exhibited by the curves a_w vs x_w (positive or negative deviations with respect to Raoult's law, cross-over point) satisfactory results are obtained over a large water concentration range with only two parameters r and c.

Abraham[43] has derived the equation for the electrolyte activity a_e corresponding to Eq. (7) by a treatment based on statistical thermodynamics[44]

$$\frac{\lambda\left(1-x_e\right)}{x_e\left(1-\lambda\right)} = \frac{r}{c} + \frac{r(c-1)}{c}\lambda \qquad (14)$$

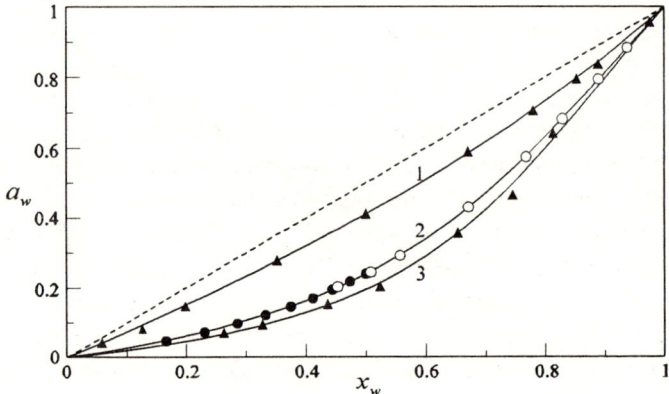

Figure 3. Water activity a_w as function of water mole fraction x_w. 1: $[C_2H_5NH_3NO_3 + H_2O]$,[18] at 298 K; $r = 1.42$, $\epsilon \approx 0$. 2: $[0.500LiNO_3–0.500KNO_3 + H_2O]$, at 393 K; N: from Ref. 25, M from Ref. 32; $r = 1.87$, $\epsilon = 2.59$ from Ref. 39. 3: $[0.606LiNO_3–0.218KNO_3–0.176NaNO_3 + H_2O]$,[40] at 473 K; $r = 2.03$, $\epsilon = 4.00$. ϵ in kJ mol^{-1} Dashed line: Raoult's law. Solid line: calculated curve from Eq. (7).

with

$$\lambda^r = a_e \qquad (15)$$

The parameters r and c are the same as in Eq. (7). In the case of an electrolyte-water system in which the electrolyte is a mixture of several components, a_e represents a sort of average value for the mixed electrolyte. Examples of curves a_e vs x_e are given in Fig. 4.

Abraham's treatment attracted criticisms from Voigt[45] who proposed an equation apparently different from Eq. (14). These criticisms were refuted by Braunstein and Ally[46] who proved that the Voigt's equation, considerably more cumbersome to apply, is in fact equivalent to the Abraham's equation and contains no new information.

Usefulness of Eqs. (7) and (14) for the prediction of thermodynamic properties of mixtures of water with single or mixed electrolytes is enhanced by the existence of additivity rules regarding r and ϵ. These rules were first proposed by Sangster, Abraham and Abraham[12] in a study on

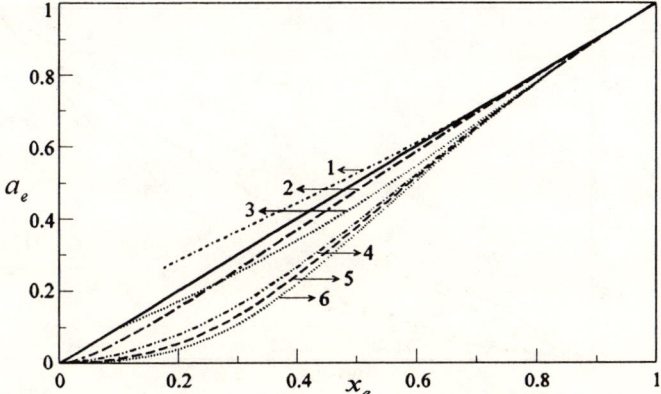

Figure 4. Electrolyte activity a_e calculated from Eq. (14) as function of electrolyte mole fraction x_e. 1: [0.515AgNO$_3$–0.485TlNO$_3$ + H$_2$O],[12] at 372 K; r = 0.49, ϵ = 1.33. 2: [C$_2$H$_5$NH$_3$NO$_3$ + H$_2$O],[18] at 298 K; r = 1.42, $\epsilon \approx$ 0. 3: [0.476AgNO$_3$–0.449TlNO$_3$– 0.075 Cd(NO$_3$)$_2$ + H$_2$O],[12] at 372 K; r = 0.72, ϵ = 2.29. 4: [0.467TlNO$_3$–0.214CsNO$_3$– 0.319 Cd(NO$_3$)$_2$ + H$_2$O],[19] at 383 K; r = 1.42, ϵ = 2.35. 5: [0.500LiNO$_3$–0.500KNO$_3$ + H$_2$O],[25,32] at 393 K; r = 1.87, ϵ = 2.59 from Ref. 39. 6: [0.606LiNO$_3$–0.218KNO$_3$– 0.176NaNO$_3$ + H$_2$O],[40] at 473 K; r = 2.03, ϵ = 4.00. ϵ in kJ mol^{-1}. Solid line: Raoult's law.

[AgNO$_3$–TlNO$_3$–Cd(NO$_3$)$_2$ + H$_2$O] systems, where X_{AgNO_3} / X_{TlNO_3} is fixed at 1.06, $X_{Cd(NO_3)_2}$ varying from 0 to 0.125, at 372 K, and subsequently confirmed with substitution of the salts NaNO$_3$, KNO$_3$, CsNO$_3$, and Ca(NO$_3$)$_2$ to Cd(NO$_3$)$_2$ at the same temperature.[13-16] In general form, these rules, also called mixing rules[40] may be expressed by

$$r = \sum_i X_{ei} r_i \quad (16)$$

$$r\epsilon = \sum_i X_{ei} (r_i \epsilon_i) \quad (17)$$

where r and ϵ are the parameters for a mixture of electrolytes, r_i and ϵ_i, the parameters for the electrolyte i.

In Figs. 5 and 6, r and the product $r\epsilon$ are plotted against $X_{M(NO_3)_n}$ for all the above-mentioned systems. On the whole, the relative positions

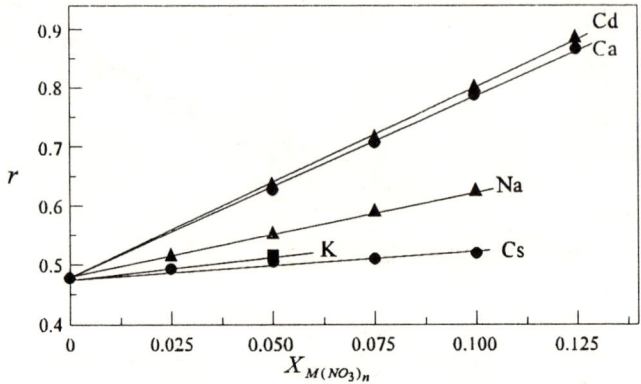

Figure 5. Structural parameter r for the systems $[AgNO_3-TlNO_3-M(NO_3)_n + H_2O]$,[12-16] where X_{AgNO_3} / X_{TlNO_3} is fixed at 1.06, with M = Na, K, Cs, Cd, and Ca, as function of the mole fraction of the added salt $X_{M(NO_3)_n}$, at 372 K.

of the straight lines are in good correlation with the hydrating power of the cations. At a given mole fraction of the added cations, the stronger the tendency of an added cation to hydrate, the higher r and $r\epsilon$ values. The latter property has suggested to express hydrating powers by means of the product $r\epsilon$.[47]

According to Abraham,[36, 47] the additivity properties of r and $r\epsilon$ are compatible with, though perhaps not uniquely required by, molten electrolyte characteristics [2, 48-55] and likely to appear when there is no chemical reaction, complex formation, hydrolysis on mixing the components. They may be the consequences of the preservation of the species, ionic or not, when the electrolytes are mixed, the lack of long range order in conjunction with the presence of structural defects or holes, and the predominance of short range interactions in the phenomenon of water adsorption.

Ally and Braunstein[40] have compared experimental values of r and $r\epsilon$ with those calculated from the additivity rules, Eqs. (16) and (17), for solutions containing binary and ternary mixed electrolytes based on the salts $LiBr$, $LiNO_3$, KNO_3, $NaNO_3$, and $Ca(NO_3)_2$, at 373 and 393 K. They concluded that the additivity rules are valid for both, r and $r\epsilon$, and apply to common anion mixtures as well as common cation mixtures. It is inter-

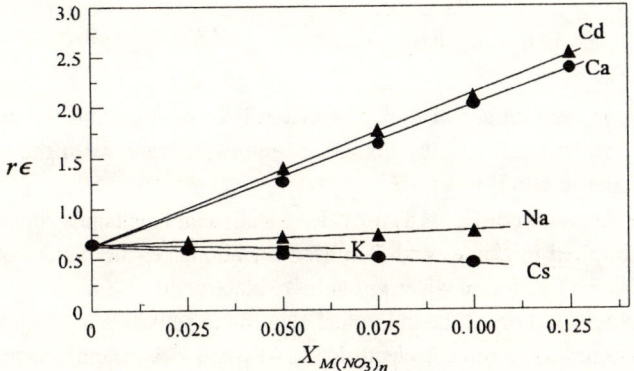

Figure 6. Product of the structural and energetic parameters, $r\epsilon$, for the systems [AgNO$_3$–TlNO$_3$–M(NO$_3$)$_n$ + H$_2$O],[12–16] where X_{AgNO_3}/X_{TlNO_3} is fixed at 1.06, with M = Na, K, Cs, Cd, and Ca, as function of the mole fraction of the added salt $X_{M(NO_3)_n}$, at 372 K. $r\epsilon$ in kJ mol^{-1}.

esting to notice that in some instances[56–58] the additivity rules have been observed even in solutions studied over limited concentration regions rich in water.

In the framework of statistical mechanics, Abraham[43, 47] has shown that in an electrolyte-water system where the electrolyte is a single one or a mixture of electrolytes whose composition is kept constant so that the system is treated as a binary system, the activities of water and electrolyte can also be expressed by

$$a_w = \frac{n_w - n_w^*}{n_w} \tag{18}$$

$$a_e = \left(\frac{rn_e - n_w^*}{rn_e}\right)^r \tag{19}$$

with the equation

$$\frac{n_w^{*2}}{(rn_e - n_w^*)(n_w - n_w^*)} = \exp\left(\frac{\epsilon}{RT}\right) = c \qquad (20)$$

where n_e is the number of moles of electrolyte, n_w, the total number of moles of water, and n_w^*, the number of moles of water with the molar binding energy E in Eq. (9).

The derivation of Eq. (18) provides a statistical mechanics support to what is implied in Stokes and Robinson equation, i.e. the water activity represents the fraction of water *unbound* to electrolyte.[2,33]

In order to express the individual electrolyte activities in electrolyte-water systems containing N electrolytes, Ally and Braunstein[59] have proposed the following generalization of Eqs. (18)–(20):

$$a_w = \frac{n_w - \sum_i^N n_{wi}^*}{n_w} \qquad (21)$$

$$a_{ei} = \frac{n_{ei}}{\sum_i^N n_{ei}} \left(\frac{r_i n_{ei} - n_{wi}^*}{r_i n_{ei}}\right)^{r_i} \qquad (22)$$

with N equations of the form:

$$\frac{n_{wi}^* \sum_i^N n_{wi}^*}{(r_i n_{ei} - n_{wi}^*)\left(n_w - \sum_i^N n_{wi}^*\right)} = \exp\left(\frac{\epsilon_i}{RT}\right) = c_i \qquad (23)$$

where n_{wi}^* is the number of moles of water linked to the electrolyte i with the molar energy E_i.

In the derivation of Eqs. (21)–(23), two basic assumptions are made. Firstly, that, in a multicomponent system, the parameters r_i and ϵ_i are constants characteristic of an electrolyte i, independent of the other elec-

Thermodynamic and Transport Properties   285

trolytes and secondly, that, on mixing water and electrolytes, it is sufficient to consider the change in the internal energy due to water adsorption. This latter assumption is a weakness in the theory, since, generally there is no ideal mixing of anhydrous electrolytes and an amount of enthalpy is involved.[50, 51]

A more rigorous approach proposed by Abraham and Abraham[60] would start with expressing the free energy of mixing all components as:

$$\Delta G_{total} = \Delta G_{e,id} + \Delta G_{e,n-id} + \Delta G_{hy} \qquad (24)$$

where $\Delta G_{e,id}$ is the ideal free energy of mixing the anhydrous electrolytes, $\Delta G_{e,n-id}$, the excess (non ideal) free energy of mixing the anhydrous electrolytes, and ΔG_{hy}, the free energy of mixing water with the anhydrous electrolytes.

Partial differentiation of ΔG_{total} with respect to n_{ei} gives the activity a_{ei} of the electrolyte i. Calculation of $\partial \Delta G_{hy}/\partial n_{ei}$ may be carried out with Eqs. (16)–(20). The additivity rules, expressed by Eqs. (16) and (17), which are not used in Ally and Braunstein approach, simplify the calculation procedure and are compatible with a possible slight influence of the electrolyte hydrating powers on one another. The result is:

$$\ln a_{ei} = \ln X_{ei} + \frac{\Delta \mu_{ei}}{RT} + r_i \ln \frac{r x_e - x_w^*}{r x_e}$$

$$+ r_i \left(\frac{r x_e}{r x_e - x_w^*} - 1 \right) - \frac{\partial n_w^*}{\partial n_{ei}} \left(\frac{r x_e}{r x_e - x_w^*} + \frac{x_w}{x_w - x_w^*} \right) \qquad (25)$$

with

$$\frac{\partial n_w^*}{\partial n_{ei}} = \frac{r_i \left(\dfrac{1}{r x_e - x_w^*} + \dfrac{\epsilon_i - \epsilon}{RT r x_e} \right)}{\dfrac{2}{x_w^*} + \dfrac{1}{r x_e - x_w^*} + \dfrac{1}{x_w - x_w^*}}, \qquad (26)$$

x_w^* obtained from Eq. (20) written in the form

$$\frac{x_w^{*2}}{(rx_e - x_w^*)(x_w - x_w^*)} = c, \quad (27)$$

and where $\Delta\mu_{ei}$ is the excess chemical potential of the electrolyte i in the anhydrous mixture of electrolytes.

The excess chemical potential $\Delta\mu_{ei}$ does not obey a simple general equation. Convenient equations have been applied only to a number of anhydrous mixtures of electrolytes, over limited composition ranges, for example, some binary mixtures of molten salts found to conform to equations of the type

$$\Delta\mu_{e1} = A X_{e2}^2 \quad (28)$$

where A is a constant introduced by regular solution theories.[50, 51, 61]

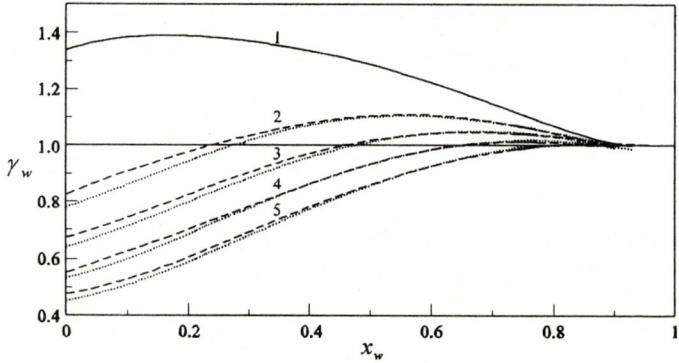

Figure 7. Water activity coefficient γ_w as function of water mole fraction x_w for [AgNO$_3$–TlNO$_3$–M(NO$_3$)$_2$ + H$_2$O] systems, with M = Cd[12] and Ca,[13] at 372 K. 1: $X_{M(NO_3)_2} = 0$, $r = 0.49$, $\epsilon = 1.33$. 2: $X_{M(NO_3)_2} = 0.050$, $r = 0.64$, $\epsilon = 1.98$ for Cd; $r = 0.63$, $\epsilon = 2.18$ for Ca. 3: $X_{M(NO_3)_2} = 0.075$, $r = 0.72$, $\epsilon = 2.29$ for Cd; $r = 0.71$, $\epsilon = 2.47$ for Ca. 4: $X_{M(NO_3)_2} = 0.100$, $r = 0.80$, $\epsilon = 2.53$ for Cd; $r = 0.79$, $\epsilon = 2.68$ for Ca. 5: $X_{M(NO_3)_2} = 0.125$, $r = 0.88$, $\epsilon = 2.69$ for Cd; $r = 0.87$, $\epsilon = 2.89$ for Ca. In all systems, X_{AgNO_3}/X_{TlNO_3} is fixed at 1.06. ϵ in kJ mol^{-1}. Dashed line: M = Cd. Dotted line: M = Ca. γ_w calculated from Eq. (35).

Application of general Margules expansions[24, 61] of $\Delta G_{e,n-id}$ would perhaps be a line of research worth exploring.

With regard to the water activity a_w, its calculation is performed by means of Eqs. (7), (8), (16) and (17) since the partial differentiation of ΔG_{total} with respect to n_w reduces to that of ΔG_{hy}.

(ii) Characteristic Features of Water Activity Coefficient Curves

Figures 7 and 8 illustrate the curves γ_w vs x_w of the systems [AgNO$_3$–Tl NO$_3$–M(NO$_3$)$_n$ + H$_2$O],[12-16] where X_{AgNO_3}/X_{TlNO_3} is fixed at 1.06, with $X_{M(NO_3)_n}$ between 0 and 0.125, M = Na, K, Cs, Cd, and Ca, at 372 K. Characteristic features of these curves have been discussed[13, 14, 41, 62] in terms of the structural parameter r and energetic parameter ϵ.

It is seen in Fig. 7 that when the strongly hydrated cations Cd^{2+} and Ca^{2+} are added to the relatively weakly hydrated cations Ag$^+$ and Tl$^+$, the

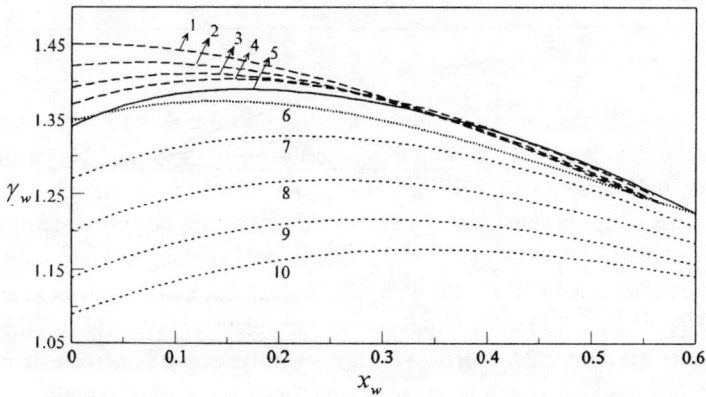

Figure 8. Water activity coefficient γ_w as function of water mole fraction x_w for [AgNO$_3$–TlNO$_3$–MNO$_3$ + H$_2$O] systems, with M = Na,[15] K,[16] and Cs,[14] at 372 K. 1: X_{CsNO_3} = 0.100, r = 0.52, ϵ = 0.89. 2: X_{CsNO_3} = 0.075, r = 0.51, ϵ = 1.00. 3: X_{CsNO_3} = 0.050, r = 0.50, ϵ = 1.10. 4: X_{CsNO_3} = 0.025, r = 0.49, ϵ = 1.21. 5: X_{MNO_3} = 0, r = 0.49, ϵ = 1.33. 6: X_{KNO_3} = 0.050, r = 0.52, ϵ = 1.13. 7: X_{NaNO_3} = 0.025, r = 0.52, ϵ = 1.30. 8: X_{NaNO_3} = 0.050, r = 0.55, ϵ = 1.27. 9: X_{NaNO_3} = 0.075, r = 0.59, ϵ = 1.23. 10: X_{NaNO_3} = 0.100, r = 0.62, = 1.20. In all systems, X_{AgNO_3}/X_{TlNO_3} is fixed at 1.06. ϵ in kJ mol^{-1}. γ_w^∞ calculated from Eq. (35).

curves γ_w vs x_w are pushed downward in a regular manner and may present a cross-over point where $\gamma_w = 1$. The similarity between the effects of Cd^{2+} and Ca^{2+} on the shapes and the situations of these curves is consistent with the similarity of their ionic radii (Cd^{2+}: 0.097 nm, Ca^{2+}: 0.099 nm)[63] and their Stokes radii in dilute aqueous solution (Cd^{2+}: 0.34 nm, Ca^{2+}: 0.31 nm).[63]

The experimental observation that there is *only one* cross-over point on a curve meets a requirement of the form of Eq. (7) which allows only one possible value such that $a_w = x_w$. The water mole fraction at this cross-over point is given by[12]

$$x_{w,cross-over\ point} = \frac{cr - 1}{c - 1} \tag{29}$$

Equation (29) implies that a necessary condition for the appearance of a cross-over point is

$$0 < \frac{cr - 1}{c - 1} < 1 \tag{30}$$

In Fig. 8, the curves γ_w vs x_w are slightly raised by addition of the cation Cs^+ to Ag^+ and Tl^+, in contrast to the effect of K^+ and Na^+ which lower the curves, like Cd^{2+} and Ca^{2+}, but much less markedly. This behavior of Cs^+ is consistent with its weakly hydrating power reflected by the fact that its Stokes radius, 0.12 nm, is smaller than its ionic radius, 0.17 nm, as usually observed with monovalent anions whose ionic radii are larger than 0.133 nm.[63] On the whole, the relative positions of all these curves are in agreement with Eq. (18) from which one expects increase of γ_w with decreasing ionic hydrating power, since a_w in this equation represents the fraction of *unbound* water.

An equation predicting the existence of an extremum value of γ_w has also been proposed.[62] The water mole fraction at this extremum is given by

$$x_{w,extremum} = f(A,B) = \left(\frac{2AB}{1+B-A} + \frac{1+B-A}{2} \right)^{-1} \tag{31}$$

with

$$A = \frac{1}{cr} \quad \text{and} \quad B = \frac{c-1}{cr} \quad (32)$$

Equation (31) implies that a necessary condition for the appearance of an extremum is

$$0 < f(A,B) < 1 \quad (33)$$

The extremum is a maximum if[41]

$$E < E_L \quad (34)$$

which means that the adsorption of water on fused electrolytes releases more energy than does water when it condenses in bulk. This is the case with most of the $AgNO_3$ and $TlNO_3$ based systems in Figs. 7 and 8. None of these systems exhibit a minimum value of γ_w. Yet, it is noticed that the maximum is removed by the presence of $CsNO_3$ at sufficiently high concentration, for which, from the additivity rules, the sign of the extrapolated value of ϵ (≈ -1.5 kJ mol^{-1}) is negative, meaning that the adsorption of water on $CsNO_3$ releases less energy than does water when it condenses in bulk ($E > E_L$), in contrast with the other electrolytes.[14]

Braunstein and Braunstein[9] have suggested application of Eq. (7) to determine the water activity coefficient at infinite dilution in electrolyte γ_w^∞. In the limit as $x_w \to 0$, Eq. (7) reduces to

$$\left(\frac{a_w}{x_w}\right)_{x_w \to 0} = \gamma_w^\infty = \frac{1}{cr} \quad (35)$$

Equation (35) leads to Henry's law constant for dissolution of water in molten electrolyte[9]

$$K_H^\infty = f_w^o \gamma_w^\infty = \frac{f_w^o}{cr} \quad (36)$$

Henry's law is strictly valid only at infinite dilution. There are cases, however, where the ratio f_w / x_w varies little over a finite concentration range. Depending on the error which can be tolerated and the concentration range of interest, a practical Henry's law constant K_H may be used[62]

$$f_w = K_H x_w \tag{37}$$

which must be distinguished from K_H^∞.

Since in the region of an extremum a function changes little, a practical Henry's law constant for water could appear if the extremum value of γ_w is present in the region of dilute solutions of water in molten electrolyte, which is the case of the systems containing the relatively weakly hydrated cations Ag^+, Tl^+, K^+, and Cs^+, as seen in Fig. 8.

Table 1

Values of the water activity coefficient at infinite dilution in molten electrolyte γ_w^∞ and the Henry's law constant for dissolution of water in molten electrolyte K_H^∞ calculated from the adsorption theory for electrolyte-water systems respectively by means of Eqs. (35) and (36)

Electrolyte	T (K)	γ_w^∞	K_H^∞ (kPa)
$[0.515AgNO_3–0.485TlNO_3]$[42, 62]	360	1.36	84
	372	1.34	127
	384	1.31	190
$[0.400AgNO_3–0.600TlNO_3]$[41]	384	1.50	218
$[0.600AgNO_3–0.400TlNO_3]$[41]	384	1.19	173
$[0.489AgNO_3–0.461TlNO_3–0.050NaNO_3]$[15]	372	1.20	113
$[0.464AgNO_3–0.436TlNO_3–0.100NaNO_3]$[15]	372	1.09	103
$[0.489AgNO_3–0.461TlNO_3–0.050KNO_3]$[16]	372	1.35	128

Table 1 (continued)

Electrolyte	T (K)	γ_w^∞	K_H^∞ (kPa)
$[0.489AgNO_3-0.461TlNO_3-0.050CsNO_3]^{14}$	372	1.39	132
$[0.464AgNO_3-0.436TlNO_3-0.100CsNO_3]^{14}$	372	1.45	138
$[0.489AgNO_3-0.461TlNO_3-0.050Ca(NO_3)_2]^{13}$	372	0.78	74
$[0.464AgNO_3-0.436TlNO_3-0.100Ca(NO_3)_2]^{13}$	372	0.54	51
$[0.489AgNO_3-0.461TlNO_3-0.050Cd(NO_3)_2]^{62}$	372	0.82	78
$[0.476AgNO_3-0.449TlNO_3-0.075Cd(NO_3)_2]^{62}$	372	0.67	63
$[0.464AgNO_3-0.436TlNO_3-0.100Cd(NO_3)_2]^{62}$	372	0.55	52
$[0.451AgNO_3-0.424TlNO_3-0.125Cd(NO_3)_2]^{62}$	372	0.47	45
$[0.467TlNO_3-0.214CsNO_3-0.319Cd(NO_3)_2]^{19}$	372	0.32	31
	383	0.34	48
$[C_2H_5NH_3NO_3]^{18}$ *	298	0.70	2
$[0.500LiNO_3-0.500KNO_3]^{25,32}$ *	393	0.24	47
$[0606LiNO_3-0.218KNO_3-0.176NaNO_3]^{40}$ *	473	0.18	256
$[0.667NH_4NO_3-0.258LiNO_3-0.075NH_4Cl]^{17}$ *	366	0.36	28

* γ_w^∞ and K_H^∞ calculated by the writers from a_w data in the given references, p_w^o, vapor pressure of pure water[64] and B, the second virial coefficient of water vapor,[65] using the equation: $f_w^o = p_w^o \exp(B p_w^o / RT)$

In Table 1, values of γ_w^∞ and K_H^∞, calculated by means of Eqs. (35) and (36), are listed for the systems shown in Figs. 2 and 3 and for other electrolyte-water systems.[13–15, 19, 41, 42, 62] Besides, in some instances, values of these constants have been proposed for dissolution of water in fused electrolytes assumed to be supercooled.[56–58, 66, 67]

(iii) Excess Properties

(a) Partial molar excess volumes

From the equation

$$\overline{V}_i^{ex} = RT\left(\frac{\partial \ln a_i}{\partial P}\right)_{T,x_i} \qquad (38)$$

where the subscript i refers to water or electrolyte and P is the pressure, and with Eqs. (7) and (14), Ally and Braunstein[40, 68] have derived equations expressing the partial molar excess volumes of water and electrolyte, $\overline{V}_w^{ex}$ and $\overline{V}_e^{ex}$ respectively, assuming arbitrarily that the pressure dependence is entirely in the energy parameter ϵ :

$$\overline{V}_w^{ex} = \epsilon' \Omega_{w\epsilon} \qquad (39)$$

$$\overline{V}_e^{ex} = \epsilon' \Omega_{e\epsilon} \qquad (40)$$

where

$$\epsilon' = \left(\frac{\partial \epsilon}{\partial P}\right)_{T,x_i} \qquad (41)$$

$$\Omega_{w\epsilon} = \frac{c(a_w - 1 + r R_e)}{c(1 - r R_e) - 2a_w(c-1) - 2} \qquad (42)$$

$$\Omega_{e\epsilon} = \frac{rc[r(\lambda - 1) + R_w]}{r[c - 2 - 2\lambda(c-1)] - cR_w} \qquad (43)$$

$$R_w = \frac{x_w}{x_e} \quad \text{and} \quad R_e = \frac{x_e}{x_w} \qquad (44)$$

Abraham and Abraham[39] assuming that the pressure dependence is not only in the energy parameter ϵ, but also in the structural parameter r, have obtained:

$$\overline{V}_w^{ex} = \epsilon' \Omega_{w\epsilon} + r' \Omega_{wr} \tag{45}$$

$$\overline{V}_e^{ex} = \epsilon' \Omega_{e\epsilon} + r' \Omega_{er} \tag{46}$$

where

$$r' = \left(\frac{\partial r}{\partial P} \right)_{T, x_i} \tag{47}$$

$$\Omega_{wr} = \frac{c R_e RT}{c(1 - r R_e) - 2 a_w (c - 1) - 2} \tag{48}$$

$$\Omega_{er} = \frac{RTr\left[(c-1)\lambda^2 - (c-2)\lambda - 1\right]}{r\lambda\left[c - 2 - 2\lambda(c-1)\right] - c\lambda R_w} + RT \ln \lambda \tag{49}$$

The functions $\Omega_{w\epsilon}$ and $\Omega_{e\epsilon}$ are the same as in Eqs. (39) and (40).

The partial molar excess volume of water at infinite dilution in molten electrolyte $\overline{V}_w^{ex\infty}$ is given by the following simple limiting form of Eq. (45):

$$\left(\overline{V}_w^{ex} \right)_{x_w \to 0} = \overline{V}_w^{ex\infty} = -\epsilon' - \frac{RTr'}{r} \tag{50}$$

The partial molar excess volumes are related to the molar excess volume of the solution V^{ex} by

$$V^{ex} = \overline{V}_e^{ex} x_e + \overline{V}_w^{ex} x_w \tag{51}$$

V^{ex} being defined by

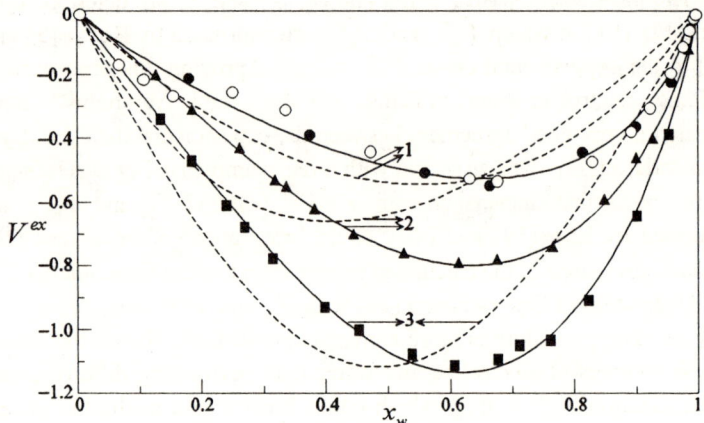

Figure 9. Molar excess volume V^{ex} as function of water mole fraction x_w. 1:[$C_2H_5NH_3NO_3$ + H_2O], at 298 K; M: from Ref. 69, N: from Ref. 70; $r' = 0.76 \times 10^{-9}$, $\epsilon' = +0.10 \times 10^{-6}$. 2:[$0.515 AgNO_3–0.485 TlNO_3 + H_2O$],[71] at 372 K; $r' = 0.58 \times 10^{-9}$, $\epsilon' = -1.99 \times 10^{-6}$. 3:[$0.451 AgNO_3–0.424 TlNO_3–0.125 Cd(NO_3)_2 + H_2O$],[71] at 372 K; $r' = 0.61 \times 10^{-9}$, $\epsilon' = +0.45 \times 10^{-6}$. Values of r' and ϵ' from Ref. 39. Solid line: calculated curve with r' and ϵ' by means of Eqs. (45), (46) and (51). Dashed line: calculated curve with r' set to zero by means of Eqs. (39), (40) and (51). r' in Pa^{-1}. ϵ' in m^3 mol^{-1}. V^{ex} in cm^3 mol^{-1}.

$$V^{ex} = V - V^{id} \qquad (52)$$

in which V^{id}, the ideal molar volume, is

$$V^{id} = V_e^o x_e + V_w^o x_w \qquad (53)$$

and V, the molar volume of the solution, is given by

$$V = \frac{x_e \sum_i X_{ei} M_{ei} + x_w M_w}{\rho} \qquad (54)$$

with V_e^o, the molar volume of the anhydrous electrolyte, V_w^o, the molar volume of pure water, M_{ei}, the molar mass of the electrolyte i, M_w, the water molar mass, and ρ, the density of the solution.

Thermodynamic and Transport Properties

The parameters r and c being known, ϵ' and r' are determined by fitting Eq. (51) in which $\overline{V}_w^{ex}$ and $\overline{V}_e^{ex}$ are expressed by Eqs. (45) and (46) to the experimental curve $V^{ex} = f(x_w)$. Applying this procedure to bridging electrolyte-water systems, Abraham and Abraham[39]* have obtained a very good agreement between the experimental data and those predicted from Eq. (51), as shown with three examples in Fig. 9. The same figure shows that dispensing with r', i.e. when $\overline{V}_w^{ex}$ and $\overline{V}_e^{ex}$ are expressed by Eqs. (39) and (40), could bring about more or less pronounced deviations of the calculated curves from the experimental data.

Utilization of Eqs. (45) and (46) with ϵ' and r' values obtained by the preceding procedure is more convenient to evaluate the partial molar excess volumes of electrolyte and water than performing differentiation of the function $V^{ex} = f(x_w)$ which could involve non negligible uncertainty.[39]

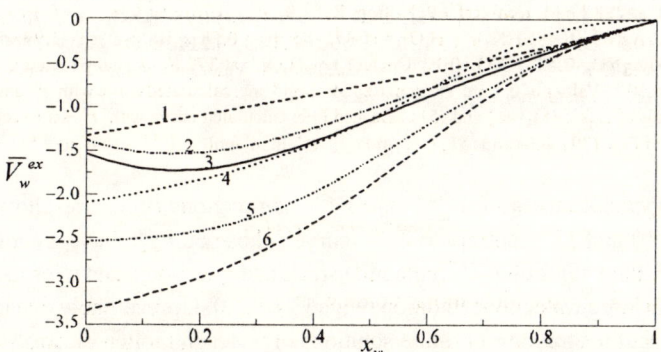

Figure 10. Partial molar excess volume of water $\overline{V}_w^{ex}$ calculated from Eq. (45) as function of water mole fraction x_w. 1: [$C_2H_5NH_3NO_3 + H_2O$],[69,70] at 298 K; $r' = 0.76 \times 10^{-9}$, $\epsilon' = +0.10 \times 10^{-6}$. 2: [$0.464AgNO_3-0.436TlNO_3-0.100CsNO_3 + H_2O$],[72] at 372 K; $r' = 0.53 \times 10^{-9}$ $\epsilon' = -1.79 \times 10^{-6}$. 3: [$0.515AgNO_3-0.485TlNO_3 + H_2O$],[71] at 372 K; $r' = 0.58 \times 10^{-9}$, $\epsilon' = -1.99 \times 10^{-6}$. 4: [$0.500LiNO_3-0.500KNO_3 + H_2O$],[39] at 393 K; $r' = 0.16 \times 10^{-9}$, $\epsilon' = +1.71 \times 10^{-6}$. 5: [$0.451AgNO_3-0.424TlNO_3-0.125Cd(NO_3)_2 + H_2O$],[71] at 372 K; $r' = 0.61 \times 10^{-9}$, $\epsilon' = +0.45 \times 10^{-6}$. 6: [$0.467TlNO_3-0.214CsNO_3-0.319Cd(NO_3)_2 + H_2O$],[19] at 383 K; $r' = 0.51 \times 10^{-9}$, $\epsilon' = +2.20 \times 10^{-6}$. Values of r' and ϵ' from Ref. 39. r' in Pa^{-1}. ϵ' in m^3 mol^{-1}. $\overline{V}_w^{ex}$ in cm^3 mol^{-1}.

---

* The chemical formula of EAN in Ref. 39 is erroneous. It should be read as $C_2H_5NH_3NO_3$ and not $N(C_2H_5)_4NO_3$.

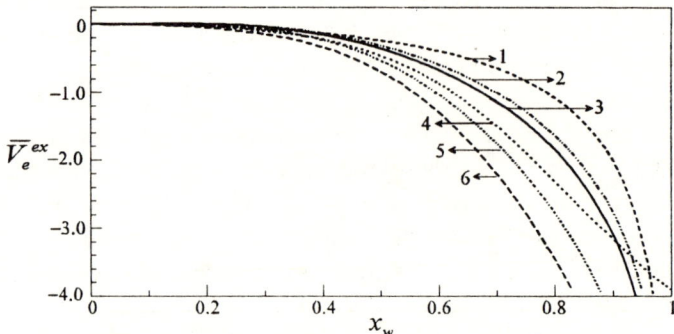

Figure 11. Partial molar excess volume of electrolyte $\overline{V}_e^{ex}$ calculated from Eq. (46) as function of water mole fraction x_w. 1: [$C_2H_5NH_3NO_3$ + H_2O],[69,70] at 298 K; $r' = 0.76 \times 10^{-9}$, $\epsilon' = +0.10 \times 10^{-6}$. 2: [0.464AgNO$_3$–0.436TlNO$_3$–0.100CsNO$_3$ + H_2O],[72] at 372 K; $r' = 0.53 \times 10^{-9}$ $\epsilon' = -1.79 \times 10^{-6}$. 3: [0.515AgNO$_3$–0.485TlNO$_3$ + H_2O],[71] at 372 K; $r' = 0.58 \times 10^{-9}$, $\epsilon' = -1.99 \times 10^{-6}$. 4: [0.500LiNO$_3$–0.500KNO$_3$ + H_2O],[39] at 393 K; $r' = 0.16 \times 10^{-9}$, $\epsilon' = +1.71 \times 10^{-6}$. 5: [0.451AgNO$_3$–0.424TlNO$_3$–0.125Cd(NO$_3$)$_2$ + H_2O],[71] at 372 K; $r' = 0.61 \times 10^{-9}$, $\epsilon' = +0.45 \times 10^{-6}$. 6: [0.467TlNO$_3$–0.214CsNO$_3$–0.319 Cd(NO$_3$)$_2$ + H_2O],[19] at 383 K; $r' = 0.51 \times 10^{-9}$, $\epsilon' = +2.20 \times 10^{-6}$. Values of r' and ϵ' from Ref. 39. r' in Pa^{-1}. ϵ' in m^3 mol^{-1}. $\overline{V}_e^{ex}$ in cm^3 mol^{-1}.

Typical curves of $\overline{V}_w^{ex}$ and $\overline{V}_e^{ex}$ as functions of x_w are shown in Figs. 10 and 11. Contrary to $\overline{V}_w^{ex}$ curves, those of $\overline{V}_e^{ex}$ have a common shape, the values of $\overline{V}_e^{ex}$ remaining small and quasi-constant over the relatively large water concentration range $x_w \approx 0 - 0.4$, presumably due to the fact that the structure of dilute solutions of water in molten electrolytes is similar to that of the anhydrous electrolytes themselves. The relative positions of the curves are in conformity with the hydrating power of the cations.

From the additivity rules, equations involving the parameters ϵ' and r' have been proposed[39]

$$r' = \sum_i X_{ei} r_i' \tag{55}$$

$$(r\epsilon)' = \epsilon r' + r\epsilon' = \sum_i X_{ei}\left(\epsilon_i r_i' + r_i \epsilon_i'\right) \tag{56}$$

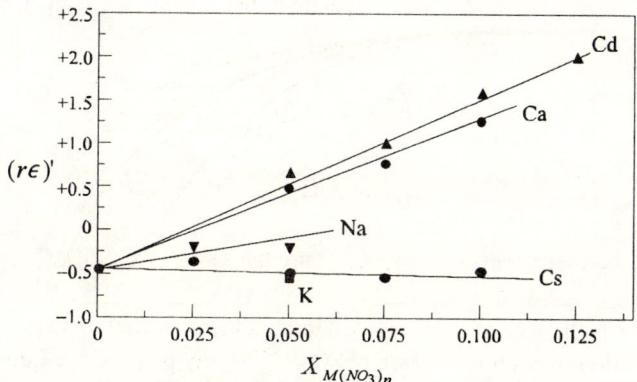

Figure 12. Derivative of the product of the structural and energetic parameters with respect to pressure P, $(r\epsilon)'$, for [$AgNO_3$–$TlNO_3$–$M(NO_3)_n$ + H_2O] systems, where X_{AgNO_3}/X_{TlNO_3} is fixed at 1.06, with M = Na, K, Cs, Cd, and Ca, as function of the mole fraction of the added salt $X_{M(NO_3)_n}$, at 372 K. $(r\epsilon)'$ in 10^{-6} m^3 mol^{-1}. (From Ref. 39, with kind permission from Monatshefte für Chemie).

Equations (55) and (56) have been tested with the same $AgNO_3$ and $TlNO_3$ based systems[12-16] whose study led to the observation of the additivity rules on the parameters r and ϵ. It is seen in Fig. 12 that $(r\epsilon)'$ obeys a linear relationship and reflects, to a certain extent, the relative cation hydrating power as does the product $r\epsilon$ illustrated in Fig 6. Contrary to r, r' remains quasi-constant at $0.58 \pm 0.05 \times 10^{-9}$ Pa^{-1} for these systems.

(b) Partial molar excess enthalpies

From the equation

$$\overline{H}_i^{ex} = R\left[\frac{\partial \ln a_i}{\partial \left(1/T\right)}\right]_{P,x_i} \tag{57}$$

and with Eqs. (7) and (14), Ally and Braunstein[40, 68] have derived equations expressing the partial molar excess enthalpies of water and electrolyte, $\overline{H}_w^{ex}$ and $\overline{H}_e^{ex}$ respectively, assuming that ϵ and r are temperature independent:

$$\overline{H}_w^{ex} = \epsilon\, \Omega_{w\epsilon} \tag{58}$$

$$\overline{H}_e^{ex} = \epsilon\, \Omega_{e\epsilon} \tag{59}$$

in which the functions $\Omega_{w\epsilon}$ and $\Omega_{e\epsilon}$ are the same* as in Eqs. (39) and (40).

These authors have also proposed the following procedure to provide crystallization data on hydrates $S.nH_2O$.[40, 68] At any point of the liquidus curve where the solid hydrate is in equilibrium with its saturated solution, $S.nH_2O(s) \rightarrow S(\ell) + n\, H_2O(\ell)$, the temperature is T and the mole fraction of the electrolyte in the liquidus phase is x_e. At the melting point T_m of the hydrate, the solid hydrate is in equilibrium with the liquid phase of the same composition in which the mole fraction of the electrolyte is x_{em}. With the approximations that the latent heat of fusion L of the hydrate and the partial molar excess enthalpies $\overline{H}_w^{ex}$ and $\overline{H}_e^{ex}$ in the hydrate melt are temperature independent, general thermodynamic development leads to the equation

$$R\ln a_e(T,x_e)a_w^n(T,x_e) - R\ln a_e(T_m,x_{em})a_w^n(T_m,x_{em}) = \left[L - \overline{H}_e^{ex}(T_m,x_{em}) - n\overline{H}_w^{ex}(T_m,x_{em})\right]\left(\frac{1}{T_m} - \frac{1}{T}\right) \tag{60}$$

where $a_e(T, x_e)$, $a_w(T, x_e)$, $a_e(T_m, x_{em})$, and $a_w(T_m, x_{em})$ are the activities of electrolyte and water at any point T of the liquidus curve, and at the melting point T_m of the hydrate.

The right-hand side and the second term of the left-hand side of Eq. (60) are evaluated knowing L, T_m, the parameters r and ϵ of the solution derived from vapor pressure data, and using Eqs. (7), (14), (58), and (59).

---

* There is an error in Ref. 40, Eq. (21), p. 228 and in Ref. 68, Eq. (7), p. 6. The parameter λ in the denominator is missing.

The two values of x_e satisfying Eq. (60) and corresponding to the two branches of the liquidus are found by trial and error using Eqs. (7) and (14). Applying this procedure, the authors have successfully predicted stable and metastable phases of the bridging system [NaOH + H_2O].*

Liquidus curves of [$MgCl_2$ + H_2O], [$CaCl_2$ + H_2O], and [$Mg(NO_3)_2$ + H_2O] systems have also been calculated by Voigt,[45] for $x_w :: 0.6 - 0.7$, on the basis of the solubility product, the linear relationship between the solubility constant and the reciprocal of the temperature, and the parameters r and ϵ obtained from vapor pressure data, taking into account, when available, the influence of temperature on r and ϵ.

(iv) Temperature Dependence of the Adsorption Parameters

Some authors [19, 42, 45] have observed linear dependences of r and ϵ or $r\epsilon$ on T. Although reasonable values of a number of thermodynamic parameters have been predicted without taking into account the influence of temperature on r and ϵ,[40, 68] such simplifications are not always appropriate.[42] For example, the partial molar excess entropy $\overline{S}_w^{ex\infty}$ and the partial molar excess enthalpy $\overline{H}_w^{ex\infty}$ of water at infinite dilution in molten electrolyte may be calculated by the following general equations

$$\overline{S}_w^{ex\infty} = -\left[\frac{\partial\left(RT\ln\gamma_w^\infty\right)}{\partial T}\right]_P \quad (61)$$

$$\overline{H}_w^{ex\infty} = R\left[\frac{\partial\ln\gamma_w^\infty}{\partial\left(1/T\right)}\right]_P \quad (62)$$

If γ_w^∞ is expressed as a function of the parameters r and ϵ, Eqs. (61) and (62) give

---

* In Ref. 40, Ally and Braunstein have noticed that if the hydrate melting points and the heats of fusion are not known, or if the hydrates melt incongruently, the liquidus curves can be calculated if some phase data are available.

$$\overline{S}_w^{ex\infty} = \left(\frac{\partial \epsilon}{\partial T}\right)_P + R\ln r + \frac{RT}{r}\left(\frac{\partial r}{\partial T}\right)_P \tag{63}$$

$$\overline{H}_w^{ex\infty} = -\epsilon + T\left(\frac{\partial \epsilon}{\partial T}\right)_P + \frac{RT^2}{r}\left(\frac{\partial r}{\partial T}\right)_P \tag{64}$$

Assuming that the parameters r and ϵ are temperature independent, Eqs. (63) and (64) reduce to

$$\overline{S}_w^{ex\infty} = R\ln r \tag{65}$$

$$\overline{H}_w^{ex\infty} = -\epsilon \tag{66}$$

For water dissolved in the molten salt mixture [0.515AgNO_3–0.485 TlNO_3],[42] data on the influence of temperature on r and ϵ and application of Eqs. (63) and (64) give for the values of $\overline{S}_w^{ex\infty}$ and $\overline{H}_w^{ex\infty}$ + 2.5 J K^{-1} mol^{-1} and + 1.8 kJ mol^{-1} respectively, compared to – 6.0 J K^{-1} mol^{-1} and – 1.3 kJ mol^{-1} with Eqs. (65) and (66).

2. Approaches Related to Regular Solution Theories

(i) Approach with Mole Fractions on an Ionized Basis

Pitzer[21,22] has proposed the use of equations applicable to regular solutions.[61] This author writes for the activities of water and electrolyte the classical equations of regular solution theories in which the ordinary mole fraction x is replaced by the mole fraction on an ionized basis y:

$$\ln a_w = \ln y_w + w_w z_e^2 \tag{67}$$

$$\ln a_e = \ln y_e + w_e z_w^2 \tag{68}$$

with the parameters z_w, z_e, and w_e expressed by

Thermodynamic and Transport Properties

$$z_w = \frac{n_w}{n_w + \nu n_e \dfrac{b_e}{b_w}} \tag{69}$$

$$z_e = \frac{\nu n_e}{\nu n_e + n_w \dfrac{b_w}{b_e}} = 1 - z_w \tag{70}$$

$$w_e = \frac{b_e}{b_w} w_w \tag{71}$$

where ν is the number of ions produced by the total dissociation of the electrolyte. The empirical parameters w_e and w_w are non-ideality parameters arising from the difference between the intermolecular attraction of unlike species as compared to the mean of the intermolecular attraction of like species, and the ratio b_w / b_e is tentatively ascribed to the ratio of the volumes of the molecules or to the ratio of the molar volumes in the liquid. For fused salt-water systems, Pitzer pointed out that it seems best to regard b_w / b_e as a freely adjustable parameter to be compared to the ratio of the volumes of species.

Equation (67) has been fitted[21] to the following bridging systems: [0.533AgNO$_3$–0.467TlNO$_3$ + H$_2$O],[7] at 372 K, with the values $w_w = +1.02$ and $b_w / b_e = 0.50$ and [0.500LiNO$_3$–0.500KNO$_3$ + H$_2$O],[9,32] at 373 and 392 K, with the values $w_w = -0.89$ and $b_w / b_e = 1.2$. The agreement between the experimental data and calculated curves has been found excellent over practically the whole water concentration range. Yet, comparison between the ratio b_w / b_e and the volume ratio (which is the ratio of molar volume of water to the average volume per ion) of the two systems, 0.82 and 0.87 respectively, is not satisfactory.[21,22]

Equation (67) has also been successfully fitted by Tripp[20] to water activities of seventeen solutions of water in mixtures of nitrates and/or nitrites containing essentially monovalent cations, Li$^+$, K$^+$, Na$^+$, Cs$^+$, Ag$^+$, and Tl$^+$, with two of them containing the divalent cations Ca^{2+} and Mg^{2+} at mole fraction 0.056. Values of the parameters lie between $w_w = -3.01$,

$b_w / b_e = 1.04$, for a system whose electrolyte has a high hydrating power, [LiNO$_3$ + H$_2$O],[9] at 393 K, and $w_w = +1.08$, $b_w / b_e = 0.58$, for a system whose electrolyte has a relatively low hydrating power, [0.464AgNO$_3$–0.436TlNO$_3$–0.100CsNO$_3$ + H$_2$O],[14] at 372 K.

Tripp has illustrated a near linear correlation between w_w and the weighted average cation charge density Q defined as

$$Q = \sum_i X_{i+} \left(\frac{z_{i+}}{r_{i+}^3} \right) \qquad (72)$$

where X_{i+}, z_{i+}, and r_{i+} are the mole fraction, the charge, and the ionic radius of the cation i in the melt, respectively.

According to this author, this correlation between Q and w_w strongly supports the concept of preferential hydration previously invoked by Abraham, Abraham, and Sangster[62] with regard to the Henry's law constant.

Besides, Tripp has brought attention to the system [AgNO$_3$–TlNO$_3$–Ca(NO$_3$)$_2$ + H$_2$O],[13] where X_{AgNO_3} / X_{TlNO_3} is fixed at 1.06, $X_{Ca(NO_3)_2}$ lying between 0.025 and 0.125, at 372 K. He pointed out the failure of Eq. (67) to correlate the water concentration dependence of water activities for this system, without stating the $X_{Ca(NO_3)_2}$ value(s) at which failure was observed.* Therefore, the writers have tested Eq. (67) noticing that with addition of Ca(NO$_3$)$_2$, the ratio b_w / b_e decreases gradually, becoming even negative at $X_{Ca(NO_3)_2} = 0.125$. At this concentration, $w_w = 0.26$ and $b_w / b_e = -0.01$, this latter value being obviously unrealistic.

In order to obtain an accurate representation of the water activity even in very dilute solutions of electrolyte in water, Pitzer[21] added to Eq. (67) an extended form of the Debye-Hückel equation previously proposed.[73, 74] This electrical contribution to the water activity coefficient is given by

---

* There is an error in Ref. 20, p. 855 concerning the evaluation of W and b, respectively designated by w_w and (b_w / b_e) in Eq. (67) in this chapter. The writers have verified that $W = 0.69$ and $b = -98$ could not correspond to the data of the system [AgNO$_3$–TlNO$_3$–Ca(NO$_3$)$_2$ + H$_2$O],[13] where the ratio X_{AgNO_3}/X_{TlNO_3} is fixed at 1.06, with $X_{Ca(NO_3)_2}$ between 0.025 and 0.125, at 372 K, whatever $X_{Ca(NO_3)_2}$. Yet, this error does not affect the comments of the author.

$$\ln \gamma_w^{el} = \frac{2\left(\dfrac{1000}{M_w}\right)^{1/2} A_\phi I_y^{3/2}}{\left(1 + d_{ca} I_y^{1/2}\right)} \quad (73)$$

where A_ϕ is the usual Debye-Hückel parameter, I_y, the ionic strength, and d_{ca}, a parameter related to the closest approach of ions.

A_ϕ is given by

$$A_\phi = \frac{1}{3}\left(\frac{2\pi N_A \rho_w}{1000}\right)^{1/2} \left(\frac{e^2}{D_w k T}\right)^{3/2} \quad (74)$$

in which ρ_w is the water density, e, the electronic charge, D_w, the dielectric constant of water, and k, the Boltzmann constant.

I_y is expressed by

$$I_y = \frac{1}{2} \sum_i z_i^2 y_i \quad (75)$$

in which y_i and z_i are the mole fraction and charge of the ionic species i, respectively.

Addition of the Debye-Hückel term to Eq. (67) by Pitzer gives the following results[21]: $w_w = +0.835$, $b_w/b_e = 0.56$ for [0.533AgNO$_3$–0.467 TlNO$_3$ + H$_2$O],[7] at 372 K, and $w_w = -1.124$, at 373 K, $w_w = -1.070$, at 392 K, $b_w/b_e = 1.0$ for [0.500LiNO$_3$–0.500KNO$_3$ + H$_2$O].[9, 32]

Subsequently, Pitzer and co-workers[22–25, 31] have proposed more general equations applicable, in principle, to multicomponent systems for calculation of the partial excess free energy, enthalpy and volume of all components. These equations comprising an ideal mixing term, with the assumption of random mixing of all particles, a general Margules term originating from short-range forces, and a third term expressing the Debye-Hückel effect, have been successfully fitted to data of many electrolyte-water systems containing essentially monovalent cations.[21, 23, 25–31, 75] However, serious difficulties are encountered with hydrate melts containing appreciable amounts of divalent cations[20, 76] and even with dilute aqueous solutions.[77] Moreover, this approach has been criticized[40, 68, 78] for its lack of predicting power, its complexity and the need of too many parameters whose evaluation may pose problems.

(ii) Approach with Mole Fractions on an Un-Ionized Basis

In a model based on the quasi-lattice concept proposed by Horsák and Sláma,[79] the water molecules are introduced into the interpenetrating cationic and anionic sublattices in the ratio of the stoichiometric coefficients of the ions, diluting these two sublattices. The excess free energy of mixing water and electrolyte, and related thermodynamic parameters, are expressed as the sum of two contributions, one arising from short-range interactions, analogous to those in regular solutions, the other arising from long-range interactions, analogous to those in ionic crystals. This model has only been tested for 1:1 electrolytes with the following equation:

$$\ln \gamma_w^{\bullet} = L_1 \frac{2 x_e^2}{(1+x_e)^2} - \frac{L_2}{6}\left[\frac{2 x_e}{(1+x_e)}\right]^{4/3} \qquad (76)$$

in which the activity coefficient $\gamma_w^{\bullet}$ is defined as

$$\gamma_w^{\bullet} = \frac{a_w}{a_w^{\bullet}} \qquad (77)$$

with

$$a_w^{\bullet} = \frac{1-x_e}{1+x_e} \qquad (78)$$

L_1 is a short-range force interaction parameter and L_2, a Coulombic long-range force interaction parameter.

Fitting of Eq. (76) over the whole water concentration range, with $L_1 = -0.55$, $L_2 = -7.85$ for the system [0.533AgNO$_3$–0.467TlNO$_3$ + H$_2$O],[7] at 372 K, and $L_1 = -2.01$, $L_2 = -1.25$ for the system [0.500LiNO$_3$– 0.500 KNO$_3$ + H$_2$O],[9,32] at 373 and 392K, is satisfactory.*

3. Surface Properties

The first ever determinations of the surface tension σ of a bridging electrolyte-water system were made by Hadded, Bahri and Letellier[80] on the [C$_2$H$_5$NH$_3$NO$_3$ + H$_2$O] system, at 298 K. Other determinations of σ were made by Abraham and co-workers on the following bridging systems, between 350 and 390 K: [0.515AgNO$_3$–0.485TlNO$_3$ + H$_2$O],[81] [0.464AgNO$_3$–0.436TlNO$_3$–0.100M(NO$_3$)$_n$+H$_2$O][81] with M = Cs, Cd, and Ca, and [0.467TlNO$_3$–0.214CsNO$_3$–0.319Cd(NO$_3$)$_2$ + H$_2$O].[19]

At a fixed water mole fraction, from anhydrous electrolyte to pure water, it was observed[19,81] that the surface tension σ is a simple linear function of the temperature T, over the explored temperature range:

$$\sigma = -m_\sigma T + b_\sigma \tag{79}$$

where m_σ and b_σ are empirical parameters.

Since σ is the surface free energy per unit area, the surface entropy S_σ and the surface enthalpy H_σ, per unit area, are expressed by the equations:

$$S_\sigma = -\frac{d\sigma}{dT} \tag{80}$$

$$H_\sigma = \sigma - T\frac{d\sigma}{dT} \tag{81}$$

Thus, S_σ and H_σ, which are respectively represented by m_σ and b_σ in Eq. (79), appear independent of the temperature for these bridging salt-

---

* The values of L_1 and L_2 in Ref. 79 are inaccurate and, consequently, the remark on the sign of L_1 is unfounded, pp. 1678 – 9. The values of L_1 and L_2 given in this chapter have been determined by the writers.

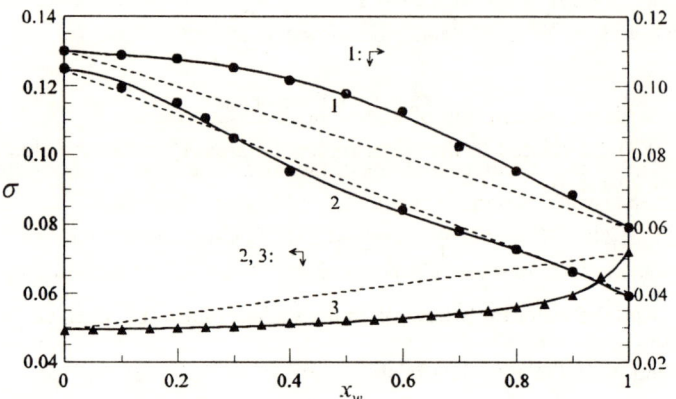

Figure 13. Surface tension σ as function of water mole fraction x_w. 1: [0.467TlNO$_3$–0.214CsNO$_3$–0.319Cd(NO$_3$)$_2$ + H$_2$O],[19] at 372 K. 2: [0.464AgNO$_3$–0.436TlNO$_3$–0.100 Cd(NO$_3$)$_2$ + H$_2$O],[81] at 372 K. 3: [C$_2$H$_5$NH$_3$NO$_3$ + H$_2$O],[80] at 298 K. σ in N m^{-1}.

water systems,[19, 81] as observed for various pure molten salts, molten salt mixtures, anhydrous or containing small amounts of water.[82, 83]

As seen in Fig. 13, isotherms σ vs x_w may exhibit positive, negative and alternate (S-shape curve) deviations from linearity, reminiscent of water activity deviations from Raoult's law.

From the surface tension and water activity data, orders of magnitude of the water mole fraction in the surface phase have been obtained[81] by application of the Guggenheim and Adam method,[84, 85] as follows. The Gibbs adsorption equation is written

$$\Gamma_{w(e)} = -\frac{1}{kT}\frac{d\sigma}{d\ln a_w} \qquad (82)$$

where $\Gamma_{w(e)}$ is the Gibbs parameter which measures the adsorption of water per unit area, relative to the electrolyte.

$\Gamma_{w(e)}$ is related to the numbers of water molecules and electrolyte entities per unit area, Γ_w and Γ_e respectively, by

Thermodynamic and Transport Properties

$$\Gamma_{w(e)} = \Gamma_w - \Gamma_e \frac{x_w}{1-x_w} \qquad (83)$$

With the assumption of a monolayer surface phase, Γ_w and Γ_e in Eq. (83) are related to the water molecular area A_w and to the electrolyte entity area A_e by

$$\Gamma_e A_e + \Gamma_w A_w = 1 \qquad (84)$$

And the water mole fraction in the surface phase x_w^s is given by

$$x_w^s = \frac{\Gamma_w}{\Gamma_w + \Gamma_e} \qquad (85)$$

The areas A_w and A_e were estimated from the particle radii.[81] The particle area of water is given by

$$A_w = \pi r_w^2 \qquad (86)$$

and taking the nitrate entities as (M^{z+}, zNO_3^-), an average value of the electrolyte entity area is expressed by

$$A_e = \pi \sum_i X_{ei} \left(r_{i+}^2 + z_{i+} r_-^2 \right) \qquad (87)$$

in which r_w is the water radius, r_-, the NO_3^- radius, r_{i+}, the cation i radius, and z_{i+}, the cation i valency.

As illustrated in Fig. 14 the curves x_w^s vs x_w show a steady enrichment in water of the surface phase as x_w increases over the whole water concentration range.[19, 81] The relative lower values of x_w^s for the system $[0.467TlNO_3-0.214CsNO_3-0.319Cd(NO_3)_2 + H_2O]^{19}$ reflect the strong tendency of the Cd^{2+} cations to attract water molecules.

Computation of the water activity in the surface phase a_w^s from the water activity in the bulk phase a_w was performed[81] by means of the Butler equation[86] written in the form

$$a_w^s = a_w \exp\left[\frac{A_w(\sigma - \sigma_w)}{kT}\right] \tag{88}$$

in which σ_w is the surface tension of pure water.

In Fig. 15, curves of a_w^s vs x_w^s show that the adsorption theory of electrolytes might be applied to the surface phase of some nitrate-water systems. However, the values of r and ϵ in the surface phase are different from those in the bulk phase, especially the value of the energetic parameter ϵ.

The related water activity coefficient in the surface phase γ_w^s is obtained by

$$\gamma_w^s = \frac{a_w^s}{x_w^s} \tag{89}$$

It has been pointed out[81] that, on the whole, the relative position of the curves γ_w^s vs x_w^s in the surface phase, for the systems [0.515AgNO$_3$ –0.485TlNO$_3$ + H$_2$O] and [0.464AgNO$_3$–0.436TlNO$_3$–0.100 M(NO$_3$)$_n$ +

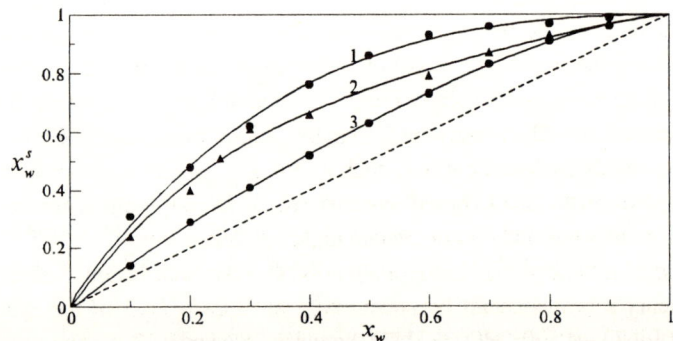

Figure 14. Water mole fraction in the surface phase x_w^s as function of water mole fraction in the bulk phase x_w, at 372 K. 1: [0.464AgNO$_3$–0.436TlNO$_3$–0.100CsNO$_3$ + H$_2$O].[81] 2: [0.515AgNO$_3$–0.485TlNO$_3$ + H$_2$O].[81] 3: [0.467TlNO$_3$–0.214CsNO$_3$–0.319Cd(NO$_3$)$_2$ + H$_2$O].[19]

H_2O] with M = Cs, Cd, and Ca, reflect, to a certain extent, the hydrating power of the cations as do the curves γ_w vs x_w in the bulk phase.

Recently, Sonowane, and Kumar[87] have proposed the following equation to represent the surface tension of the above-mentioned nitrate-water systems, at 373 K, over the whole water concentration range:

$$\sigma = \sigma_w x_w + \sigma_e x_e + RT x_w x_e \left(\frac{1}{A_w} - \frac{1}{A_e}\right)(\delta_p + \delta_m x_e) \quad (90)$$

where σ_e is the surface tension of pure electrolyte and δ_p, δ_m are two empirical constants.

It must be underlined that there is a weakness in the approach of these authors. In effect, from the outset of the derivation of Eq. (90), the Butler equation, Eq. (88), is written with substitution of the mole fractions to the activities which is inconsistent with the fact that these systems are not ideal mixtures.

4. Other Approaches and Observations

(i) Water Vapor Pressure

Tripp and Braunstein[1, 2, 20, 32, 88–90] have observed that in a number of hydrate melts the water vapor pressure and the water activity are linear in the water mole ratio R_H (= moles of water / moles of electrolyte), in the region where water is the solute. Tripp[20] pointed out that this property holds only in systems which show negative deviations from Raoult's law and contain highly hydrated cations, such as Li^+, Cd^{2+}... but does not hold in the $AgNO_3$ and $TlNO_3$ based systems which show positive deviations from Raoult's law and do not contain highly hydrated cations.

Besides, Tripp[89] has brought attention to a constant isotopic effect on the Henry's law constant by substitution of D_2O to H_2O in the system [$0.500LiNO_3$–$0.500KNO_3$ + H_2O], between 383 and 423 K. The ratio of the Henry's law constants K_H, for water, and K_D, for deuterium oxide, is $K_H / K_D \approx 1.04$.

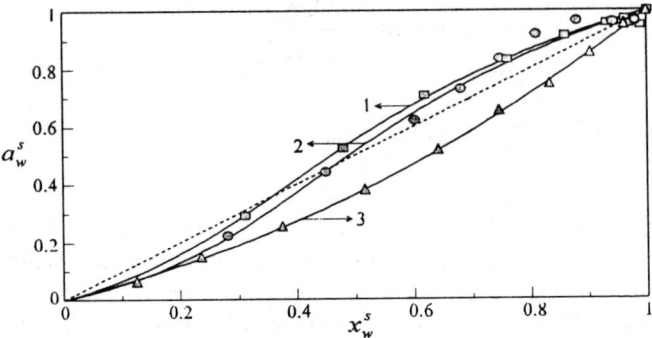

Figure 15. Water activity in the surface phase a_w^s as function of water mole fraction in the surface phase x_w^s, at 372 K. 1: [0.464AgNO$_3$–0.436TlNO$_3$–0.100 CsNO$_3$ + H$_2$O][81]; $r = 0.5$, $\epsilon = 4$. 2: [0.515AgNO$_3$–0.485TlNO$_3$ + H$_2$O][81]; $r = 0.6$, $\epsilon = 4$. 3: [0.467TlNO$_3$– 0.214 CsNO$_3$–0.319Cd(NO$_3$)$_2$ + H$_2$O][19]; $r = 1.5$, $\epsilon = 1$. ϵ in kJ mol^{-1}. Dashed line: Raoult's law. Solid line: calculated curve from Eq. (7).

(ii) Molar Volume

Sacchetto and Kodejš[91, 92] have proposed an equation for the dependence of the molar volume of hydrate melts on molar composition which is based on the calculation of the distribution of water molecules between sites near ions and sites near water molecules. This distribution is obtained by a statistical thermodynamic treatment of a quasi-lattice model. Supposing that in an electrolyte-water system the effective molar volume of water in the neighborhood of the ions is the molar volume of water at infinite dilution in the molten electrolyte, V_w^∞, and that the effective molar volume of water at sites near the water molecules is the molar volume of pure water, V_w^o, they obtained

$$V = V_e^o x_e + \left[\frac{\beta \theta x_e}{1 + x_e(\beta\theta - 1)} V_w^\infty + \frac{1 - x_e}{1 + x_e(\beta\theta - 1)} V_w^o \right] (1 - x_e) \quad (91)$$

The parameters β and θ are given by

$$\beta = \exp\left(-\frac{\Delta H_{tr}}{Z_e RT}\right) \tag{92}$$

and

$$\theta = \frac{Z_e}{Z_w} \tag{93}$$

where ΔH_{tr} is the enthalpy of transfer of water from its pure state to infinite dilution in the molten electrolyte. The parameters Z_e and Z_w are interpreted as average quasi-lattice coordination numbers of the molten electrolyte and pure liquid water, respectively.

Equation (91) was successfully fitted to the data of the bridging systems [$C_2H_5NH_3NO_3$ + H_2O], at 298 K, by Hadded et al.,[69] [0.667 NH_4NO_3–0.258 $LiNO_3$–0.075NH_4Cl + H_2O],[93] at 366 K, by Pacák,[94] but failed[95] with the data of the system [0.515$AgNO_3$–0.485$TlNO_3$ + H_2O],[71] at 372 K.

(iii) Substitution of an Organic Substance to Water

Kodejš and Sacchetto[75, 95] have measured vapor pressures of the system [0.5$AgNO_3$–0.5$TlNO_3$ + $(CH_3)_2SO$], between 360 and 390 K, and compared their results with those obtained with the system [0.515 $AgNO_3$–0.485$TlNO_3$ + H_2O],[12] at 372 K. They pointed out[75] that the substitution of dimethyl sulfoxide (DMSO) to water intensifies the ion-solvent interaction in a manner analogous to the addition of an appreciable amount of divalent cation, such as Cd^{2+} or Ca^{2+}, to the aqueous system, i.e. positive deviations from Raoult's law become markedly negative over the whole concentration range. They attributed the difference in behavior of these two systems mainly to the stronger cation-dipole interaction energy in DMSO[66, 75], the dipole moments of the two molecules being[64] μ_{H_2O} = 1.85 D and $\mu_{(CH_3)_2SO}$ = 3.96 D, respectively.

It must be noticed that the difference in behavior of these two systems is simply reflected in the adsorption parameters, $r = 0.49$, $\epsilon = 1.33$ kJ mol^{-1}, whence $r\epsilon = 0.65$ kJ mol^{-1} for water,[12] at 372 K, and $r = 1.13$, $\epsilon = 5.5$ kJ mol^{-1}, whence $r\epsilon = 6.22$ kJ mol^{-1} for DMSO,[75] at 383 K, suggesting the use of the product $r\epsilon$ as a criterion of solvation aptitude of various solvents with reference to a selected electrolyte, by analogy with the use of the same product $r\epsilon$ as a parameter expressing the hydrating power of various electrolytes.[47]

Upon substituting DMSO to water in the bridging systems containing the electrolytes [0.667NH$_4$NO$_3$–0.258 LiNO$_3$–0.075NH$_4$Cl],[94] at 377 K, and [0.5AgNO$_3$–0.5TlNO$_3$],[95] at 383 K, the excess molar volume of the solution V^{ex} becomes more important in both cases while Eq. (91) can still be fitted to the first system experimental data[94] but fails with the second system.[95]

III. TRANSPORT PROPERTIES

The first ever determinations of viscosity and electrical conductivity of a bridging electrolyte-water system were made in 1921 by Rabinowitsch[96] on the system [0.500AgNO$_3$–0.500TlNO$_3$ + H$_2$O], at 373 K, although there were few measurements over the concentration range where water is the solute. Further measurements of viscosity and electrical conductivity isotherms were carried out, more than 37 years later, on the following systems: [LiClO$_3$ + H$_2$O],[97] at 405 K, [0.667NH$_4$NO$_3$–0.258 LiNO$_3$–0.075NH$_4$Cl + H$_2$O],[93] at 366 K, [AgNO$_3$–TlNO$_3$–M(NO$_3$)$_n$ + H$_2$O],[71, 72, 98] where X_{AgNO_3} / X_{TlNO_3} is fixed at 1.06, with $X_{M(NO_3)_n}$ between 0 and 0.125, M = Na, K, Cs, Cd, and Ca, at 372 K, [0.467TlNO$_3$ –0.214CsNO$_3$–0.319Cd(NO$_3$)$_2$ + H$_2$O],[99] at 383 K, [C$_2$H$_5$NH$_3$NO$_3$ + H$_2$O],[100] at 298 K, [KF.2HF + H$_2$O],[101] at 358 K. Electrical conductivity isotherms have been measured on the systems [AgNO$_3$ + H$_2$O],[102] at 495 K, and [NH$_4$NO$_3$ + H$_2$O],[102] at 453 K.

The influence of temperature on both, viscosity and electrical conductivity, have been studied on the following systems: [0.410LiNO$_3$– 0.590KNO$_3$ + H$_2$O],[103] between 413 and 463 K, [0.515AgNO$_3$–0.485 TlNO$_3$ + H$_2$O],[104] and [0.464AgNO$_3$–0.436TlNO$_3$–0.100M(NO$_3$)$_2$ + H$_2$O],[105, 106] with M = Cd and Ca, between 350 and 380 K, [0.467TlNO$_3$ –0.214CsNO$_3$–0.319Cd(NO$_3$)$_2$ + H$_2$O],[19] between 355 and 400 K, [H$_2$SO$_4$ + H$_2$O],[107] between 200 and 300 K, and, on the electrical conductivity alone on [KF.2HF + H$_2$O],[101] between 345 and 365 K.

1. Activation Energy for Viscous Flow

The temperature dependence of viscosity η at constant pressure is frequently represented by an Arrhenius type equation[52, 108]:

$$\eta = A_\eta \exp\left(\frac{E_\eta^*}{RT}\right) \tag{94}$$

where the preexponential term A_η and the activation energy E_η^* for viscous flow are two empirical parameters, often found practically constant over limited temperature ranges.

For the bridging system [0.410LiNO$_3$–0.590KNO$_3$ + H$_2$O], Claes, Michielsen, and Glibert[103] found a linear dependence of the logarithm of the viscosity, $\ln \eta$, on the reciprocal of the temperature, $1/T$, whatever the water mole fraction, allowing to extract constant values of E_η^* valid over the explored temperature range, i.e. between 413 and 463 K.

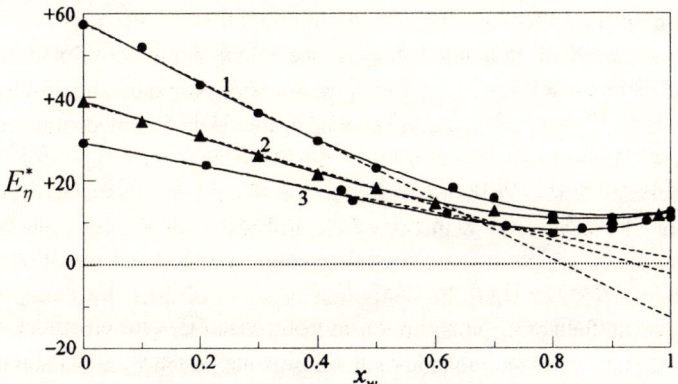

Figure 16. Activation energy for viscous flow E_η^* in Eq. (94) as function of water mole fraction x_w. 1: [0.467TlNO$_3$–0.214CsNO$_3$–0.319Cd(NO$_3$)$_2$ + H$_2$O],[19] between 355 and 400 K. 2: [0.464AgNO$_3$–0.436TlNO$_3$–0.100Cd(NO$_3$)$_2$ + H$_2$O],[105] between 350 and 380 K. 3: [0.410LiNO$_3$–0.590KNO$_3$ + H$_2$O],[103] between 413 and 463 K. Dashed line: linear relationship Eq. (95). E_η^* in kJ mol^{-1}.

Abraham and co-workers have also applied Eq. (94) and extracted values of activation energy E_η^* for the systems: [0.515AgNO$_3$– 0.485 TlNO$_3$ + H$_2$O],[104] [0.464AgNO$_3$–0.436 TlNO$_3$–0.100M(NO$_3$)$_2$ + H$_2$O],[105] with M = Cd and Ca, between 350 and 380 K, and [0.467TlNO$_3$– 0.214 CsNO$_3$–0.319Cd(NO$_3$)$_2$+H$_2$O],[19] between 355 and 400 K. Some curves E_η^* vs x_w are shown in Fig. 16.

Claes, Michielsen, and Glibert,[103] observing a minimal value of E_η^* at $x_w \approx 0.8$, concluded that this change in the activation energy suggests a transition from a system exclusively composed of hydrated ions to aqueous solutions where free water is also found.

For the five above-mentioned bridging systems, Abraham, Ziogas, and Abraham[105] have pointed out the existence of a linear, or quasi-linear, relationship between E_η^* and x_w over the range $x_w \approx 0$–0.5, which may be expressed by

$$E_\eta^* = E_{\eta e}^* x_e + E_{\eta w}^{*/} x_w \tag{95}$$

where $E_{\eta e}^*$ is the activation energy of the anhydrous electrolyte and $E_{\eta w}^{*/}$, an empirical parameter, is the apparent activation energy of water, for viscous flow.

The empirical parameter $E_{\eta w}^{*/}$ is one among a number of other empirical parameters called *apparent parameters* by Abraham and co-workers[37, 72, 104, 105] used to take into account in the formulations of the properties of the electrolyte-water systems the behavior of water perturbed by the presence of ions, which is linked to the electrolyte in a manner reminiscent of the adsorption theory of electrolytes, over the concentration range $x_w \approx 0$–0.5. In this state of water, there would be perturbations of hydrogen bonds or even the disruption of some of these by the strong electrostatic field of the ions, similar to the effect of temperature and pressure on pure water and aqueous solutions as suggested from Raman and infrared spectral investigations[109, 110] and recently from neutron diffraction.[111] Thus, the apparent energy of activation of water $E_{\eta w}^{*/}$ could be seen as the energy of activation pure water would exhibit if it underwent structural modifications analogous to those brought about by the presence of the electrolyte.

It is seen in Fig. 16 that $E^{*/}_{\eta w}$ may be negative. Concerning these negative values, a similarity between the behavior of perturbed water in these very ionic melts and pure water in its gaseous state has been underlined.[105] As a matter of fact, application of Eq. (94) to water vapor viscosity data[64, 112] between 373 and 673 K, leads to $E^{*}_{\eta w} = -4$ kJ mol^{-1}. These results suggest some connection of Eq. (95) with the picture of Eyring significant structure theory of liquids[113] which combines solid-like parts with gas-like parts, although this Eyring theory was not meant to deal with hydrates specifically.

For the bridging system [KF.2HF + H$_2$O], studied at 358 K by Dumont, Qian, and Conway[101] at 358 K, a linear relationship between ln η and x_w, over the very large range $x_w \approx 0$–0.88, has been attributed to a linear relationship between E^{*}_{η} and x_w, i.e. formally Eq. (95). This has been correlated by the authors with the formation of hydration shells which are supposed to be complete at $x_w \approx 0.88$, in accordance with electrical conductivity data.

2. Transition State Theory of Viscosity

Eyring[114, 115] has proposed the following equation for the viscosity η of a simple fluid

$$\eta = \frac{h}{\upsilon_h} \exp\left(\frac{\Delta G^{*}_{\eta}}{RT}\right) \qquad (96)$$

where h is the Planck constant, υ_h, the volume of a hole considered to be close to that of a flow unit, and ΔG^{*}_{η}, the molar free energy of acti-vation for viscous flow.

Generally, υ_h is related to the molar volume of the fluid V by

$$\upsilon_h = \frac{V}{N_A} \qquad (97)$$

and Eq. (96) is often written

$$\eta = \frac{h N_A}{V} \exp\left(\frac{\Delta G_\eta^*}{RT}\right) \qquad (98)$$

ΔG_η^* is related to the enthalpy of activation, ΔH_η^*, and the entropy of activation, ΔS_η^*, for viscous flow by

$$\Delta H_\eta^* = \Delta G_\eta^* - T\left(\frac{\partial \Delta G_\eta^*}{\partial T}\right) \qquad (99)$$

and

$$\Delta S_\eta^* = -\left(\frac{\partial \Delta G_\eta^*}{\partial T}\right) \qquad (100)$$

In this viscous flow process theory, a particle i, or flow unit, passes from one equilibrium position where it oscillates with a characteristic frequency v_i to another equilibrium position over a free energy barrier ΔG_η^*. This passage requires the availability of a hole, or empty site, in the quasi-crystalline lattice, in which the particle will jump, its vibration being transformed into a translation movement. In molten salts, it has been considered[116-118] that ΔH_η^* originates essentially from the formation of a hole and the jump, or movement, of the flow unit into this hole, whereas the physical meaning of ΔS_η^* has been less clearly stated.

As a matter of fact, holes are parts of liquid structures and their existence has often been invoked in various studies of molten salts.[52, 53, 104, 116-129] Similarities between holes in liquids and defects in solid crystals, such as Schottky (ion pair or single ion vacancy) and Frenkel (dislocated interstitial cation) defects, have been put forward in investigations on molten halides and nitrates.[123-125, 128] Besides, studies on water have led to theories making use of such concepts as hole, cavity, defect, interstitial position.[129-133] From these theories and other arguments, it has been inferred[104] that pure water resembles pure nitrates with respect to hole mechanisms, so that the existence of these mechanisms might also be highly probable in mixtures of water and molten nitrates, and possibly other electrolytes.

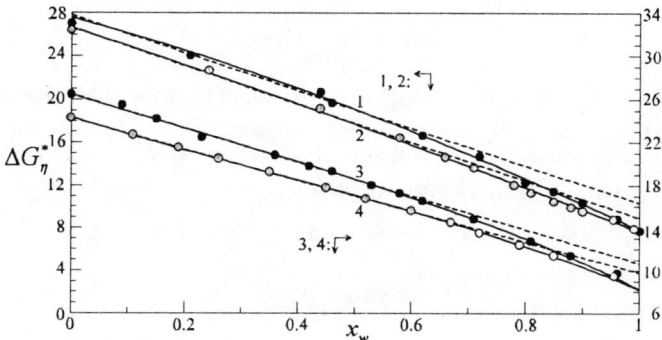

Figure 17. Free energy of activation for viscous flow ΔG_η^* in Eq. (98) as function of water mole fraction x_w. 1: [0.410LiNO$_3$–0.590KNO$_3$ + H$_2$O],[103] at 423 K. 2: [LiClO$_3$ + H$_2$O],[97] at 405 K. 3: [0.464AgNO$_3$–0.436TlNO$_3$–0.100Cd(NO$_3$)$_2$ + H$_2$O],[71] at 372 K. 4: [0.515 AgNO$_3$–0.485TlNO$_3$ + H$_2$O],[71] at 372 K. Dashed line: linear relationship Eq. (101). ΔG_η^* in kJ mol^{-1}.

Although Eyring equation was proposed for a simple fluid and other approaches are used when cooperative phenomena have to be considered in liquids of relatively high viscosity,[134] interesting results have been obtained by applying Eyring equation to hydrate melts despite the complexity of their liquid structures. In practical situation, υ_h in Eq.(96) must be obviously viewed as an average parameter, like ΔG_η^*, and, in hydrates, it will depend on the nature of the flow units originating from the electrolyte and water. Species involved in the viscous flow mechanisms are not known and one might suppose the existence of ion pairs, complex ion clusters... Abraham and co-workers[19, 37, 98, 99, 105] have applied Eyring equation with υ_h expressed by Eq. (97) to the following systems: [AgNO$_3$–TlNO$_3$–M(NO$_3$)$_n$ + H$_2$O],[71, 72, 98] where X_{AgNO_3} / X_{TlNO_3} is fixed at 1.06, $X_{M(NO_3)_n}$ being between 0 and 0.125, with M = Na, K, Cs, Cd, and Ca, [0.467TlNO$_3$–0.214CsNO$_3$–0.319Cd(NO$_3$)$_2$ + H$_2$O],[19, 99] [0.500AgNO$_3$–0.500TlNO$_3$ + H$_2$O],[96] [LiClO$_3$ + H$_2$O],[97] [0.667NH$_4$NO$_3$–0.258LiNO$_3$–0.075NH$_4$Cl + H$_2$O],[93] and [0.410LiNO$_3$–0.590KNO$_3$ + H$_2$O].[103] Curves of ΔG_η^* vs x_w are illustrated in Fig. 17 for some of these systems.

Over the range $x_w \approx 0$–0.5, ΔG_η^* is a linear, or quasi-linear, function of x_w, which may be written

$$\Delta G_\eta^* = \Delta G_{\eta e}^* x_e + \Delta G_{\eta w}^{*/} x_w \qquad (101)$$

where $\Delta G_{\eta e}^*$ is the free energy of activation of the anhydrous electrolyte and $\Delta G_{\eta w}^{*/}$, the apparent free energy of activation assigned to the perturbed water.*

For the $AgNO_3$ and $TlNO_3$ based systems,[37, 71, 72, 98] $\Delta G_{\eta w}^{*/}$ reflects the perturbing power of the cations through an additivity rule expressed by

$$\Delta G_{\eta w}^{*/} = \sum_i X_{ei} \, \Delta G_{\eta wi}^{*/} \qquad (102)$$

where $\Delta G_{\eta w}^{*/}$ is the parameter for a mixture of electrolytes and $\Delta G_{\eta wi}^{*/}$, the parameter for the component i.

As seen in Fig. 18, at any given value of the added nitrate mole fraction, the trend of the $\Delta G_{\eta w}^{*/}$ values

$$Cs^+ \approx K^+ < (Ag^+ + Tl^+) \approx Na^+ < Ca^{2+} < Cd^{2+}$$

is consistent, on the whole, with the characterization of ions in dilute aqueous solutions as water structure breakers, such as Cs^+, and water structure formers, such as Cd^{2+}.[135-139]

If the evaluation of ΔG_η^* is made assuming the flow units to be the water molecules and the ions, the average volume of a hole υ_h in Eq.(96) is then

$$\upsilon_h = \frac{V}{N_A \left[x_e \sum_i X_{ei} (\nu_{i+} + \nu_{i-}) + x_w \right]} \qquad (103)$$

---

* Concerning the system [$0.667 NH_4NO_3 - 0.258 LiNO_3 - 0.075 NH_4Cl + H_2O$],[93] at 366 K, it was observed in Ref. 37 that ΔG_η^* does not obey the linear relationship expressed by Eq. (101). This is all the more surprising that, for the same system, the free energy for electrical conductance obeys a linear relationship analogous to Eq. (101), as will be seen in Section III.6, and that, moreover the writers have verified that the system [$KF.2HF + H_2O$][101] obeys Eq. (101) although it does not contain any nitrate. In our opinion, the experimental data provided by courtesy of Dr Claes on the viscosity of the system [$0.667 NH_4NO_3 - 0.258\ LiNO_3 - 0.075 NH_4Cl + H_2O$] could be unreliable given the fact that the times of drain of solutions very rich in electrolyte through the capillary tube in the method of measurement are much too long in comparison to those of the solutions rich in water: for example, $\Delta t = 12745$ s at $x_w = 0$ compared to $\Delta t = 241$ s at $x_w = 0.975$.

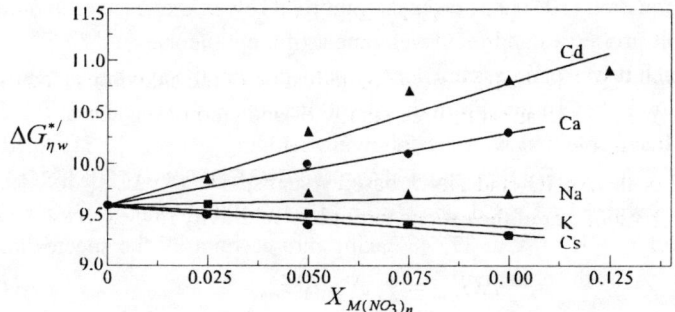

Figure 18. Apparent free energy of activation for viscous flow assigned to the perturbed water $\Delta G^{*/}_{\eta w}$ for the systems [AgNO$_3$–TlNO$_3$–M(NO$_3$)$_n$ + H$_2$O],[71, 72, 98] where X_{AgNO_3}/X_{TlNO_3} is fixed at 1.06, with M = Na, K, Cs, Cd, and Ca, as function of the mole fraction of the added salt $X_{M(NO_3)_n}$, at 372 K. $\Delta G^{*/}_{\eta w}$ in kJ mol^{-1}.

where ν_{i+} and ν_{i-} are the numbers of positive and negative ions respectively produced by the total dissociation of one mole of the electrolyte i and V is given by Eq. (54).

For the system [0.467TlNO$_3$–0.214CsNO$_3$–0.319Cd(NO$_3$)$_2$ + H$_2$O], the two alternatives in the evaluation of υ_h lead to an appreciable difference in ΔG^*_η only for the solutions of water in electrolyte, at worst 10% for the anhydrous electrolyte, and the free energy ΔG^*_η is still linear in x_w when υ_h is expressed by Eq. (103).[19]

In molten carbonates,[116] the molar free energy of activation for viscous flow ΔG^*_η has been compared to the molar free energy of hole formation ΔG_σ, evaluated by means of the following equation used in Fürth hole theory of liquids[140]

$$\Delta G_\sigma = 4\pi\,\sigma\,N_A\,r_h^2 \tag{104}$$

where r_h is the radius of a hole.

In the Fürth theory, outside the holes, a liquid is considered as a continuum with normal surface tension and the sizes of the holes obey a statistical distribution law,[140] so that they are not comparable to holes in a quasi-lattice.[117]

Equation (104) has played an important role in the discussion of molten salt properties and the development of their theories[52, 53, 116, 118, 121, 123] in which it was often assumed, or concluded, that the average hole size is close to that of an ion. From this point of view, average values of $\Delta \overline{G}_\sigma$ have been obtained with the following bridging systems [0.515AgNO$_3$–0.485TlNO$_3$ + H$_2$O],[81] [0.464AgNO$_3$–0.436TlNO$_3$–0.100M(NO$_3$)$_n$ + H$_2$O],[81] with M = Cs, Cd, and Ca, and [0.467TlNO$_3$–0.214CsNO$_3$–0.319 Cd(NO$_3$)$_2$ + H$_2$O],[19] at 372 K, taking into account all the species in the solutions, by the equation

$$\Delta \overline{G}_\sigma = 4\pi\sigma N_A \overline{r_h^2} \qquad (105)$$

where the parameter $\overline{r_h^2}$ is the weighted average of the square particle radii expressed by

$$\overline{r_h^2} = \frac{x_e \left[\sum_i X_{ei} \left(r_{i+}^2 + z_{i+} r_-^2 \right) \right] + x_w r_w^2}{x_e \left[\sum_i X_{ei} \left(1 + z_{i+} \right) \right] + x_w} \qquad (106)$$

For the AgNO$_3$ and TlNO$_3$ based systems, it was observed that $\Delta \overline{G}_\sigma$ is close to ΔG_η^* from fused electrolyte to water, and inferred that indeed hole formation is an essential step of the viscous flow mechanisms.[81] For the system [0.467TlNO$_3$–0.214CsNO$_3$–0.319Cd(NO$_3$)$_2$ + H$_2$O],[19] over the range $x_w \approx 0.4$–1, $\Delta \overline{G}_\sigma$ is also close to ΔG_η^*, but their difference is greater for the dilute solutions of water in the molten salt, probably due to the high degree of quasi-crystalline character of the mixture containing an important amount of Cd^{2+} cations. The presence of divalent cations strengthens the structure of the liquid which will become more crystal-like due to intensification of the cationic electrical field, particularly near the pure fused salt end of the concentration range at relatively low temperature, and, therefore, the approximation of a liquid continuum could perhaps be less appropriate, holes resembling more to sort of Schottky or Frenkel defects.[19]

In the transition state theory, the enthalpy of activation ΔH_η^* in Eq. (99) is identified to the energy of activation E_η^* in Eq. (94)

$$\Delta H_\eta^* \cong E_\eta^* \tag{107}$$

and the preexponential term A_η in Eq. (94) is interpreted as

$$A_\eta = \frac{hN_A}{V} \exp\left(-\frac{\Delta S_\eta^*}{R}\right) \tag{108}$$

Generally and rigorously, ΔS_η^*, like ΔH_η^*, is valid only at a specified temperature. For the bridging system [0.464AgNO$_3$–0.436TlNO$_3$–0.100Cd(NO$_3$)$_2$ + H$_2$O],[105] ΔG_η^* has been plotted against T, at different water mole fractions, showing that ΔS_η^* varies perceptibly with T, especially at low water concentrations. Nevertheless, this dependence of ΔS_η^* on T seems sufficiently weak to warrant extraction of meaningful average values of ΔS_η^* from data on A_η.

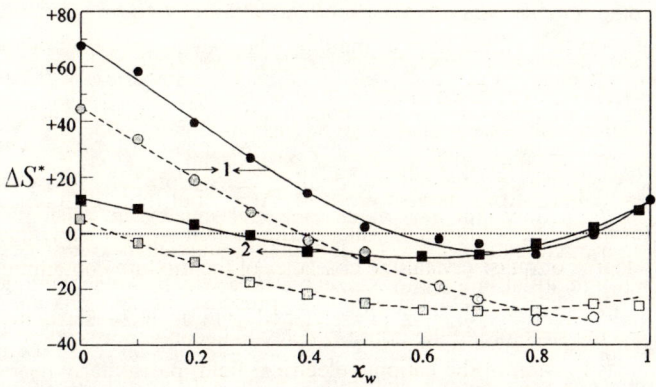

Figure 19. Entropy of activation for viscosity, ΔS_η^*, and entropy of activation for equivalent electrical conductance, ΔS_Λ^*, as functions of water mole fraction x_w. 1: [0.467TlNO$_3$–0.214CsNO$_3$–0.319Cd(NO$_3$)$_2$ + H$_2$O],[19] between 355 and 400 K. 2: [0.515AgNO$_3$–0.485 TlNO$_3$ + H$_2$O],[104] between 350 and 380 K. Solid line: ΔS_η^*. Dashed line: ΔS_Λ^*. ΔS^* in J K^{-1} mol^{-1}.

Values of ΔS_η^* have also been extracted[19, 37, 105] for other bridging nitrate-water systems: [0.515AgNO$_3$–0.485TlNO$_3$+H$_2$O],[104] [0.410LiNO$_3$ – 0.590KNO$_3$ + H$_2$O],[103] [0.464AgNO$_3$–0.436TlNO$_3$–0.100Ca(NO$_3$)$_2$ + H$_2$O],[105] and [0.467TlNO$_3$–0.214CsNO$_3$–0.319Cd(NO$_3$)$_2$ + H$_2$O].[19] Two examples of curves ΔS_η^* vs x_w are given in Fig. 19.

The curves ΔS_η^* vs x_w have been discussed[37] considering that the composition range can be divided, at the most, into four regions defined by the algebraic sign of ΔS_η^* and $\partial \Delta S_\eta^* / \partial x_w$, bounded by the pure components, the two values of x_w at which ΔS_η^* is zero, and the value of x_w at which ΔS_η^* is minimum.

In region 1, ΔS_η^* is positive while $\partial \Delta S_\eta^* / \partial x_w$ is negative. In region 2, ΔS_η^* and $\partial \Delta S_\eta^* / \partial x_w$ are both negative. In region 3, ΔS_η^* is negative while $\partial \Delta S_\eta^* / \partial x_w$ is positive. In region 4, ΔS_η^* and $\partial \Delta S_\eta^* / \partial x_w$ are both positive.

The changes in the sign of ΔS_η^* and the existence of a minimum indicate that there is no unique and simple mechanism of passage of a flow unit from one equilibrium position to another over a potential energy barrier. Several simultaneous competing mechanisms occur in different regions of the liquid, all accompanied by local structural rearrangements to which different local entropy variations are associated.

Some regions would have a relatively high degree of order, close to that of a crystal, where hole formation may obey a sort of Schottky type local mechanism. In this case, one could expect a lowering of the degree of order, and therefore a positive contribution to the entropy of activa-tion. Other regions would have a relatively low degree of order, with interstices of various sizes but too small to be occupied by a flow unit. In these parts, a flow unit may push back and squeeze up other particles yielding a modification of the local structure equivalent to a coalescence of small interstitial holes into a sizeable hole which can accommodate a flow unit. This phenomenon is associated with an increase of the local degree of order and it follows that the passage of a flow unit is accompanied by a negative contribution to the entropy of activation.

From this point of view, ΔS_η^* may be expressed by[37]

$$\Delta S_\eta^* = P_{\eta+} \Delta S_{\eta+}^* + P_{\eta-} \Delta S_{\eta-}^* \tag{109}$$

where $\Delta S^*_{\eta+}$ and $\Delta S^*_{\eta-}$ are the average positive and negative contributions to the entropy of activation, respectively. $P_{\eta+}$ and $P_{\eta-}$ are the probabilities of occurrence of a passage with a positive and negative contribution to the entropy of activation, respectively.

These four functions are expected to be generally dependent upon the temperature and the composition of the mixture in an intricate manner. Near the pure electrolyte which has a quasi-lattice structure, processes involving hole formation analogous to that of Schottky defects may predominate. As x_w increases, although the structure of the mixture is similar to that of the molten electrolyte, the electrolyte-water mixture structure becomes more disordered so that the overall entropy of activation decreases. Likewise, near pure water, since the water molecules are connected with each other in a water-like quasi-lattice structure, processes involving hole formation analogous to that of Schottky defects may predominate. As the electrolyte is added to water, the structure becomes more disordered and the overall entropy of activation decreases. Whence the appearance of a minimum in ΔS^*_η.

Over the range $x_w \approx 0$–0.4, ΔS^*_η is a linear, or quasi-linear, function of x_w expressed by[105]

$$\Delta S^*_\eta = \Delta S^*_{\eta e}\, x_e + \Delta S^{*/}_{\eta w}\, x_w \qquad (110)$$

where $\Delta S^*_{\eta e}$ is the entropy of activation of the anhydrous electrolyte and $\Delta S^{*/}_{\eta w}$, the apparent entropy of activation assigned to the perturbed water, for viscous flow.

In Tables 2 and 3 are reported the values of $\Delta H^*_{\eta e}$, $\Delta S^*_{\eta e}$, $\Delta G^*_{\eta e}$, $\Delta H^{*/}_{\eta w}$, $\Delta S^{*/}_{\eta w}$, and $\Delta G^{*/}_{\eta w}$, for the five nitrate-water systems[19, 103-105] whose viscosity has been studied as function of temperature. The negative values of $\Delta S^{*/}_{\eta w}$, like $\Delta H^{*/}_{\eta w}$, indicate that water at low concentration in these ionic systems behaves presumably as a disordered substance in contrast to ordinary pure water for which $\Delta S^*_{\eta w}$ and $\Delta H^*_{\eta w}$ are positive,[105] as seen in Figs. 16 and 19. These negative values of $\Delta S^{*/}_{\eta w}$ are consistent with the remark made in Section III.1 concerning an analogy between the perturbed water and the water vapor, since a gaseous state is characterized by a low degree of order. Although $\Delta H^{*/}_{\eta w}$ is negative, $\Delta S^{*/}_{\eta w}$ is so negative that $\Delta G^{*/}_{\eta w}$ is positive, like $\Delta G^*_{\eta w}$ for pure water (see Fig. 17).

Table 2
Enthalpy of activation $\Delta H^*_{\eta e}$, entropy of activation $\Delta S^*_{\eta e}$, and free energy of activation $\Delta G^*_{\eta e}$, of anhydrous electrolytes, for viscous flow

Electrolyte	$\Delta H^*_{\eta e}$ (kJ mol^{-1})	$\Delta S^*_{\eta e}$ (J K^{-1} mol^{-1})	$\Delta G^*_{\eta e}$ (kJ mol^{-1})
[0.515AgNO$_3$–0.485TlNO$_3$][104, a]	29	11	24.3
[0.464AgNO$_3$–0.436TlNO$_3$–0.100Cd(NO$_3$)$_2$][105, a]	39	34	26.6
[0.464AgNO$_3$–0.436TlNO$_3$–0.100Ca(NO$_3$)$_2$][105, a]	42	39	27.5
[0.467TlNO$_3$–0.214CsNO$_3$–0.319Cd(NO$_3$)$_2$][19, b]	58	68	32.4
[0.410LiNO$_3$–0.590KNO$_3$][103, c]	29	5	27.0

[a] $T = 372$ K for $\Delta G^*_{\eta e}$; 350 K $\leq T \leq$ 380 K for $\Delta H^*_{\eta e}$ and $\Delta S^*_{\eta e}$.
[b] $T = 372$ K for $\Delta G^*_{\eta e}$; 355 K $\leq T \leq$ 400 K for $\Delta H^*_{\eta e}$ and $\Delta S^*_{\eta e}$.
[c] $T = 423$ K for $\Delta G^*_{\eta e}$; 413 K $\leq T \leq$ 463 K for $\Delta H^*_{\eta e}$ and $\Delta S^*_{\eta e}$.

Table 3
Enthalpy of activation $\Delta H^{*/}_{\eta w}$, entropy of activation $\Delta S^{*/}_{\eta w}$, and free energy of activation $\Delta G^{*/}_{\eta w}$, assigned to the perturbed water, for viscous flow

Electrolyte	$\Delta H^{*/}_{\eta w}$ (kJ mol^{-1})	$\Delta S^{*/}_{\eta w}$ (J K^{-1} mol^{-1})	$\Delta G^{*/}_{\eta w}$ (kJ mol^{-1})
[0.515AgNO$_3$–0.485TlNO$_3$][104, a]	≈ -2	≈ -32	9.8
[0.464AgNO$_3$–0.436TlNO$_3$–0.100Cd(NO$_3$)$_2$][105, a]	≈ -3	≈ -38	10.6
[0.464AgNO$_3$–0.436TlNO$_3$–0.100Ca(NO$_3$)$_2$][105, a]	≈ -7	≈ -47	10.1
[0.467TlNO$_3$–0.214CsNO$_3$–0.319Cd(NO$_3$)$_2$][19, b]	≈ -13	≈ -65	11.1
[0.410LiNO$_3$–0.590KNO$_3$][103, c]	$\approx +1$	≈ -23	10.8

[a] $T = 372$ K for $\Delta G^{*/}_{\eta w}$; 350 K $\leq T \leq$ 380 K for $\Delta H^{*/}_{\eta w}$ and $\Delta S^{*/}_{\eta w}$.
[b] $T = 372$ K for $\Delta G^{*/}_{\eta w}$; 355 K $\leq T \leq$ 400 K for $\Delta H^{*/}_{\eta w}$ and $\Delta S^{*/}_{\eta w}$.
[c] $T = 423$ K for $\Delta G^{*/}_{\eta w}$; 413 K $\leq T \leq$ 463 K for $\Delta H^{*/}_{\eta w}$ and $\Delta S^{*/}_{\eta w}$.

3. Free Volume for Viscous Flow

In a study on the viscosity and the diffusivity of molecules in liquids, Hildebrand[141] has suggested that the ideas of quasi-lattice structures and activated processes could, in fact, be unrealistic and that the equation of Batschinski,[142] based on the available free volume, could be a better approach. This equation, proposed for non-associated liquids, as long ago as 1913, expresses, at a fixed composition, the reciprocal of the viscosity η, i.e. the fluidity ϕ, as an indirect function of the temperature, through the difference between the specific volume υ and a constant υ_ϕ, similar to the van der Waals co-volume

$$\phi = C_\phi \left(\upsilon - \upsilon_\phi \right) \tag{111}$$

where C_ϕ is another constant characteristic of the liquid.

Equation (111) may also be written

$$\phi = C_\phi \, \upsilon_{f\phi} \tag{112}$$

in which $\upsilon_{f\phi}$ is the specific free volume defined by

$$\upsilon_{f\phi} = \upsilon - \upsilon_\phi \tag{113}$$

If the specific volume υ is replaced by the molar volume V, Eqs. (111)–(113) are transformed into[141]

$$\phi = B_\phi \left(V - V_\phi \right) \tag{114}$$

$$\phi = B_\phi \, V_{f\phi} \tag{115}$$

$$V_{f\phi} = V - V_\phi \tag{116}$$

with the parameters B_ϕ and C_ϕ related by

$$B_\phi = \frac{C_\phi}{M} \tag{117}$$

where M is the molar mass, $V_{f\phi}$, the molar free volume of the liquid, V_ϕ and B_ϕ, two empirical constants.*

The constant V_ϕ has been designated by several expressions[52, 141]: molar co-volume, molar intrinsic volume, molar incompressible volume, molar volume at which the fluidity is zero. It has been found, in some instances, close to the molar volume of the solid at the melting point[141, 143] or close to the molar volume of the liquid in equilibrium with the solid at the melting point.[104]

The other empirical constant B_ϕ may be considered as a measure of the tendency of the fluidity to increase with the free volume so that it could depend primarily upon the interactions between the particles whereas V_ϕ would primarily depend upon the size and form of the particles, reflecting steric effects.[104]

Abraham and Abraham,[98, 144] in a derivation of the Batschinski equation from the Eyring equation, have proposed an interpretation of the physical meaning of B_ϕ based on a quantum concept of hole. In the 1930s, Eyring[114, 115] considered the holes as playing the same part in a liquid as molecules do in a gas. He thought that a liquid may be regarded as made up of holes moving about in matter just as a gas consists of molecules moving about in empty space. In a study published in 1941, Fürth[140] viewed the holes in a liquid as the counterparts of the clusters in a dense gas, formed and destroyed by the action of the statistical fluctuations, interacting with each other, and performing a kind of Brownian motion. The phenomenon of viscous flow was then explained by Fürth in a way similar to that in a gas. He assumed that the Brownian motion of the holes produces a transfer of momentum between adjacent layers of the moving liquid. These analogies between holes in matter and particles of matter in space have suggested to Abraham, Chevillot, and Brenet[145] that a hole in a liquid exhibits, in connection with its surroundings, wavelike properties.

---

* There is an error in Ref. 104, Eq. (19), p. 1480. The equation $B_\phi = 1/MC_\phi$ must be replaced by $B_\phi = C_\phi/M$, which is Eq. (117) in the present chapter.

A frequency v_h and a quantum of energy hv_h are associated to the hole. The moving hole together with the induced perturbation of the surrounding quasi-lattice is called the lacunon or fluctuating hole.[37, 98, 144, 145]

In an ideal picture of a simple liquid, ΔG_η^* in Eq. (96) has been related[37] to an equilibrium constant of hole formation K_h,

$$\Delta G_\eta^* = -RT\ln K_h \tag{118}$$

with

$$K_h = \frac{N_h}{N} \tag{119}$$

N being the number of flow units and N_h, the number of lacunons in a given volume of liquid taken as the molar volume.

From Eqs. (96), (118) and (119), one can write

$$\phi = \frac{v_h}{h}\frac{N_h}{N} \tag{120}$$

The ratio of the total volume of the lacunons ($v_h N_h$) to the free volume is assumed to be constant over the temperature range where the Batschinski equation is valid

$$v_h N_h = \xi V_{f\phi} \tag{121}$$

with ξ, the proportionality constant.

Equations (115), (120) and (121) lead to the identification of B_ϕ as

$$B_\phi = \frac{\xi}{hN} \tag{122}$$

The Bastchinski equation has been applied to various molten electrolytes[52, 143, 146] over large temperature ranges and subsequently, by Abraham and co-workers[104, 105, 144] to the following systems: [0.515AgNO$_3$–0.485 TlNO$_3$ + H$_2$O],[104] [0.410LiNO$_3$–0.590KNO$_3$ + H$_2$O],[103] [0.464AgNO$_3$–

0.436TlNO_3–$0.100\text{M(NO}_3)_2 + \text{H}_2\text{O}]$,[105] with M = Cd and Ca, and [0.467 TlNO_3–0.214CsNO_3–$0.319\text{Cd(NO}_3)_2 + \text{H}_2\text{O}]$.[19]

From fitting of Eq. (114), the constants V_ϕ and B_ϕ are determined as functions of the water mole fraction. From V_ϕ, values of $V_{f\phi}$ are calculated, at a fixed temperature, by means of Eq. (116) in which V is given by Eq. (54). From B_ϕ, values of ξ are determined by means of Eq. (122) where N is taken equal to N_A, assuming the flow units to be electro-lyte entities ($M^{z+}, z\text{NO}_3^-$) and H_2O molecules. From the values of $V_{f\phi}$ and ξ, those of ($\upsilon_h N_h$) are evaluated by means of Eq. (121).

It is seen in Fig. 20, showing some curves $V_{f\phi}$ vs x_w and ($\upsilon_h N_h$) vs x_w, that the fluctuating holes involved in the viscous flow process represent presumably only a small part of the free volume in the fused electrolytes, contrary to water where about all the free volume would appear in the form of holes which could be one reason why those fused nitrates are much more viscous than water.[98, 144]

It should be noticed that the preceding development which simultaneously relates Batschinski equation to Eyring equation through the lacunon concept and identify the first Batschinski constant, shows that the two

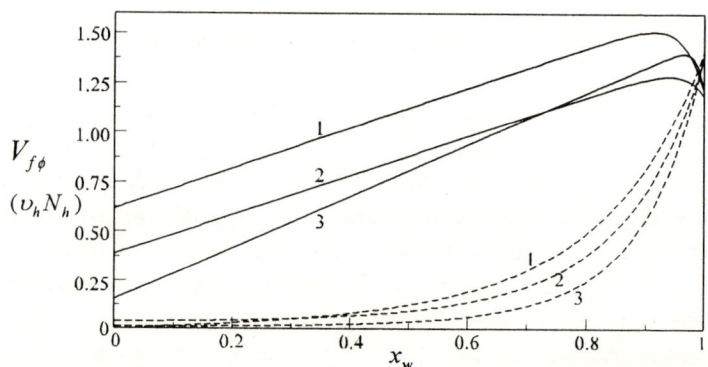

Figure 20. Molar free volume for fluidity $V_{f\phi}$ and total volume of the lacunons in one mole of solution ($\upsilon_h N_h$) as functions of water mole fraction x_w, at 372 K. 1: [0.515AgNO_3–$0.485\text{TlNO}_3 + \text{H}_2\text{O}]$.[104] 2: [$0.464\text{AgNO}_3$–$0.436\text{TlNO}_3$–$0.100\text{Ca(NO}_3)_2 + \text{H}_2\text{O}]$.[105] 3: [$0.467$ TlNO_3–0.214CsNO_3–$0.319\text{Cd(NO}_3)_2 + \text{H}_2\text{O}]$.[19] Solid line: $V_{f\phi}$. Dashed line: ($\upsilon_h N_h$). $V_{f\phi}$ and ($\upsilon_h N_h$) in $\text{cm}^3 \text{ mol}^{-1}$.

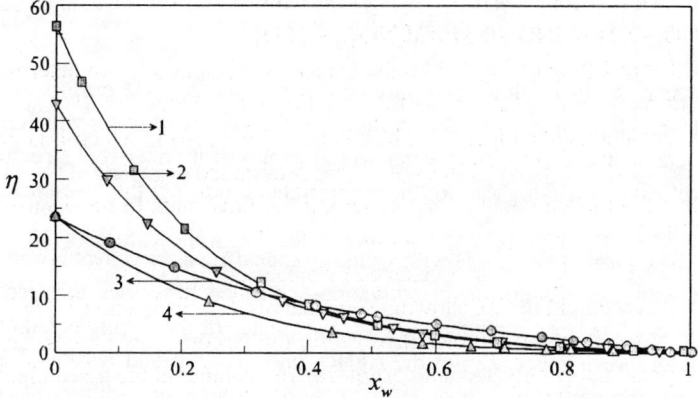

Figure 21. Isotherm of viscosity η as function of water mole fraction x_w. 1: [0.464AgNO$_3$–0.436TlNO$_3$–0.100Ca(NO$_3$)$_2$ + H$_2$O],[98] at 372 K. 2: [0.464AgNO$_3$–0.436TlNO$_3$–0.100 Cd(NO$_3$)$_2$ + H$_2$O],[71] at 372 K. 3: [KF.2HF + H$_2$O],[101] at 358 K. 4: [LiClO$_3$ + H$_2$O],[97] at 405 K. η in 10^{-3} Pa s.

approaches in fact support each other and even merge and there is no reason to give more importance to one approach than to the other.

4. Equation for Fluidity with Apparent Parameter

In a study on the systems [AgNO$_3$–TlNO$_3$–Cd(NO$_3$)$_2$ + H$_2$O], where X_{AgNO_3} / X_{TlNO_3} is fixed at 1.06, $X_{Cd(NO_3)_2}$ varying from 0 to 0.125, at 372 K, Abraham et al.[71] found that the significant decrease which occurs when water is added to the electrolyte, especially for solutions of water in molten electrolytes, may be expressed by a linear relationship between $\ln \eta$, or $\ln \phi$, and x_w over the range $x_w \approx 0$–0.6. This decrease of the viscosity with increasing water concentration has also been illustrated with other bridging systems[72,93,97,100,101,103] and some curves η vs x_w are shown in Fig. 21.

Abraham and Abraham[72] put the relationship between $\ln \phi$ and x_w in the following form, containing only one empirical parameter:

$$\phi = \left(\phi'_w\right)^{x_w} \left(\phi_e\right)^{x_e} \quad (123)$$

in which ϕ_e is the fluidity of pure electrolyte and ϕ'_w, the empirical parameter, is the apparent fluidity assigned to the perturbed water which may be seen as the fluidity pure water would exhibit if it underwent structural modifications analogous to those brought about by the presence of electrolytes.

The form of Eq. (123) with x_w and x_e appearing as exponent is consistent with the assumption of activation-controlled processes for viscous flow and has been considered as a consequence of the Eyring equation.[37]

As shown in Fig. 22, for the $AgNO_3$ and $TlNO_3$ based systems,[98] ϕ'_w reflects the perturbing power of the cations through an additive rule expressed by

$$\phi'_w = \sum_i X_{ei} \phi'_{wi} \quad (124)$$

where ϕ'_w is the parameter for a mixture of electrolytes and ϕ'_{wi}, the parameter for the component i.

Moreover, the fluidity from pure molten electrolyte to pure water may be expressed by the following two empirical parameter equation[72]:

$$\phi = \left(\phi'_w\right)^{x_w} \left(\phi'_e\right)^{x_e} + \left(\Delta\phi_w\right)^{x_w} \left(\Delta\phi_e\right)^{x_e} \quad (125)$$

where $\Delta\phi_w$ and $\Delta\phi_e$, called the excess fluidity of water and the excess fluidity of electrolyte, respectively, are defined by

$$\Delta\phi_w = \phi_w - \phi'_w \quad (126)$$

$$\Delta\phi_e = \phi_e - \phi'_e \quad (127)$$

with ϕ_w, the fluidity of pure water and ϕ'_e, the apparent fluidity of electrolyte.

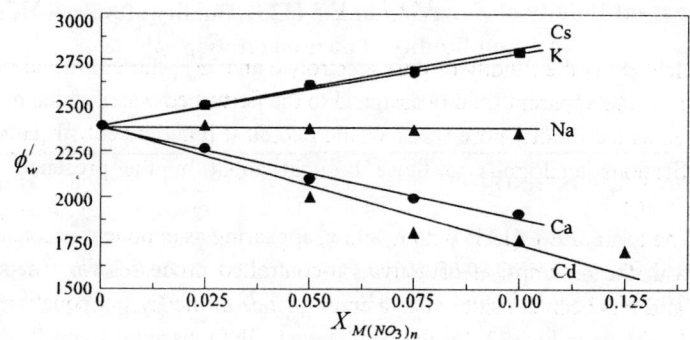

Figure 22. Apparent fluidity assigned to perturbed water, ϕ'_w in Eq. (123), for the systems [AgNO$_3$–TlNO$_3$–M(NO$_3$)$_n$ + H$_2$O],[98] where X_{AgNO_3} / X_{TlNO_3} is fixed at 1.06, with M = Na, K, Cs, Cd, and Ca, as function of the mole fraction of the added salt, $X_{M(NO_3)_n}$ at 372 K. ϕ'_w in Pa^{-1} s^{-1}.

Similarly to the apparent fluidity of water, the apparent fluidity of electrolyte is defined as the fluidity the electrolyte would exhibit if it underwent structural modifications analogous to those brought about by the presence of water at low concentration, for example, the degree of defects or holes.

It was found[37, 72, 98, 104, 105] that ϕ'_e is very close to ϕ_e, whereas ϕ'_w and ϕ_w are quite different. Examples of ϕ'_w values are given in Table 4 together with ϕ_w and ϕ_e for various hydrate melts.[19, 72, 96, 98, 101, 103–105]

It must be underlined that Eq. (125) is consistent with the cybotactic theory[38] and the significant structure theory[113] adapted to feature a picture of electrolyte-water systems comprising submicroscopic parts such as pure electrolyte, electrolyte with small amounts of water, ordinary water, perturbed water, ion pairs, complex ion clusters ... whose relative quantities vary with the water mole fraction. The fact that Eq. (125) reduces to Eq. (123) for the solutions of water in electrolyte is in good agreement with a picture comprising above all pure electrolyte and perturbed water submicroscopic parts while for solutions rich in water, other submicroscopic parts have to be taken into account.

Table 4
Apparent fluidity of water ϕ'_w in Eq. (123), fluidity of pure water ϕ_w, and fluidity of pure electrolyte ϕ_e

Electrolyte	T (K)	ϕ'_w (Pa^{-1}s^{-1})	ϕ_w [a] (Pa^{-1}s^{-1})	ϕ_e (Pa^{-1}s^{-1})
$[0.515AgNO_3–0.485TlNO_3]$[72, 104]	358	2420	3000	31.0
	372	2390	3490	44.4
$[0.489AgNO_3–0.461TlNO_3–0.050NaNO_3]$[98]	372	2370	3490	41.8
$[0.464AgNO_3–0.436TlNO_3–0.100NaNO_3]$[98]	372	2345	3490	39.5
$[0.489AgNO_3–0.461TlNO_3–0.050KNO_3]$[98]	372	2610	3490	41.2
$[0.464AgNO_3–0.436TlNO_3–0.100KNO_3]$[98]	372	2790	3490	37.9
$[0.489AgNO_3–0.461TlNO_3–0.050CsNO_3]$[98]	372	2610	3490	40.4
$[0.464AgNO_3–0.436TlNO_3–0.100CsNO_3]$[72]	372	2800	3490	37.0
$[0.489AgNO_3–0.461TlNO_3–0.050Ca(NO_3)_2]$[98]	372	2100	3490	28.5
$[0.464AgNO_3–0.436TlNO_3–0.100Ca(NO_3)_2]$[98,105]	372	1920	3490	17.7
	381	1830	3855	24.8
$[0.489AgNO_3–0.461TlNO_3–0.050Cd(NO_3)_2]$[98]	372	1990	3490	32.7
$[0.464AgNO_3–0.436TlNO_3–0.100Cd(NO_3)_2]$[72,105]	372	1760	3490	23.3
	381	1640	3855	31.6
$[0.467TlNO_3–0.214CsNO_3–0.319Cd(NO_3)_2]$[19]	365	1490	3240	3.0
	388	1170	4120	9.5
$[0.500AgNO_3–0.500TlNO_3]$[96, b]	373	2700	3550	47.1
$[LiClO_3]$[97, b]	405	3870	4790	42.4
$[0.410LiNO_3–0.590KNO_3]$[103, b]	423	3000	5510	51.8
$[KF.2HF]$[101, b]	358	570	3000	42.5

[a] ϕ_w data from Ref. 147.
[b] ϕ'_w evaluated by the writers.

If an analysis is made with the mole fraction on an ionized basis,[98] the experimental data cannot be represented over the whole concentration range by an equation similar to Eq. (125). However, in dilute aqueous solutions, i.e. over the range $y_w \approx 0.55-1$, which corresponds to $x_w \approx 0.75-1$, the following equation is obeyed:

$$\phi = \left(\phi_w\right)^{y_w} \left(\phi_e''\right)^{y_e} \qquad (128)$$

with ϕ_e'', an empirical parameter.

The parameter ϕ_e'' plays formally a role similar to that of ϕ_w' in Eq. (123). Arguing from this similarity, ϕ_e'' has been defined as the fluidity that the electrolyte would exhibit if it underwent structural modifications analogous to those brought about by the presence of water at high concentration, in contrast to ϕ_e' which is defined in the concentration region rich in electrolyte.

It was observed that the perturbed salt behaves in dilute aqueous solutions as a liquid whose fluidity ϕ_e'' is higher than that of the anhydrous salt ϕ_e whereas, as seen in Table 4, in very concentrated solutions, the perturbed water behaves as a liquid whose fluidity ϕ_w' is lower than that of ordinary water ϕ_w. This is exemplified with the following systems: [0.515AgNO$_3$–0.485TlNO$_3$],[72] for which $\phi_e'' \approx 125$ Pa^{-1} s^{-1} and ϕ_e = 44.4 Pa^{-1} s^{-1}, at 372 K, [0.464AgNO$_3$–0.436TlNO$_3$–0.100Cd(NO$_3$)$_2$],[72] for which $\phi_e'' \approx 75$ Pa^{-1} s^{-1} and ϕ_e = 23.3 Pa^{-1} s^{-1}, at 372 K, and [0.467TlNO$_3$–0.214CsNO$_3$–0.319Cd(NO$_3$)$_2$],[19] for which $\phi_e'' \approx 26$ Pa^{-1} s^{-1} and ϕ_e = 7.8 Pa^{-1} s^{-1}, at 383 K.

The shifting of the linear relationship between the logarithm of the fluidity and the concentration towards the dilute aqueous solution range when the ordinary mole fraction is replaced by the mole fraction on an ionized basis and the fact that ϕ_e'' is greater than ϕ_e may be ascribed, at least partially, to the isolation of the ions at high dilution in the solutions. It has been noticed that a relatively high temperature seems required for a simple equation, such as Eq. (128), to be valid.

5. Activation Energy for Electrical Conductance

The temperature dependence of electrical conductance at constant pressure has often been represented by the following Arrhenius type equations[52, 108]

$$\kappa = A_\kappa \exp\left(-\frac{E_\kappa^*}{RT}\right) \qquad (129)$$

$$\Lambda = A_\Lambda \exp\left(-\frac{E_\Lambda^*}{RT}\right) \qquad (130)$$

where the preexponential terms A_κ, A_Λ, and the activation energies E_κ^*, E_Λ^*, for the electrical conductivity κ and the equivalent electrical conductance Λ, are taken as constants over given temperature ranges.

The equivalent electrical conductance Λ is defined by

$$\Lambda = \frac{\kappa}{C_{eq}} \qquad (131)$$

with C_{eq}, the electrolyte concentration in equivalent per unit of volume.

For a solution in which the electrolyte is a mixture of several components, C_{eq} is related to the molar volume V by

$$C_{eq} = \frac{x_e \sum_i X_{ei} n_{ei}}{V} \qquad (132)$$

where n_{ei} is the electrochemical valency defined by

$$n_{ei} = v_{i+} z_{i+} = v_{i-} z_{i-} \qquad (133)$$

v_{i+} and v_{i-} being the number of positive and negative ions respectively produced by the total dissociation of one mole of the electrolyte i, z_{i+}, the cation i valency, and z_{i-}, the anion i valency.

For a mixture of electrolytes, the equivalent electrical conductance Λ in Eq. (131) may be seen as the equivalent conductance of a virtual single electrolyte which would consist of anions and cations carrying the same value of electrical charge and having the same mobility[37, 106]:

$$\Lambda = 2F\bar{u}_i \qquad (134)$$

where F is the Faraday constant and $\bar{u}_i$, the average of the mobilities of all ions, is defined by

$$\bar{u}_i = \frac{\sum_i X_{ei}\left(v_{i+}\,u_{i+} + v_{i-}\,u_{i-}\right)}{\sum_i X_{ei}\left(v_{i+} + v_{i-}\right)} \qquad (135)$$

u_{i+} and u_{i-} being the cationic and anionic mobilities, respectively.

The activation energies E_κ^* and E_Λ^* are related by the following equation[148]:

$$E_\Lambda^* = E_\kappa^* + RT^2\alpha \qquad (136)$$

where α is the coefficient of thermal expansion of the solution given by

$$\alpha = \frac{1}{V}\left(\frac{\partial V}{\partial T}\right)_P \qquad (137)$$

For the bridging system [0.410LiNO$_3$–0.590KNO$_3$ + H$_2$O], Claes, Michielsen, and Glibert[103] found a linear relationship between ln Λ and $1/T$, whatever the water mole fraction, and extracted constant values of E_Λ^* valid over the explored temperature range, i.e. between 413 and 463 K.

Equation (130) has also been applied to the following bridging systems: [0.515AgNO$_3$–0.485TlNO$_3$ + H$_2$O],[104] [0.464AgNO$_3$–0.436TlNO$_3$–0.100M(NO$_3$)$_2$ + H$_2$O],[106] with M = Cd and Ca, [0.467TlNO$_3$–0.214 CsNO$_3$–0.319Cd(NO$_3$)$_2$ + H$_2$O],[19] [KF.2HF + H$_2$O][101] and solutions of water in LiClO$_3$.[149] Two curves E_Λ^* vs x_w are represented in Fig. 23.

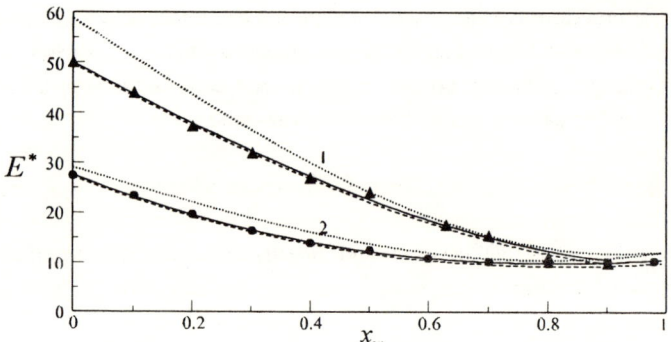

Figure 23. Activation energy for viscosity, E_η^*, activation energy for equivalent electrical conductance, E_Λ^*, and activation energy for electrical conductivity, E_κ^*, as functions of water mole fraction x_w. 1: [0.467TlNO$_3$–0.214CsNO$_3$–0.319Cd(NO$_3$)$_2$ + H$_2$O],[19] between 355 and 400 K. 2: [0.515AgNO$_3$–0.485TlNO$_3$ + H$_2$O],[104] between 350 and 380 K. Dotted line: E_η^* in Eq. (94). Solid line: E_Λ^* in Eq. (130). Dashed ligne: E_κ^* in Eq. (129). E^* in kJ mol^{-1}.

Dumont, Qian, and Conway[101] have observed for the system [KF.2HF + H$_2$O]* a linearity between E_Λ^* and x_w over the very large water concentration range $x_w \approx 0$–0.88.

By analogy with Eq. (95) for viscosity, this relationship may be expressed by

$$E_\Lambda^* = E_{\Lambda e}^* x_e + E_{\Lambda \infty}^{*/} x_w \tag{138}$$

where $E_{\Lambda e}^*$ is the activation energy for the equivalent electrical conductance of the anhydrous electrolyte and $E_{\Lambda \infty}^{*/}$, the apparent activation energy for the equivalent electrical conductance of the electrolyte at infinite dilution in the perturbed water.

---

* In Ref. 101, p. 157, in Fig. 9, values of $\Delta E_{activation}$, which is $-E_\Lambda^*$ in Eq. (130) of this chapter, are erroneous. Calculations made by the writers with the experimental data provided by courtesy of Dr Dumont give values of E_Λ^* between 7 and 22 kJ mol^{-1} and not between 0.1 and 0.4 kJ mol^{-1}. Besides, there is obviously an error concerning the experimental value corresponding to $x_w = 0$ which is far apart from the straight line; the experimental value is $E_\Lambda^* = 13.3$ kJ mol^{-1} while the extrapolated value is $E_\Lambda^* = 22.4$ kJ mol^{-1}.

This equation is also valid for nitrate-water systems[19, 103, 104, 106] over the concentration range $x_w \approx 0-0.5$.

Equation (138) for equivalent electrical conductance, like Eqs. (95) and (123) for viscous flow, is consistent with a picture of hydrate melts inspired by the cybotactic theory[38] and the significant structure theory.[113] Obedience to Eqs. (95), (123) and (138) suggests that H_2O is not involved in chemical reactions with ions produced by the electrolyte dissociation. With regard to the salt $KF \cdot 2HF$ made up of K^+ and $(HF)_2F^-$ ions, Dumont, Qian, and Conway[101] concluded from thermodynamic investigations that water added to the electrolyte is ionized to a negligible extent.

For solutions of water in $LiClO_3$, Campbell and Williams[149] have compared E_κ^* in Eq. (129) and E_Λ^* in Eq. (130). Since α in Eq. (136) is small and T is relatively low, E_κ^* and E_Λ^* were found very close to one another. It can be verified that it is also the case for systems from molten electrolyte to dilute aqueous solution,[19, 101, 103, 104, 106] as illustrated in Fig. 23. Besides, the ratio E_η^* / E_Λ^*, usually close to unity for nitrate melts,[52] remains quasi-constant at 1.1 ! 0.1 as water is added to the nitrate mixtures over the whole concentration range, suggesting analogous hole mechanisms for both, the electrical conductance and the viscous flow, in dilute aqueous solutions as well as in molten nitrates.[104]

6. Transition State Theory of Electrical Conductance

The transition state theory of the mobility of the hydrogen ion in hydroxylic solvent, proposed by Stearn and Eyring[150] has been adapted to the ionic mobility of interstitial ions in liquid silicates by Bockris et al.[151] Further applications of this approach were made to molten chlorides[152, 153] and nitrates.[154] Abraham and Abraham[37] have applied the transition state theory to bridging systems, from pure electrolyte to dilute aqueous solutions, extending the equation proposed by Bockris et al.[151] to take into account the presence of water and various ions, giving:

$$\Lambda = \frac{4}{3} \frac{F^2}{hN_A} E_e d^2 z \exp\left(-\frac{\Delta G_\Lambda^*}{RT}\right) \qquad (139)$$

where ΔG_Λ^* represents the free energy of activation for the conduction process of the electrolyte whose Λ is related to $\bar{u}_i$ by Eq. (134).

The parameter z is an average value of the number of charges carried by all the anions and cations:

$$z = \frac{2\sum_i X_{ei} n_{ei}}{\sum_i X_{ei}\left(v_{i+} + v_{i-}\right)} \quad (140)$$

The parameter d is something like an average of half the migration distances across the potential barriers taken as equal to the radius that corresponds to the average of all the particle volumes, including water:

$$d = \left(\frac{1}{x_e \sum_i X_{ei}\left(v_{i+} + v_{i-}\right) + x_w}\right)^{1/3} \left(\frac{3}{4\pi}\frac{V}{N_A}\right)^{1/3} \quad (141)$$

The effective electrical field resulting from a unit applied field strength E_e is expressed by

$$E_e = \left[1 - \left(x_e \frac{V_e^o\left(D_e - 1\right)}{V\left(D_e + 2\right)}\right) + x_w \frac{V_w^o\left(D_w - 1\right)}{V\left(D_w + 2\right)}\right]^{-1} \quad (142)$$

where D_e and D_w are the dielectric constants of electrolyte and water, respectively.

The free energy of activation ΔG_Λ^* is related to the enthalpy of activation ΔH_Λ^* and the entropy of activation ΔS_Λ^*, for the electrical conduction process by

$$\Delta H_\Lambda^* = \Delta G_\Lambda^* - T\left(\frac{\partial \Delta G_\Lambda^*}{\partial T}\right) \quad (143)$$

and

$$\Delta S_\Lambda^* = -\left(\frac{\partial \Delta G_\Lambda^*}{\partial T}\right) \quad (144)$$

Values of ΔG_Λ^* in Eq. (139) have been determined[19, 37, 98, 99, 106] for the following systems: [0.515AgNO$_3$–0.485TlNO$_3$ + H$_2$O],[72, 104] [0.464 AgNO$_3$–0.436TlNO$_3$–0.100M(NO$_3$)$_n$ + H$_2$O],[72, 98, 106] with M = K, Cs, Cd, and Ca, [0.467TlNO$_3$–0.214CsNO$_3$–0.319Cd(NO$_3$)$_2$ + H$_2$O],[19, 99] [0.500 AgNO$_3$–0.500TlNO$_3$ + H$_2$O],[96] [LiClO$_3$ + H$_2$O],[97] [AgNO$_3$ + H$_2$O],[102] [NH$_4$NO$_3$ + H$_2$O],[102] [0.667NH$_4$NO$_3$–0.258LiNO$_3$–0.075NH$_4$Cl + H$_2$O],[93] [0.410LiNO$_3$–0.590KNO$_3$ + H$_2$O],[103] and, in this chapter, for the system [KF.2HF + H$_2$O].[101] In Fig. 24, ΔG_Λ^* vs x_w curves are given for three systems.

A linear, or quasi-linear, relationship is observed between ΔG_Λ^* and x_w for all the above-mentioned systems, over the range $x_w \approx 0$–0.5, which may be expressed by the following equation[37, 98, 99]:

$$\Delta G_\Lambda^* = \Delta G_{\Lambda e}^* x_e + \Delta G_{\Lambda \infty}^{*\prime} x_w \quad (145)$$

where $\Delta G_{\Lambda e}^*$ is the free energy of activation for the equivalent electrical conductance of the anhydrous electrolyte and $\Delta G_{\Lambda \infty}^{*\prime}$, the apparent free energy of activation for the equivalent electrical conductance of the electrolyte at infinite dilution in the perturbed water.

It must be underscored that $\Delta G_{\Lambda e}^*$ and $\Delta G_{\eta e}^*$ appear close to one another for some nitrate systems,[19, 37, 98, 99] indicating similarities in viscous flow and electrical conductance mechanisms in fused electrolytes. However, as water is added to the anhydrous electrolyte, the difference between the free energies of activation ΔG_Λ^* and ΔG_η^* increases, as seen in Fig. 24. This has been discussed[19, 37, 106] in investigations regarding the influence of temperature on viscosity and electrical conductance on the following systems: [0.515AgNO$_3$–0.485TlNO$_3$ + H$_2$O],[104] [0.464AgNO$_3$–0.436TlNO$_3$–0.100 M(NO$_3$)$_2$ + H$_2$O],[105, 106] with M = Cd and Ca, [0.467 TlNO$_3$–0.214CsNO$_3$–0.319Cd(NO$_3$)$_2$ + H$_2$O],[19] and [0.410LiNO$_3$–0.590 KNO$_3$ + H$_2$O].[103] Since for these systems ΔH_Λ^* (or E_Λ^*) is near ΔH_η^* (or

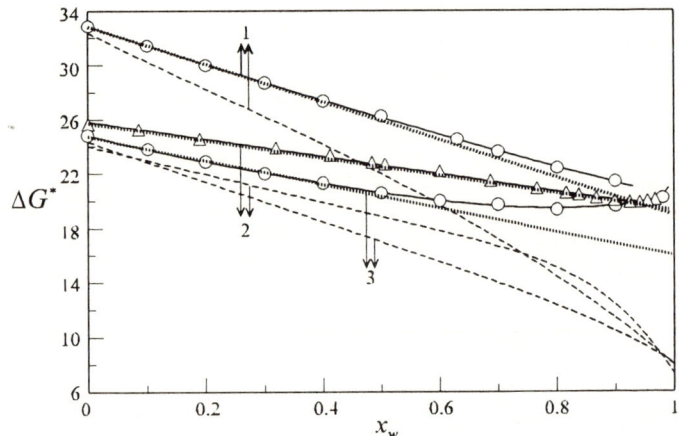

Figure 24. Free energy of activation for equivalent electrical conductance ΔG_Λ^* and free energy of activation for viscous flow ΔG_η^* as functions of water mole fraction x_w. 1: [0.467TlNO$_3$–0.214CsNO$_3$–0.319Cd(NO$_3$)$_2$ + H$_2$O],[19] at 372 K. 2: [KF.2HF + H$_2$O],[101] at 358 K. 3: [0.515AgNO$_3$–0.485 TlNO$_3$ + H$_2$O][71,72] at 372 K. Solid line: ΔG_Λ^* in Eq. (139). Dashed line: ΔG_η^* in Eq. (98). Dotted line: linear relationship Eq. (145). ΔG^* in kJ mol^{-1}.

E_η^*), as seen in Fig. 23, the difference ($\Delta G_\Lambda^* - \Delta G_\eta^*$) must be looked for in an increase of the difference between ΔS_Λ^* and ΔS_η^*.

In Fig. 19, it is observed that ($\Delta S_\eta^* - \Delta S_\Lambda^*$) becomes larger as water is added to the electrolyte, in particular for x_w ‰ 0.5, in connection with the fact that ΔS_Λ^* does not exhibit a marked minimum contrary to ΔS_η^*. For the electrical conductance, the unique participating component is the electrolyte itself over the whole concentration range whereas for viscous flow, electrolyte and water are both participating components. For solutions rich in water, with a water-like quasi-lattice structure, although the ions interact with the water molecules, they are, less than water itself, structural parts of the quasi-lattice. Consequently, they move under the influence of an electric field in obedience to predominant mechanisms in which neighboring water molecules are squeezed up. This is equivalent to coalescence of small holes into larger holes. Whence an overall entropy of activation for the electrical conductance which decreases with increasing water concentration, even in dilute aqueous solutions.

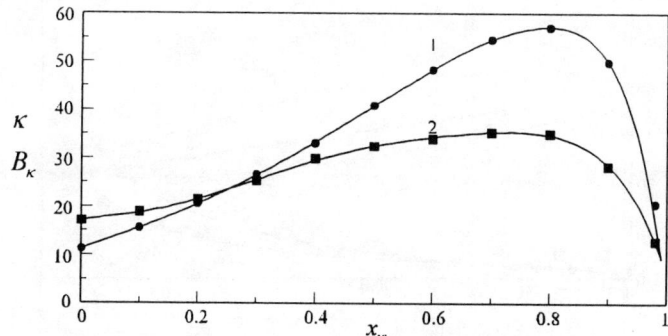

Figure 25. Electrical conductivity κ and parameter B_κ in Eq. (146) as functions of water mole fraction x_w for the system [0.515AgNO$_3$–0.485TlNO$_3$ + H$_2$O].[104] 1: κ, at 372 K. 2: B_κ, between 350 and 380 K. κ in S m^{-1}. B_κ in 10^6 S mol m^{-4}.

7. Free Volume for Electrical Conductance

Abraham and Abraham[104] have extended utilization of Batschinski type equations to the electrical conductance and, by analogy with the fluidity, wrote the following equations:

$$\kappa = B_\kappa \left(V - V_\kappa \right) \tag{146}$$

$$\kappa = B_\kappa V_{f\kappa} \tag{147}$$

$$V_{f\kappa} = V - V_\kappa \tag{148}$$

and

$$\Lambda = B_\Lambda \left(V - V_\Lambda \right) \tag{149}$$

$$\Lambda = B_\Lambda V_{f\Lambda} \tag{150}$$

$$V_{f\Lambda} = V - V_\Lambda \tag{151}$$

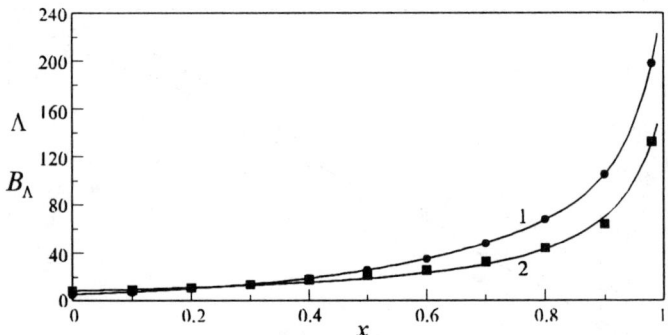

Figure 26. Equivalent electrical conductance Λ and parameter B_Λ in Eq. (149) as functions of water mole fraction x_w for the system [0.515AgNO$_3$–0.485TlNO$_3$ + H$_2$O].[104] 1: Λ, at 372 K. 2: B_Λ, between 350 and 380 K. Λ in 10^{-4} S m^2 eq^{-1}. B_Λ in 10^2 S mol m^{-1} eq^{-1}.

in which V_κ and V_Λ are two empirical constants called molar co-volumes, B_κ and B_Λ are other constants measuring the tendency of the electrical conductance to increase with the molar free volumes $V_{f\kappa}$ and $V_{f\Lambda}$, for the electrical conductivity κ and equivalent electrical conductance Λ, respectively.

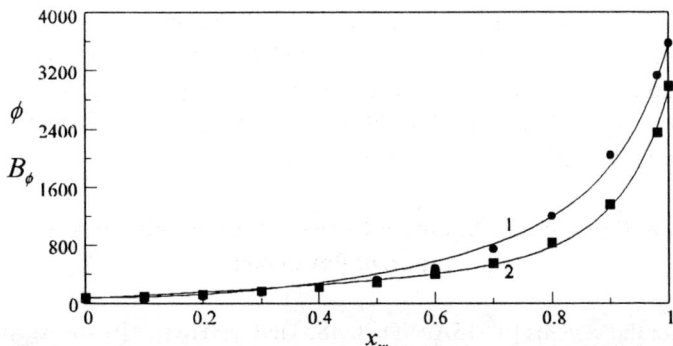

Figure 27. Fluidity ϕ and parameter B_ϕ in Eq. (114) as functions of water mole fraction x_w for the system [0.515AgNO$_3$–0.485TlNO$_3$ + H$_2$O].[104] 1: ϕ, at 372 K. 2: B_ϕ, between 350 and 380 K. ϕ in Pa^{-1} s^{-1}. B_ϕ in 10^6 mol Pa^{-1} s^{-1} m^{-3}.

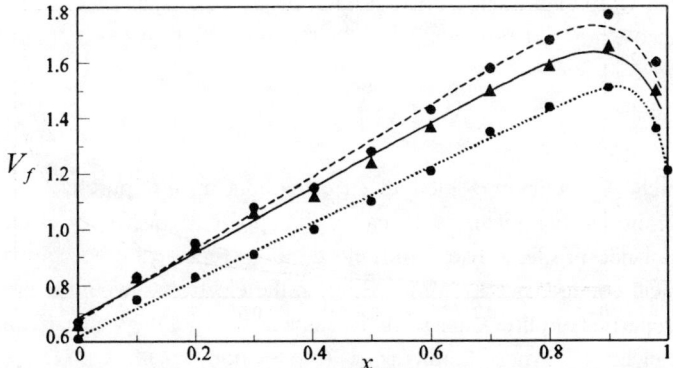

Figure 28. Molar free volume for electrical conductivity $V_{f\kappa}$, molar free volume for equivalent electrical conductance $V_{f\Lambda}$, and molar free volume for fluidity $V_{f\phi}$, as functions of water mole fraction x_w for the system $[0.515AgNO_3-0.485TlNO_3 + H_2O]$,[104] at 372 K. Dashed line: $V_{f\kappa}$. Solid line: $V_{f\Lambda}$. Dotted line: $V_{f\phi}$. V_f in cm³ mol⁻¹.

Equations (146)–(151) have been applied[104] to the systems $[0.515 AgNO_3-0.485TlNO_3 + H_2O]$[104] and $[0.410LiNO_3-0.590KNO_3 + H_2O]$,[103] from fused salts to dilute aqueous solutions. As illustrated in Figs. 25–27 with the system $[0.515AgNO_3-0.485TlNO_3 + H_2O]$, B_κ and B_Λ as well as B_ϕ in the Batschinski equation for fluidity, Eq. (114), behave respectively like the functions κ, Λ, and ϕ when plotted against x_w.

The molar free volumes $V_{f\kappa}$, $V_{f\Lambda}$, and $V_{f\phi}$ are close together, over the whole water concentration range for the system $[0.515AgNO_3-0.485 TlNO_3 + H_2O]$,[104] as seen in Fig. 28, and, over the range $x_w \approx 0$–0.8 for the system $[0.410LiNO_3-0.590KNO_3 + H_2O]$.[103]

8. Equation for Equivalent Electrical Conductance with Apparent Parameter

For the systems $[0.515AgNO_3-0.485TlNO_3 + H_2O]$,[72] $[0.464AgNO_3-0.436TlNO_3-0.100M(NO_3)_n + H_2O]$,[72] with M = Cs and Cd, and $[0.410 LiNO_3-0.590KNO_3 + H_2O]$,[103] Abraham and Abraham[72] have proposed the following equation, with only one empirical parameter, to represent at fixed temperature experimental data on the equivalent electrical conductance Λ as function of x_w, over the range $x_w \approx 0$–0.6:

$$\Lambda = \left(\Lambda'_\infty\right)^{x_w} \left(\Lambda_e\right)^{x_e} \tag{152}$$

in which Λ_e is the equivalent electrical conductance of pure electrolyte. The empirical parameter Λ'_∞, called the apparent equivalent electrical conductance of electrolyte at infinite dilution, represents the equivalent electrical conductance at infinite dilution the electrolyte would exhibit if the properties of pure water were the same as those of the water dissolved in the electrolyte whose fluidity is ϕ'_w (see Section III.4).

Equation (152) has also been applied to other hydrate melts and various values of Λ'_∞ and Λ_e are listed in Table 5. For the AgNO$_3$ and TlNO$_3$ based systems, it must be pointed out that the order of increasing Λ'_∞, like ϕ'_w in Fig. 22, is the same as that of increasing water structure breaking power of the ions in dilute aqueous solutions,[135–139] and the order of decreasing $\Delta G^{*\prime}_{\eta w}$ in Fig. 18.

The common exponential form of the curves Λ vs x_w and ϕ vs x_w, from pure molten electrolyte to dilute aqueous solutions (see Figs. 26 and 27) has suggested to represent Λ by the following two empirical parameter equation,[72] similar to Eq. (125) which expresses ϕ:

$$\Lambda = \left(\Lambda'_\infty\right)^{x_w} \left(\Lambda'_e\right)^{x_e} + \left(\Delta\Lambda_\infty\right)^{x_w} \left(\Delta\Lambda_e\right)^{x_e} \tag{153}$$

where $\Delta\Lambda_\infty$ is the excess equivalent electrical conductance of the electrolyte at infinite dilution in water and $\Delta\Lambda_e$, the excess equivalent electrical conductance of the anhydrous electrolyte defined by

$$\Delta\Lambda_\infty = \Lambda_\infty - \Lambda'_\infty \tag{154}$$

$$\Delta\Lambda_e = \Lambda_e - \Lambda'_e \tag{155}$$

with Λ'_e, the apparent equivalent electrical conductance of the anhydrous electrolyte and Λ_∞, the equivalent electrical conductance of the electrolyte at infinite dilution in water obtained by extrapolation of Eq. (153).

Table 5
Apparent equivalent electrical conductance at infinite dilution in perturbed water, Λ'_∞ in Eq. (152), and equivalent electrical conductance of pure electrolyte, Λ_e

Electrolyte	T (K)	$\Lambda'_\infty \times 10^4$ (S m^2 eq^{-1})	$\Lambda_e \times 10^4$ (S m^2 eq^{-1})
[0.515AgNO$_3$–0.485TlNO$_3$][72, 104]	348	123	3.0
	372	124	5.2
[0.464AgNO$_3$–0.436TlNO$_3$–0.100KNO$_3$][98]	372	140	4.6
[0.464AgNO$_3$–0.436TlNO$_3$–0.100CsNO$_3$][72]	372	145	4.4
[0.464AgNO$_3$–0.436TlNO$_3$–0.100Ca(NO$_3$)$_2$][105]	372	110	2.3
	381	105	3.1
[0.464AgNO$_3$–0.436TlNO$_3$–0.100Cd(NO$_3$)$_2$][105]	372	96	2.9
	381	90	3.8
[0.467TlNO$_3$–0.214CsNO$_3$–0.319Cd(NO$_3$)$_2$][19]	365	63	0.4
	388	58	1.1
[0.500AgNO$_3$–0.500TlNO$_3$][96]*	373	126	5.4
[AgNO$_3$][102]*	495	127	31.0
[NH$_4$NO$_3$][102]*	453	150	24.1
[LiClO$_3$][97]*	405	39	5.1
[0.410LiNO$_3$–0.590KNO$_3$][103]*	423	153	4.7
[0.667NH$_4$NO$_3$–0.258LiNO$_3$–0.075NH$_4$Cl][93]*	366	90	4.4
[KF.2HF][101]*	358	70	6.3

* Λ'_∞ evaluated by the writers.

Since the equivalent electrical conductance of a solution is an homogeneous function of degree one of the ionic mobilities, the physical meaning of the above-mentioned parameters proceeds from their relation to these ionic mobilities expressed by Eqs. (134) and (135). Equations (154) and (155) correspond to the excess mobility of the ion i at infinite dilution in water $\Delta u_{i\infty}$, and the excess mobility of the ion i in the anhydrous electrolyte Δu_i, defined as follows

$$\Delta u_{i\infty} = u_{i\infty} - u'_{i\infty} \tag{156}$$

$$\Delta u_i = u_i - u'_i \tag{157}$$

where u_i and $u_{i\infty}$ are the mobilities of the ionic species i in the anhydrous electrolyte and at infinite dilution in water, respectively. The parameter $u'_{i\infty}$ is the mobility at infinite dilution the ionic species i would exhibit if the structure and properties of pure water were the same as that of the perturbed water and, similarly, the parameter u'_i is the mobility the ionic species i would exhibit in pure electrolyte if its structure and properties were the same as that of the electrolyte modified by the presence of water.

It has been observed[98] that, contrary to the fluidity, the logarithm of the equivalent electrical conductance $\ln \Lambda$ is not a linear function of the water mole fraction on an ionized basis y_w over the dilute aqueous solution concentration range where extensions of the Debye, Hückel and Onsager theories are expected to apply. Consequently, there is no equation similar to Eq. (128) for fluidity. However, a linear relation between $\ln \Lambda$ and y_w is perceptible over an intermediate concentration range.

9. Relation Between Viscosity and Electrical Conductance

Abraham and Abraham[104] have found that for the bridging systems [0.515 AgNO$_3$–0.485TlNO$_3$ + H$_2$O][104] and [0.410LiNO$_3$–0.590KNO$_3$ + H$_2$O],[103] over the range $x_w \approx 0$– 0.7, the parameters B_κ in Eq. (146) and B_ϕ in Eq. (114) are related in the following simple manner:

Thermodynamic and Transport Properties 347

$$\frac{B_\kappa}{B_\phi} = m_B\, x_e \qquad (158)$$

where m_B is the proportionality constant.

Since over the range $x_w \approx 0\text{–}0.7$, the ratio $V_{f\kappa}/V_{f\phi}$ is a quasi-constant at a given temperature and the ratio B_κ / B_ϕ obeys Eq. (158), the following equation can be deduced from Eqs. (115) and (147):

$$\frac{\kappa}{\phi} = \eta\kappa = \eta_e \kappa_e\, x_e \qquad (159)$$

in which κ_e and η_e are the electrical conductivity and the viscosity of the pure electrolyte, respectively.

This equation is also applicable[72, 99, 104, 129, 144] to the bridging systems [0.464AgNO$_3$–0.436TlNO$_3$–0.100M(NO$_3$)$_n$ + H$_2$O],[72, 129] with M = K, Cs, Cd, and Ca, [0.467TlNO$_3$–0.214CsNO$_3$–0.319Cd(NO$_3$)$_2$ + H$_2$O],[99] [0.500 AgNO$_3$–0.500TlNO$_3$ + H$_2$O],[96] but fails when tested with the systems [LiClO$_3$ + H$_2$O][97] and [KF.2HF + H$_2$O].[101]

The product of the viscosity and the equivalent electrical conductance, referred to as the Walden product $\eta\Lambda$, has often been considered in structural discussions on hydrate melts.[21, 22, 72, 93, 100, 101, 129] Some curves $\eta\Lambda$ vs x_w are shown in Fig. 29.

The first determinations of the Walden product over the whole water concentration range were made with the system [0.667NH$_4$NO$_3$–0.258 LiNO$_3$–0.075NH$_4$Cl + H$_2$O], at 366 K, by Glibert et al.[93] who pointed out that $\eta\Lambda$ decreases rapidly over the range $x_w \approx 0\text{–}0.2$ and remains approximately constant between 0.2 and 1 which is, according to these authors, surprising since at high electrolyte concentration the relaxation and electrophoretic effects should have considerable importance.*

For the system [LiClO$_3$ + H$_2$O],[97] at 405 K, the existence of a shallow minimum leading to a quasi-constancy of the Walden product over the range $x_w \approx 0.65\text{–}0.85$, seen in Fig. 29, has suggested to Pitzer[21, 22] full ion-

---

* In our opinion, viscosity measurements of the system [0.667NH$_4$NO$_3$–0.258LiNO$_3$– 0.075 NH$_4$Cl + H$_2$O] should be redone before conducting any discussion implying viscosity data on this system (see note in Section III.2).

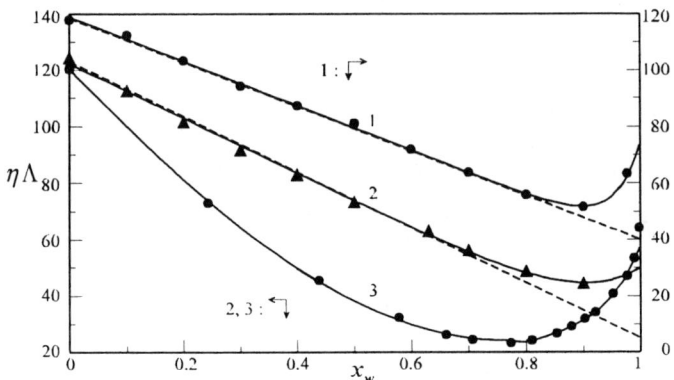

Figure 29. Walden product $\eta\Lambda$ as function of water mole fraction x_w. 1: [0.515AgNO$_3$–0.485TlNO$_3$ + H$_2$O],[104] at 372 K. 2: [0.467TlNO$_3$–0.214CsNO$_3$–0.319Cd(NO$_3$)$_2$ + H$_2$O],[19] at 372 K. 3: [LiClO$_3$ + H$_2$O],[97] at 405 K. $\eta\Lambda$ in 10^{-7} C^2 m^{-1} eq^{-1}.

ization for hydrate melts as well as for dilute aqueous solutions in which the electrostatic effects are responsible for a decrease in equivalent electrical conductance from its limiting value following theoretical equations for fully ionized solute. [155, 156]

For the bridging nitrate-water systems[72, 96, 99, 103, 129] obeying Eq. (159), Abraham and Abraham[72] have found that the Walden product is a linear function of x_w over a large concentration range, as illustrated in Fig. 29 with two systems.

Using the concept of the apparent molar volume of water,[157] V'_w, the molar volume of the solution may be written as follows:

$$V = V_e^o x_e + V'_w x_w \tag{160}$$

Combining Eqs. (131), (132), (159) and (160) gives the following equation to express the Walden product:

$$\eta\Lambda = \eta_e \Lambda_e \left(x_e + \frac{V'_w}{V_e^o} x_w \right) \tag{161}$$

For these nitrate-water systems, as a first approximation, the apparent volume V_w^I being nearly independent of the composition over the range $x_w \approx 0$–0.7,[17, 71, 72] $\eta\Lambda$ is a linear function of x_w over this concentration range. It has been numerically tested[72] that a minimum value of the Walden product may be viewed as a consequence of Eqs. (125), (153) and the values of the empirical parameters in these equations.

10. Other Approaches and Observations

(i) Logarithmic Equation for Equivalent Electrical Conductance

Campbell et al.[102] have reported for the systems $[AgNO_3 + H_2O]$, at 495 K, and $[NH_4NO_3 + H_2O]$, at 453 K, a linear relationship between Λ and the logarithm of the molarity of the solution, $\ln C$, valid from pure electrolyte to dilute aqueous solutions, expressed by:

$$\Lambda = \Lambda_e - m_\Lambda \left(\ln C_e - \ln C\right) \qquad (162)$$

where C_e is the molarity of the pure electrolyte and m_Λ, a proportionality constant.

It has been noticed[129] that this empirical equation is also valid for the $AgNO_3$ and $TlNO_3$ based systems, at 372 K, but over more limited concentration ranges, especially in presence of divalent cations. For example, for the system $[0.464 AgNO_3 – 0.436 TlNO_3 – 0.100 Cd(NO_3)_2 + H_2O]$, Eq. (162) is valid over the limited range $x_w \approx 0$–0.4.

(ii) Power-Law Equations for Viscosity and Electrical Conductivity

Concerning the system $[H_2SO_4 + H_2O]$, studied in the subambient region, between 200 and 300 K, Das et al.[107] have found that a power-law type equation[158-160] based on the mode coupling theory of relaxation in supercooled liquids[161] is better than a Vogel-Tammann-Fulcher type equation[108] to describe the temperature dependence of the electrical conductivity κ and the viscosity η.

The power-law equation for κ [158, 159] is written

$$\kappa = \kappa_o \left(\frac{T}{T_{01}} - 1\right)^{\beta_1} \tag{163}$$

and for η [160]

$$\eta = \eta_o \left(\frac{T}{T_{02}} - 1\right)^{-\beta_2} \tag{164}$$

where κ_o, T_{o1}, β_1, η_o, T_{o2}, and β_2 are empirical parameters.

For a given composition, the computed value T_{o1} is expected to be very close to T_{o2}, since it represents the ideal glass transition temperature which should be independent of the transport property.[162] Das et al.[107] found that the values of T_{o1} and T_{o2} are in fact close to one another over the explored concentration range $x_w \approx 0$–0.6. Their investigations have also highlighted correlations between the viscosity-concentration curves and the phase diagram of the system.[163]

(iii) Substitution of an Organic Substance to Water

Pacák[94] has measured the viscosity of the system [0.667NH_4NO_3–0.258$LiNO_3$–0.075NH_4Cl + $(CH_3)_2SO$], at 369 K. Comparison of the curve η vs x_{DMSO} with the curve η vs x_w for the system [0.667NH_4NO_3–0.258$LiNO_3$–0.075NH_4Cl + H_2O],[93] at 366 K, shows[94] that the viscosity increases considerably when DMSO is substituted to water. Moreover, contrary to the hydrates, the curve with DMSO exhibits a maximum at $x_e \approx 0.8$ which, according to Pacák, cannot be described by any currently known equation, but could be related to the solubility curve of the system and to solvate formation. This maximum is not specific to this system but has been also observed with the system [$Zn(NO_3)_2$ + $(CH_3)_2SO$],[164] between 298 and 333 K, however at $x_e \approx 0.2$–0.3.

Ziogas and Papanastasiou[165, 166] have measured the viscosity of the systems [0.515$AgNO_3$–0.485$TlNO_3$ + $(CH_3)_2SO$][165] and [0.515$AgNO_3$–

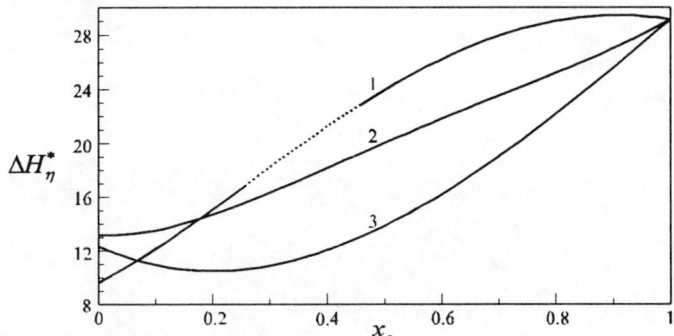

Figure 30. Enthalpy of activation for viscosity ΔH_η^* as function of electrolyte mole fraction x_e. 1: [0.515AgNO$_3$–0.485 TlNO$_3$ + (CH$_3$)$_2$SO],[165] between 376 and 394 K. 2: [0.515AgNO$_3$–0.485TlNO$_3$ + 0.500(CH$_3$)$_2$SO–0.500H$_2$O],[166] between 376 and 394 K. 3: [0.515AgNO$_3$–0.485TlNO$_3$ + H$_2$O],[104] between 350 and 380 K. Dotted line in curve 1: concentration range where a miscibility gap occurs. ΔH_η^* in kJ mol^{-1}.

0.485TlNO$_3$ + 0.500(CH$_3$)$_2$SO–0.500H$_2$O],[166] between 376 and 394 K, focusing their discussion on the influence of composition upon the activation parameters ΔH_η^* and ΔS_η^* in the transition state theory, Eqs. (98)–(100), in connection with the assumption of hole mechanisms in the viscous flow process.

As evidenced in Figs. 30 and 31, for the hydrate melts [0.515AgNO$_3$–0.485TlNO$_3$ + H$_2$O],[104] i.e. $x_e \approx 0.2$–1, replacement of water by DMSO increases significantly ΔH_η^* and ΔS_η^* due to stronger cation-dipole interactions in DMSO solutions than in aqueous solutions,[165, 166] also invoked in vapor pressure investigations[75, 94] (see Section II.4). When water is totally replaced by DMSO, for solutions rich in electrolyte, the simple linear relationships ΔH_η^* vs x_e and ΔS_η^* vs x_e, Eqs. (95) and (110), do not hold and the pronounced minimum in the curve ΔS_η^* vs x_e disappears.

The low values of ΔS_η^* in the concentration range rich in DMSO has been attributed by Ziogas and Papanastasiou[165] to the existence in DMSO of monomers, dimers and chain-like molecules[167] contributing to a relatively low degree of order in contrast to the quasi-lattice structure of nitrate-water systems rich in water as well as systems rich in nitrates.

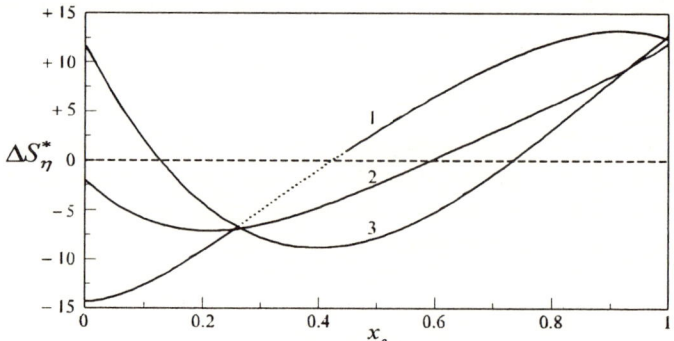

Figure 31. Entropy of activation for viscosity ΔS_η^* as function of electrolyte mole fraction x_e. 1: [0.515AgNO$_3$–0.485 TlNO$_3$ + (CH$_3$)$_2$SO],[165] between 376 and 394 K. 2: [0.515 AgNO$_3$–0.485TlNO$_3$ + 0.500(CH$_3$)$_2$SO–0.500H$_2$O],[166] between 376 and 394 K. 3: [0.515 AgNO$_3$–0.485TlNO$_3$ + H$_2$O],[104] between 350 and 380 K. Dotted line in curve 1: concentration range where a miscibility gap occurs. ΔS_η^* in J K^{-1} mol^{-1}.

Claes, Mestdagh, and Glibert[168] have studied the effect of substituting various organic substances to water on the electrical conductivity of the system [0.667NH$_4$NO$_3$–0.258LiNO$_3$–0.075NH$_4$Cl + H$_2$O][93] at 366 K: ethylene glycol (EG), (CH$_2$OH)$_2$, at 403 K, formamide (F), HCONH$_2$, at 363 and 403 K, dimethyl formamide (DMF), HCON(CH$_3$)$_2$, at 403 K and dimethyl sulfoxide (DMSO), (CH$_3$)$_2$SO, at 403 K. The most striking difference between the behavior of the electrical conductivity isotherm of the aqueous system and the isotherms of the organic systems is that from pure electrolyte to dilute solutions κ is less than 20 S m^{-1} with the organic substances, while, with water, it increases dramatically from about 10 to the maximum value 65 S m^{-1} at $x_w \approx 0.84$, in a manner analogous to that of the system [0.515AgNO$_3$– 0.485TlNO$_3$ + H$_2$O][104] illustrated in Fig. 25. The disappearance of the pronounced maximum in the curve κ vs x_w induced by the substitution of an organic substance to water is also observed with the system [C$_2$H$_5$NH$_3$NO$_3$ + H$_2$O],[100] at 298 K, when water is replaced by acetonitrile (AN), CH$_3$CN.

Occurrence of a pronounced maximum, often observed in aqueous solutions at a mixture composition where the liquid medium is relatively depleted of electrical charge carriers,[169] must obviously be ascribed to much higher mobilities of these charge carriers in water compared to the investigated organic substances. In fact, Λ increases much more rapidly with x_w than with the mole fraction of these organic substances.[100, 168] From their study, Claes, Mestdagh, and Glibert[168] concluded that the dimensions and shapes on the non-electrolyte molecules are factors more influential than the dielectric constants and acidity properties in the behavior of the isotherms κ vs x_e.

IV. REFERENCES

1. J. Braunstein, *Inorg. Chim. Acta Rev.* **2** (1968) 19.
2. J. Braunstein, "Statistical Thermodynamics of Molten Salts and Concentrated Aqueous Electrolytes," in: *Ionic Interactions: From Dilute Solutions to Fused Salts*, S. Petrucci, ed., Vol. 1, Chapter 4, Academic Press, New York, 1971.
3. R. M. Fuoss, *Chem. Rev.* **17** (1935) 27.
4. C. A. Kraus, *J. Phys. Chem.* **58** (1954) 673.
5. R. M. Fuoss and L. Onsager, *J. Phys. Chem.* **61** (1957) 668.
6. R. H. Stokes and R. Mills, "Concentrated Solutions and Molten Salts," in: *Viscosity of Electrolytes and Related Properties. The International Encyclopedia of Physical Chemistry and Chemical Physics*, E. A. Guggenheim, J. E. Mayer, and F. C. Tompkins, eds., Vol. 3, Chapter 7, Pergamon Press, New York. 1965.
7. M.-C. Trudelle, M. Abraham, and J. Sangster, *Can. J. Chem.* **55** (1977) 1713.
8. P. Claes and J. Glibert, "Water Concentration Dependence of Transport Properties in Ionic Melts," in: *Ionic Liquids*, D. Inman and D. G. Lovering, eds., Chapter 14, Plenum Press, New York, 1981.
9. H. Braunstein and J. Braunstein, *J. Chem. Thermodynamics* **3** (1971) 419.
10. A. N. Campbell, J. B. Fishman, G. Rutherford, T. P. Schaefer, and L. Ross, *Can. J. Chem.* **34** (1956) 151.
11. R. A. Robinson and R. H. Stokes, *Electrolyte Solutions*, 2nd ed., Butterworths, London, 1965.
12. J. M. Sangster, M.-C. Abraham, and M. Abraham, *Can. J. Chem.* **56** (1978) 348.
13. J. Sangster, M.-C. Abraham, and M. Abraham, *J. Chem. Thermodynamics* **11** (1979) 619.
14. M.-C. Abraham, M. Abraham, and J. Sangster, *J. Solution Chem.* **8** (1979) 647.
15. M.-C. Abraham, M. Abraham, and J. Sangster, *J. Chem. Eng. Data* **25** (1980) 331.
16. J. Sangster, M.-C. Abraham, and M. Abraham, *J. Chem. Thermodynamics* **14** (1982) 599.
17. M. Michielsen, G. Y. Loix, J. Glibert, and P. Claes, *J. Chim. Phys.* **79** (1982) 247.
18. M. Biquard, P. Letellier, and M. Fromon, *Can. J. Chem.* **63** (1985) 3587.

19. M. Abraham, M.-C. Abraham, I. Ziogas, and Z. Kodejš, *J. Am. Chem. Soc.* **115** (1993) 4300.
20. T. B. Tripp, *J. Electrochem. Soc.* **134** (1987) 848.
21. K. S. Pitzer, *J. Am. Chem. Soc.* **102** (1980) 2902.
22. K. S. Pitzer, *Phys. Chem. Earth* **13-14** (1981) 249.
23. J. M. Simonson, Thermodynamic Properties of Very Concentrated Electrolyte Solutions, Ph. D. Thesis, University of California, Berkeley, 1983.
24. K. S. Pitzer and J. M. Simonson, *J. Phys. Chem.* **90** (1986) 3005.
25. J. M. Simonson and K. S. Pitzer, *J. Phys. Chem.* **90** (1986) 3009.
26. O. Weres and L. Tsao, *J. Phys. Chem.* **90** (1986) 3014.
27. J. C. Barry, J. Richter, and E. Stich, *Ber. Bunsenges. Phys. Chem.* **92** (1988) 1118.
28. S. L. Clegg and P. Brimblecombe, *J. Phys. Chem.* **94** (1990) 5369.
29. E. Boßmann, J. Richter, and A. Stark, *Ber. Bunsenges. Phys. Chem.* **97** (1993) 240.
30. J. Geerlings, J. Richter, L. Rørmark, and H. A. Øye, *Ber. Bunsenges. Phys. Chem.* **101** (1997) 1129.
31. S. V. Petrenko and K. S. Pitzer, *J. Phys. Chem. B* **101** (1997) 3589.
32. T. B. Tripp and J. Braunstein, *J. Phys. Chem.* **73** (1969) 1984.
33. R. H. Stokes and R. A. Robinson, *J. Am. Chem. Soc.* **70** (1948) 1870.
34. S. Brunauer, P. M. Emmet, and E. Teller, *J. Am. Chem. Soc.* **60** (1938) 309.
35. R. B. Anderson, *J. Am. Chem. Soc.* **68** (1946) 686.
36. M. Abraham, *J. Chim. Phys.* **81** (1984) 207.
37. M. Abraham and M.-C. Abraham, *Electrochim. Acta* **33** (1988) 967.
38. G. W. Stewart and R. M. Morrow, *Phys. Rev.* **30** (1927) 232.
39. M.-C. Abraham and M. Abraham, *Monatsh. Chem.* **128** (1997) 805.
40. M. R. Ally and J. Braunstein, *Fluid Phase Equilibria* **87** (1993) 213.
41. M.-C. Abraham, M. Abraham, and J. Sangster, *J. Chim. Phys.* **76** (1979) 125.
42. M.-C. Abraham, M. Abraham, and J. Sangster, *Can. J. Chem.* **58** (1980) 1480.
43. M. Abraham, *J. Chim. Phys.* **78** (1981) 57.
44. T. L. Hill, *J. Chem. Phys.* **14** (1946) 263.
45. W. Voigt, *Monatsh. Chem.* **124** (1993) 839.
46. J. Braunstein and M. R. Ally, *Monatsh. Chem.* **127** (1996) 269.
47. M. Abraham, *Electrochim. Acta* **26** (1981) 1397.
48. Y. K. Delimarskii and B. F. Markov, *Electrochemistry of Fused Salts*, Sigma Press, Washington, 1961.
49. M. Blander, "Thermodynamic Properties of Molten Salt Solutions," in: *Molten Salt Chemistry*, M. Blander, ed., Chapter 3, Interscience Publishers, New York, 1964.
50. J. Lumsden, *Thermodynamics of Molten Salt Mixtures*, Academic Press, New York, 1966.
51. G. J. Janz, *Molten Salts Handbook*, Academic Press, New York, 1967.
52. H. Bloom, *The Chemistry of Molten Salts*, W. A. Benjamin Inc., New York, 1967.
53. J. O'M. Bockris and A. K. N. Reddy, "Ionic Liquids," in: *Modern Electrochemistry*, Vol.1, Chapter 6, Plenum Press, New York, 1970.
54. J. Lumsden, "Polarization Energy in Ionic Melts," in: *Ionic Liquids*, D. Inman and D. G. Lovering, eds., Chapter 20, Plenum Press, New York, 1981.

55. G. N. Papatheodorou, "Structure and Thermodynamics of Molten Salts,"in : *Comprehensive Treatise of Electrochemistry,* B. E. Conway, J. O'M. Bockris, and E. Yeager, eds., Vol. 5, Chapter 5, Plenum Press, New York, 1983.
56. Z. Kodejš and G. A. Sacchetto, *J. Chem. Faraday Trans. 1* **82** (1986) 1853.
57. V. Brendler and W. Voigt, *J. Solution Chem.* **24** (1995) 917.
58. V. Brendler and W. Voigt, *J. Solution Chem.* **25** (1996) 83.
59. M. R. Ally and J. Braunstein, *J. Chem. Thermodynamics* **30** (1998) 49.
60. M. Abraham and M.-C. Abraham, *Electrochim. Acta* **46** (2000) 137.
61. K. S. Pitzer, *Thermodynamics,* 3rd ed., McGraw Hill, New York, 1995.
62. M.-C. Abraham, M. Abraham, and J. M. Sangster, *Can. J. Chem.* **56** (1978) 635.
63. E. Darmois and G. Darmois, *Electrochimie Théorique,* Masson et Cie, Paris, 1960.
64. *Handbook of Chemistry and Physics,* 61st ed., C. R. C. Press, Boca Raton, 1980-1981.
65. A. Apelblat, *AIChE Journal* **39** (1993) 918.
66. G. A. Sacchetto and Z. Kodejš, *J. Chem. Soc., Faraday Trans 1* **84** (1988) 2885.
67. G. A. Sacchetto and Z. Kodejš, *J. Chem. Thermodynamics* **21** (1989) 585.
68. M. R. Ally and J. Braunstein, United States Patent, Number: 5,294,357, Mar. 15, 1994.
69. M. Hadded, M. Biquard, P. Letellier, and R. Schaal, *Can. J. Chem.* **63** (1985) 565.
70. M. Allen, D. F. Evans, and R. Lumry, *J. Solution Chem.* **14** (1985) 549.
71. M.-C. Abraham, M. Abraham, A. Combey, and J. Sangster, *J. Chem. Eng. Data* **28** (1983) 259.
72. M. Abraham and M.-C. Abraham, *Electrochim. Acta* **31** (1986) 821.
73. K. S. Pitzer, *J. Phys. Chem.* **77** (1973) 268.
74. K. S. Pitzer, *Acc. Chem. Res.* **10** (1977) 371.
75. Z. Kodejš and G. A. Sacchetto, *J. Chem. Soc. Faraday Trans.* **88** (1992) 2187.
76. H. H. Emons, W. Voigt, and W. F. Wollny, *J. Electroanal. Chem.* **180** (1984) 57.
77. K. H. Khoo, K. R. Fernando, R. J. Fereday, and C.-Y. Chan, *J. Solution Chem.* **24** (1995) 1039.
78. M. R. Ally and J. Braunstein, *Fluid Phase Equilibria* **120** (1996) 131.
79. I. Horsák and I. Sláma, *Collect. Czech. Chem. Commun.* **52** (1987) 1672.
80. M. Hadded, H. Bahri, and P. Letellier, *J. Chim. Phys.* **83** (1986) 419.
81. M. Abraham, M.-C. Abraham, and I. Ziogas, *J. Am. Chem. Soc.* **113** (1991) 8583.
82. A. N. Campbell and D. F. Williams, *Can. J. Chem.* **42** (1964) 1778.
83. H. Bloom, F. G. Davis, and D. W. James, *Trans. Faraday Soc.* **56** (1960) 1179.
84. E. A. Guggenheim and N. K. Adam, *Proc. Roy. Soc.* **A139** (1933) 218.
85. E. A. Guggenheim, *Mixtures,* Clarendon Press, Oxford, 1952.
86. J. A. V. Butler, *Proc. Roy. Soc.* **A135** (1932) 348.
87. P. D. Sonawane and A. Kumar, *Fluid Phase Equilibria* **157** (1999) 17.
88. T. B. Tripp and J. Braunstein, *Chem. Commun. Com.* **1244** (1968) 144.
89. T. B. Tripp, *J. Chem. Thermodynamics* **7** (1975) 263.
90. T. B. Tripp, "Vapor Pressures of Aqueous Melts at 130 ° Containing LiNO$_3$–KNO$_3$," in: *Molten Salts,* J. P. Pemsler, J. Braunstein, D. R. Morris, K. Nobe, and N. E. Richards, eds., p. 560, The Electrochemical Society Proceeding Series, Princeton, NJ, 1976.
91. G. A. Sacchetto and Z. Kodejš, *J. Chem. Soc., Faraday Trans 1* **78** (1982) 3519.
92. Z. Kodejš and G. A. Sacchetto, *J. Chem. Soc., Faraday Trans 1* **78** (1982) 3529.
93. J. Glibert, G. Y. Loix, R. Culot, and P. Claes, *Electrochim. Acta* **23** (1978) 1205.
94. P. Pacák, *J. Solution Chem.* **22** (1993) 839.

95. G. A. Sacchetto and Z. Kodejš, "Studies on the Behaviour of Hydrous Melts: the Non-Aqueous $(Ag,Tl)NO_3$ + DMSO Liquid System from Thermodynamic and Volumetric Measurements," in: *Molten Salt Chemistry and Technology,* M. Chemla and D. Devilliers, eds., Vol. 73-75, p. 227, Trans Tech Publications, Materials Science Forum, Switzerland. 1991.
96. A. J. Rabinowitsch, *Z. Phys. Chem.* **99** (1921) 417.
97. A. N. Campbell and W. G. Paterson, *Can. J. Chem.* **36** (1958) 1004.
98. M. Abraham and M.-C. Abraham, *J. Phys. Chem.* **94** (1990) 900.
99. I. Ziogas, M.-C. Abraham, and M. Abraham, *Electrochim. Acta* **37** (1992) 349.
100. G. Perron, A. Hardy, J.-C. Justice, and J. E. Desnoyers, *J. Solution Chem.* **22** (1993) 1159.
101. H. Dumont, S. Y. Qian, and B. E. Conway, *J. Molec. Liq.* **73** (1997) 147.
102. A. N. Campbell, E. M. Kartzmark, M. E. Bednas, and J. T. Herron, *Can. J. Chem.* **32** (1954) 1051.
103. P. Claes, M. Michielsen, and J. Glibert, *Electrochim. Acta* **28** (1983) 429.
104. M. Abraham and M.-C. Abraham, *Electrochim. Acta* **32** (1987) 1475.
105. M. Abraham, I. Ziogas, and M.-C. Abraham, *J. Solution Chem.* **19** (1990) 693.
106. M. Abraham, M.-C. Abraham, and I. Ziogas, *Electrochim. Acta* **41** (1996) 903.
107. A. Das, S. Dev, H. Shangpliang, K. L. Nonglait, and K. Ismail, *J. Phys. Chem.* B **101** (1997) 4166.
108. C. T. Moynihan, "Mass Transport in Fused Salts," in: *Ionic Interactions: From Dilute Solutions to Fused Salts,* S. Petrucci, ed., Vol. 1, Chapter 5, Academic Press, New York, 1971.
109. G. E. Walrafen, "Raman and Infrared Spectral Investigations of Water Structure," in: *Water, a Comprehensive Treatise,* F. Franks, ed., Vol. 1, Chapter 5, Plenum Press, New York, 1972.
110. W. A. P. Luck, *Disc. Faraday Soc.* **43** (1967) 115.
111. R. Leberman and A. K. Soper, *Nature* **378** (1995) 364.
112. K. Sigwart, in: *Properties of Ordinary Water-Substance,* Compiled by N. E.Dorsey, p. 64, Reinold, New-York, 1940.
113. M. S. Jhon and H. Eyring,"The Significant Structure Theory of Liquids,"in: *Physical Chemistry, an Advance Treatise,* D. Henderson, ed., Vol. VIIIA, Chapter 5, Academic Press, New York, 1971.
114. H. Eyring, *J. Chem. Phys.* **4** (1936) 283.
115. S. Glasstone, K. J. Laidler, and H. Eyring, *The Theory of Rate Processes,* McGraw-Hill, New York, 1941.
116. G. J. Janz and F. Saegusa, *J. Electrochem. Soc.* **110** (1963) 452.
117. J. O'M. Bockris and S. R. Richards, *J. Phys. Chem.* **69** (1965) 671.
118. T. Emi and J. O'M. Bockris, *J. Phys. Chem.* **74** (1970) 159.
119. J. Zarzycki, *J. Phys. Rad.* **18** (1957) 65.
120. J. O'M. Bockris and N. E. Richards, *Proc. Roy. Soc.* **A241** (1957) 44.
121. J. O'M. Bockris, E. H. Crook, H. Bloom, and N. E. Richards, *Proc. Roy. Soc.* **A255** (1960) 558.
122. M. Abraham and J. Brenet, *C. R. Acad. Sci.* **251** (1960) 2921.
123. G. J. Janz and M. R. Lorenz, *J. Electrochem. Soc.* **108** (1961) 1052.

124. M. Abraham, Etude sur les Potentiels d'Electrodes dans les Sels Fondus, Thèse de Doctorat, Université de Strasbourg, France, 1963.
125. B. Cleaver, *Nature* **207** (1965) 1291.
126. P. Cerisier, *Rev. Int. Hautes Tempér. et Réfract.* **8** (1971) 133.
127. P. Cerisier and B. Blin, *Ultrasonics* **5** (1982) 130.
128. M. Abraham and M.-C. Abraham, *J. Chim. Phys.* **83** (1986) 115.
129. M.-C. Abraham, Pression de Vapeur d'Eau, Densité, Viscosité et Conductivité Electrique de Solutions Electrolytiques Fortement Ioniques, Thèse de Doctorat, Université de Montréal, Canada, 1987.
130. T. L. Hill, *J. Chem. Phys.* **28** (1958) 1179.
131. A. H. Narten and H. A. Levy, "Liquid Water: Scattering of X-Rays," in: *Water, a Comprehensive Treatise,* F. Franks, ed., Vol. 1, Chapter 8, Plenum Press, New York, 1972.
132. A. Ben-Naim, "Application of Statistical Mechanics in the Study of Liquid Water," in: *Water, a Comprehensive Treatise,* F. Franks, ed., Vol. 1, Chapter 11, Plenum Press, New York, 1972.
133. H. S. Frank, "Structural Models," in: *Water, a Comprehensive Treatise,* F. Franks, ed., Vol. 1, Chapter 14, Plenum Press, New York, 1972.
134. M. Spiro and F. King, "Transport Properties in Concentrated Aqueous Electrolyte Solutions," in: *Ionic Liquids,* D. Inman and D. G. Lovering, eds., Chapter 5, Plenum Press, New York, 1981.
135. H. S. Frank and M. W. Evans *J. Chem. Phys.* **13** (1945) 507.
136. H. S. Frank and W. Y. Wen, *Discuss. Faraday Soc.* **24** (1957) 133.
137. M. Kaminsky, *Discuss. Faraday Soc.* **24** (1957) 171.
138. R. L. Kay and D. F. Evans, *J. Phys. Chem.* **70** (1966) 2325.
139. R. E. Verrall, "Infrared Spectroscopy of Aqueous Electrolyte Solutions," in: *Water, a Comprehensive Treatise,* F. Franks, ed., Vol. 3, Chapter 5, Plenum Press, New York, 1973.
140. R. Fürth, *Proc. R. Soc. Camb. Phil. Soc.* **27** (1941) 252.
141. J. H. Hildebrand, *Science* **174** (1971) 490.
142. A. J. Batschinski, *Z. Phys. Chem.* **84** (1913) 643.
143. R. P. Chhabra and R. J. Hunter, *Rheol. Acta* **20** (1981) 203.
144. M. Abraham, "Some Recent Contributions of Molten Salts Studies to the Understanding of Very Concentrated Aqueous Solutions," in: *Molten Salts: Ninth International Symposium. Proceedings,* C. L. Hussey, D. S. Newman, G. Mamantov, and Y. Ito, eds., Volume 94-13, p. 83, The Electrochemical Society, Pennington, NJ, 1994.
145. M. Abraham, J. P. Chevillot, and J. Brenet, *C. R. Acad. Sci.* **256** (1963) 2129.
146. S. Zuca, *Rev. Roum. Chim.* **15** (1970) 1277.
147. J. Kestin, M. Sokolov, and W. A. Wakeham, *J. Phys. Chem. Ref. Data* **7** (1978) 941.
148. R. L. Martin, *J. Chem. Soc.* **3** (1954) 3246.
149. A. N. Campbell and D. F. Williams, *Can. J. Chem.* **42** (1964) 1984.
150. A. E. Stearn and H. Eyring, *J. Chem. Phys.* **5** (1937) 113.
151. J. O'M. Bockris, J. A. Kitchener, S. Ignatowicz, and J. W. Tomlinson, *Trans. Faraday Soc.* **48** (1952) 75.
152. E. R. Van Artsdalen and I. S. Yaffe, *J. Phys. Chem.* **59** (1955) 118.
153. E. R. Van Artsdalen and I. S. Yaffe, *J. Phys. Chem.* **60** (1956) 1125.
154. S. K. Jain and H. C. Gaur, *Indian J. Chem.* **15** (1977) 645.

155. P. Debye and E. Hückel, *Physik Z.* **24** (1923) 305.
156. L. Onsager, *Physik Z.* **28** (1927) 277.
157. Z. Kodejš and I. Sláma, *Collect. Czech. Chem. Commun.* **45** (1980) 17.
158. R. J. Speedy, *J. Phys. Chem.* **87** (1983) 320.
159. R. J. Speedy, J. A. Ballance, and B. D. Cornish, *J. Phys. Chem.* **87** (1983) 325.
160. S. S. N. Murthy, *J. Chem. Soc., Faraday Trans. 2* **85** (1989) 581.
161. W. Götze, "Aspects of Structural Glass Transitions," in: *Liquids, Freezing and Glass Transition*, J. P. Hansen, D. Levesque, and J. Zinn-Justin, eds., Part 1, Course 5, North-Holland, Amsterdam, 1991.
162. C. A. Angell and R. D. Bressel, *J. Phys. Chem.* **76** (1972) 3244.
163. W. F. Giauque, E. W. Hornung, J. E. Kunzler, and T. R. Rubin, *J. Am. Chem. Soc.* **82** (1960) 62.
164. P. Pacák, Z. Kodejš, and H. Špalková, *Z. Physik. Chem. Neue Folge* **142** (1984) 157.
165. I. I. Ziogas and G. Papanastasiou, *Electrochim. Acta* **39** (1994) 2517.
166. I. I. Ziogas and G. Papanastasiou, *Electrochim. Acta* **42** (1997) 107.
167. H. H. Szmant, "Chemistry of DMSO," in: *Dimethyl Sulfoxide*, S. W. Jacob, E. E. Rosenbaum, and D. C. Wood, eds., Vol. 1, Chapter 1, Marcel Dekker Inc., New York, 1971.
168. P. Claes, C. Mestdagh, and J. Glibert, *Bull. Soc. Chim. Belg.* **102** (1993) 305.
169. P. Claes, G. Y. Loix, and J. Glibert, *Electrochim. Acta* **28** (1983) 421.

Index

Active anodes, electrochemical oxidation of organics, aqueous media, 91–108; see also Organics electrochemical oxidation
Active energy, for viscous flow, bridging electrolyte-water systems, 313–315
Adsorption theory of electrolytes, 276–300
  component activities, 276–287
  excess properties, 292–299
  temperature dependence, 299–300
  water activity coefficient curves, 287–291
Alkaline fuel cell (AFC), numerical modeling, 223–225
Analytical modeling (fuel cells), 156–169
  differential equations in, 166–169
  by division in hypothetical sub-cells, 160–166
  numerical modeling compared, 156–159
Apparent parameter equations electrical conductance, bridging electrolyte-water systems, 343–346

Apparent parameter equations (cont.)
  fluidity, bridging electrolyte-water systems, 329–333
Aqueous media, electrochemical oxidation of organics
  using active anodes, 91–108; see also Organics electrochemical oxidation
  using non-active anodes, 108–126; see also Organics electrochemical oxidation

Boron doped diamond based anode (BDD, electrochemical oxidation) of organics, 109–126
  in potential region before O_2 evolution, 110–113
  in potential region of O_2 evolution, 113–116
  p-Si/BDD anode modeling, 116–126
Braunstein classification, 271
Bridging electrolyte-water systems, 271–358
  overview, 271–273
  thermodynamic properties, 273–312

Bridging electrolyte-water
systems (*cont.*)
thermodynamic properties
(*cont.*)
adsorption theory, 276–300
component activities,
276–287
excess properties, 292–
299
temperature dependence,
299–300
water activity coefficient
curves, 287–291
molar volume, 310–311
organic substance
substitution, 311–312
regular solution theories,
300–305
mole fractions on ionized
basis, 300–304
mole fractions on un-
ionized basis, 304–
305
surface properties, 305–309
water vapor pressure, 309–
310
transport properties, 312–353
electrical conductance
activation energy for,
334–337
with apparent parameter,
equation for, 343–
346
free volume for, 341–343
logarithmic equation for
equivalent, 349
power-law equation for
viscosity and, 349–
350

Bridging electrolyte-water
systems (*cont.*)
transport properties (*cont.*)
electrical conductance
(*cont.*)
substitution of organic
substance to water,
350–353
transition state theory of,
337–341
viscosity relation with,
346–349
fluidity with apparent
parameter, equation for,
329–333
generally, 312
viscosity, transition state
theory of, 315–324
viscous flow
active energy for, 313–315
free volume for, 325–329

Carnot engine, fuel cells
compared, 132–136
Carnot engine/fuel cell
combination, 136–146
with overall FC reaction with
$^aS < 0$ (irreversible),
138–141
with overall FC reaction with
$^aS < 0$ (reversible), 136–
138
with overall FC reaction with
$^aS > 0$ (irreversible),
145–146
with overall FC reaction with
$^aS > 0$ (reversible), 144
with overall FC reaction with
$^aS = 0$ (irreversible), 143

Index

Carnot engine/fuel cell combination (*cont.*)
  with overall FC reaction with $^aS = 0$ (reversible), 141–142
Catalysis of outer-sphere heterogeneous charge-transfer reactions: *see* Outer-sphere heterogeneous charge-transfer reactions catalysis
Charge-transfer reactions: *see* Outer-sphere heterogeneous charge-transfer reactions catalysis
Coal-biomass fuel cell systems, 233–234
Constant load resistance, fuel cells, 192–193
Corrosion
  galvanic, titanium, 69–72
  general, titanium, 58–69
  pitting, titanium, 72–75
Corrosion conditions, hydrogen entry, IPZ analyses, permeation data under, 46–48

Dead-end mode fuel cell (2 in / 1 out), 217–218
Delayed fracture, hydrogen degradation, 5
Devanathan-Strachurski electrochemical permeation cell: *see* Electrochemical hydrogen permeation cell
Differential equations, fuel cells, analytical modeling, 166–169

Direct carbon fuel cell systems (DCFC), 239–244
Direct methanol fuel cell (DMFC), numerical modeling, 228–229
Disk shaped fuel cell geometry, 210–212

Electrochemical hydrogen permeation cell
  advantages of technique, 22
  cell composition, 10–11
  difficulties of technique, 20–21
  experimental procedure, 12–13
  generally, 9–10
  hydrogen concentration, 17–18
  hydrogen diffusivity, 14–16
  metal membrane preparation, 12
  quantities obtained, 13
  solution preparation, 12
  trapping effect, 18–20
Electrochemical incineration (ECI), organics, 87, 108, 110
Electrochemical oxidation of organics: *see* Organics electrochemical oxidation
Electrolyte-water systems: *see* Bridging electrolyte-water systems
Embrittlement, of titanium, hydrogen, 75–83
Equipartition principle, fuel cells, 187–189
  entropy production in non-isothermal fuel cell, 189
  general formulation, 187–188
Eyring equation, transition state theory of viscosity, bridging electrolyte-water systems, 317

Ferrous/ferric reaction, outer-sphere heterogeneous charge-transfer reactions catalysis, 256–266
Fluidity with apparent parameter, equation for, bridging electrolyte-water systems, 329–333
Free volume, for viscous flow, bridging electrolyte-water systems, 325–329
Fuel cells, 131–251
   analytical modeling, 156–169
     differential equations in, 166–169
     by division in hypothetical sub-cells, 160–166
     numerical modeling compared, 156–159
   analytical models compared (110 cm^2 MCFC), 197–209
     applications to other fuel cell types, 208–209
     experimental results, 197–198
     extended model compared with experimental results, 204–208
     simple and extended models, 198–202
     simple model compared with experimental results, 202–204
   Carnot engine combined, 136–146
     Nernst loss minimization, 179
     with overall FC reaction with $^aS < 0$ (reversible), 136–138

Fuel cells (*cont.*)
   Carnot engine combined (*cont.*)
     with overall FC reaction with $^aS < 0$ (irreversible), 138–141
     with overall FC reaction with $^aS > 0$ (irreversible), 145–146
     with overall FC reaction with $^aS > 0$ (reversible), 144
     with overall FC reaction with $^aS = 0$ (irreversible), 143
     with overall FC reaction with $^aS = 0$ (reversible), 141–142
   Carnot engine compared, 132–136
   configuration and geometries, 209–223
     dead-end mode fuel cell (2 in / 1 out), 217–218
     disk shaped geometry, 210–212
     generally, 209–210
     MCFC with separate CO_2 supply (3 in / 3 out), 218–223
     single-chamber fuel cell (1 in / 1 out), 216
     tubular geometry, 212–216
   equipartition principle, 187–189
     entropy production in non-isothermal fuel cell, 189
     general formulation, 187–188

Fuel cells (*cont.*)
  further analysis, 190–196
    constant load resistance, 192–193
    galvanostatic control, 193–194
    heat effects, 194–196
    maximum power versus maximum efficiency, 190–191
    potentiostatic control, 191–192
  irreversible losses, 183–185
  multistage oxidation, 185–187
  Nernst law, 146–155
    molten carbonate fuel cell (MCFC) with diluted fuel gas, 150–151
    pure hydrogen fuel (MCFC or SOFC), 151–155
  Nernst loss, 170–182
    decreased OCV by gas recycling minimization, 175
    FC-CE systems, 179–181
    generally, 170–174
    initial dip modeling of Nernst potential, 181–182
    lower fuel utilization minimization, 174–175
    lower operating temperature minimization, 174
    V_{eq} (u) relation change minimization, 175–177
    V_{eq} (x) relation change minimization, 177–178
  new developments and applications, 232–244

Fuel cells (*cont.*)
  new developments and applications (*cont.*)
    coal-biomass systems, 233–234
    direct carbon systems (DCFC), 239–244
    indirect carbon systems (IDCFC), 234–238
    trigeneration systems, 232–233
  numerical modeling, 223–232
    alkaline fuel cell (AFC), 223–225
    direct methanol fuel cell (DMFC), 228–229
    molten carbonate fuel cell (MCFC), 230–231
    phosphoric acid fuel cell (PAFC), 229
    polymer electrolyte membrane fuel cell (PEMFC), 225–228
    solid oxide fuel cell (SOFC), 231–232
  overview, 131–132
Fürth theory, transition state theory of viscosity, bridging electrolyte-water systems, 319

Galvanic corrosion, titanium, 69–72
Galvanostatic control, fuel cells, 193–194
General corrosion, titanium, 58–69
Generalized IPZ analyses (hydrogen entry), 32–40

Heterogeneous charge-transfer reactions: see Outer-sphere heterogeneous charge-transfer reactions catalysis
HMTA concentrations (IPZ analyses)
   new form, 44–45, 47
   original form, 27, 28
Hot filament chemical vapor deposition technique (HF-CVD), 113
Hydrogen absorption reaction (HAR) mechanisms
   hydrogen degradation, 5, 9
   hydrogen entry, 2, 4
   IPZ analysis
     generalized, 32
     new, 40
Hydrogen blistering, hydrogen degradation, 5
Hydrogen concentration, electrochemical hydrogen permeation cell, 17–18
Hydrogen degradation, hydrogen entry, 5–9
Hydrogen diffusivity, electrochemical hydrogen permeation cell, 14–16
Hydrogen embrittlement, of titanium, 75–83
Hydrogen entry, 1–54
   Devanathan-Strachurski electrochemical permeation cell, 9–22
   electrochemical hydrogen permeation cell
     advantages of technique, 22
     cell composition, 10–11
     difficulties of technique, 20–21

Hydrogen entry (*cont.*)
   electrochemical hydrogen permeation cell (*cont.*)
     experimental procedure, 12–13
     generally, 9–10
     hydrogen concentration, 17–18
     hydrogen diffusivity, 14–16
     metal membrane preparation, 12
     quantities obtained, 13
     solution preparation, 12
     trapping effect, 18–20
   hydrogen degradation, 5–9
   hydrogen evolution reaction (HER) mechanisms, 3–4
   IPZ analyses, 22–48
     corrosion conditions, permeation data under, 46–48
     generalized, 32–40
     2_H calculation and rate constants of HER from polarization data, 30–32
     membrane thickness effect, 29–30
     new, 40–46
     original form, 22–29
   overview, 1–3
Hydrogen evolution reaction (HER) mechanisms
   hydrogen degradation, 5–9
   hydrogen entry, 3–4
   IPZ analyses, original form, 22–23
   IPZ analysis
     generalized, 32

Index

Hydrogen evolution reaction (HER) mechanisms (*cont.*)
  IPZ analysis (*cont.*)
    2_H calculation and rate constants of HER from polarization data, 30–32
    new, 40
Hydrogen induced cracking (HIC), hydrogen degradation, 5, 7

Indirect carbon fuel cell systems (IDCFC), 234–238
IPZA analyses (hydrogen entry), 42–44
IPZ analyses (hydrogen entry), 22–48
  generalized, 32–40
  2_H calculation and rate constants of HER from polarization data, 30–32
  membrane thickness effect, 29–30
  new, 40–46
  original form, 22–29
Iyer-Pickering-Zamenzadah (IPZ) analysis, 9

Membrane thickness effect, IPZ analyses (hydrogen entry), 29–30
Molar volume, bridging electrolyte-water systems, 310–311
Molten carbonate fuel cell (MCFC): *see also* Fuel cells
  with diluted fuel gas, Nernst law, 150–151

Molten carbonate fuel cell (MCFC) (*cont.*)
  numerical modeling, 230–231
  with separate CO_2 supply (3 in / 3 out), 218–223
Multistage oxidation, fuel cells, 185–187

Nernst law (fuel cells), 146–155
  derivation of, 146–150
  molten carbonate fuel cell (MCFC) with diluted fuel gas, 150–151
  pure hydrogen fuel (MCFC or SOFC), 151–155
Nernst loss (fuel cells), 170–182
  decreased OCV by gas recycling minimization, 175
  FC-CE system minimization, 179–181
  generally, 170–174
  initial dip modeling of Nernst potential, 181–182
  lower fuel utilization minimization, 174–175
  lower operating temperature minimization, 174
  $V_{eq}(u)$ relation change minimization, 175–177
  $V_{eq}(x)$ relation change minimization, 177–178
Numerical modeling, fuel cells, 223–232
  alkaline fuel cell (AFC), 223–225
  analytical modeling compared, 156–159
  direct methanol fuel cell (DMFC), 228–229

Numerical modeling, fuel cells (*cont.*)
    molten carbonate fuel cell (MCFC), 230–231
    phosphoric acid fuel cell (PAFC), 229
    polymer electrolyte membrane fuel cell (PEMFC), 225–228
    solid oxide fuel cell (SOFC), 231–232

Open cell voltage (OCV) calculation, Nernst law, fuel cells, 146

Organics electrochemical oxidation, 87–130
    aqueous media using active anodes, 91–108
    generally, 91–93
    kinetic model of oxidation on Ti/IrO$_2$ anodes, 103–108
    Ti/IrO$_2$ anodes in potential region before O$_2$ evolution, 93–96
    Ti/IrO$_2$ anodes in potential region of O$_2$ evolution, 96–102
    aqueous media using non-active anodes, 108–126
    generally, 108–110
    p-Si/BDD anode in potential region before O$_2$ evolution, 110–113
    p-Si/BDD anode in potential region of O$_2$ evolution, 113–116
    p-Si/BDD anode modeling, 116–126

Organics electrochemical oxidation (*cont.*)
    overview, 87–91
    Organic substance substitution, bridging electrolyte-water systems, 311–312, 350–353
    Outer-sphere heterogeneous charge-transfer reactions
        catalysis, 253–269
        ferrous/ferric reaction, 256–266
        discussed, 264–266
        Gerischer, 257
        Hung and Nagy, 260–264
        Johnson and Resnick, 258
        Nagy et al., 260
        Smith and Wadden, 259
        Weber et al., 258–259
        observation summary, 254–256
        overview, 253–254
    Oxidation of organics: *see* Organics electrochemical oxidation

Paladium, electrochemical hydrogen permeation cell, 12
Phosphoric acid fuel cell (PAFC), numerical modeling, 229
Pitting corrosion, titanium, 72–75
Polymer electrolyte membrane fuel cell (PEMFC), numerical modeling, 225–228
Potentiostatic control, fuel cells, 191–192
p-Si/BDD anode: *see* Boron doped diamond based anode (BDD)

Index

Raoult's law, 274–276
Reversible losses: *see* Nernst loss (fuel cells)

Saturated calomel electrode (SCE), electrochemical hydrogen permeation cell, cell composition, 11
SEM studies, titanium, hydrogen embrittlement of, 82
p-Si/BDD anode: *see* Boron doped diamond based anode (BDD)
Single-chamber fuel cell (1 in / 1 out), 216
Solid oxide fuel cell (SOFC)
  numerical modeling, 231–232
  pure hydrogen fuel, 151–155; *see also* Fuel cells
Step wise cracking (SWC), hydrogen degradation, 5
Stress corrosion cracking (SCC), hydrogen degradation, 5
Stress oriented hydrogen induced cracking (SOHIC), hydrogen degradation, 5
Sulfide hydrogen embrittlement, hydrogen degradation, 5

TEM studies, titanium, hydrogen embrittlement of, 82
Ti-H phase diagram, titanium, hydrogen embrittlement of, 75–77
Ti/IrO$_2$ anodes
  kinetic model of oxidation on, organics, 103–108
  in potential region before O$_2$ evolution, organics, 93–96

Ti/IrO$_2$ anodes (*cont.*)
  in potential region of O$_2$ evolution, organics, 96–102
Titanium
  galvanic corrosion, 69–72
  general corrosion, 58–69
  hydrogen embrittlement of, 75–83
  overview, 55–58
  pitting corrosion, 72–75
Transition state theory
  of electrical conductance, bridging electrolyte-water systems, 337–341
  of viscosity, bridging electrolyte-water systems, 315–324
Transport properties, bridging electrolyte-water systems, 312–353; *see also* Bridging electrolyte-water systems
Trapping effect, electrochemical hydrogen permeation cell, 18–20
Trigeneration fuel cell systems, 232–233
Tubular fuel cell geometry, 212–216

Viscosity
  electrical conductance and, bridging electrolyte-water systems, 346–349
  power-law equation for, and electrical conductance, bridging electrolyte-water systems, 349–350

Vicosity (*cont.*)
  transition state theory of, bridging electrolyte-water systems, 315–324
Viscous flow
  active energy for, bridging electrolyte-water systems, 313–315
  free volume for, bridging electrolyte-water systems, 325–329

Water activity coefficient curves, bridging electrolyte-water systems, adsorption theory, 287–291
Water vapor pressure, bridging electrolyte-water systems, 309–310

Zinc, titanium and, galvanic corrosion, 71, 72

William F. Maag Library
Youngstown State University